W0049893

РОСТ КРИСТАЛЛОВ

ROST KRISTALLOV

GROWTH OF CRYSTALS

VOLUME 18

Growth of Crystals

Volume 18

Edited by

E. I. Givargizov and S. A. Grinberg

Shubnikov Institute of Crystallography
Russian Academy of Sciences
Moscow, Russia

Translated by

Dennis W. Wester

CONSULTANTS BUREAU • NEW YORK AND LONDON

The Library of Congress cataloged the first volume of this title as follows:

Growth of crystals. v. [1]
   New York, Consultants Bureau, 1958 –
     v. illus., diagrs. 28 cm.
   Vols. 1,3 –   constitute reports of 1st –   Conference on Crystal
Growth, 1956 –   v. 2 contains interim reports between the 1st and
2nd Conference on Crystal Growth, Institute of Crystallography,
Academy of Sciences, USSR.
   "Authorized translation from the Russian" (varies slightly)
   Editors: 1958 –   A. V. Shubnikov and N. N. Sheftal'.
   1. Crystal – Growth. I. Shubnikov, Aleksei Vasil'evich, ed. II.
Sheftal', N. N., ed. III. Consultants Bureau Enterprises, inc., New
York, IV. Soveshchanie po rostu kristallov. V. Akademiia nauk
SSSR. Institut kristallografii.
QD921.R633   548.5   58-1212

ISBN 0-306-18118-5

The original Russian text was published for the Institute of Crystallography of the
Russian Academy of Sciences by Nauka Press in Moscow in 1989

© 1992 Consultants Bureau, New York
A Division of Plenum Publishing Corporation
233 Spring Street, New York, N.Y. 10013

All rights reserved

No part of this book may be reproduced, stored in a retrieval system, or transmitted
in any form or by any means, electronic, mechanical, photocopying, microfilming,
recording, or otherwise, without written permission from the Publisher

PREFACE

This 18th volume of the series includes invited papers from the Seventh All-Union Conference on the Growth of Crystals and the Symposium on Molecular-Beam Epitaxy that were held in Moscow in November, 1988. In choosing papers, the Program Committee of the conference gave priority to studies in rapidly emerging areas of the growth and preparation of crystals and crystalline films. The qualifications of the authors were also considered. This ensured that the material was of a high standard and that the problems discussed covered a wide range. These are the same criteria that, we hope, are typical of the volumes of this series.

The articles of the present volume are divided into four sections:

I. Processes on the growth surface.
II. Molecular-beam epitaxy.
III. Growth of crystals and films from solutions and fluxes.
IV. Growth of crystals from the melt.

Following tradition, the series opens with three theoretical articles. These examine problems applicable to various crystallization media: instability of the crystallization front (for a more general case than before and for a comparatively complicated system, a solution), adsorption and migration of atoms and molecules (the analysis is made on a quantum-chemical level), and the kinetics of step and dislocation growth in the presence of surface anisotropy as well as impurity adsorption (several earlier known methods are summarized). The next two articles are experimental and methodical. One of these is concerned with high resolution electron microscopy (as applied to epitaxial layers of semiconductors, superconductors, and other practically important materials). The other discusses the recently perfected method of reflective electron microscopy (demonstrating its successes in the cases of sublimation and epitaxy of silicon).

The second section contains four articles and illustrates the achievements of the Novosibirsk school in molecular-beam epitaxy. This is one of the newest and most important practical methods of contemporary microelectronics for preparing films. The theme of these articles encompasses basic problems and basic materials such as crystallization of elemental semiconductors (silicon), $A^{III}B^V$ semiconductors (primarily Ga and In arsenides, their solid solutions, and Ga–Al arsenide), and $A^{II}B^{VI}$ compounds (Cd–Hg telluride). The articles are practically oriented. Thus, fundamental rules of epitaxial crystallization are analyzed using the most modern study methods.

The third section includes several articles on the growth of crystals and films from condensed media, i.e., solutions and fluxes. The section opens with an article in which the kinetics of crystallization from solution in the presence of impurities, a very complicated case, is studied by careful experimentation (using interference measurements of the growth rates). This is followed by a review article on the growth of single crystals of complex oxide systems from a flux. Problems with the growth of such important single crystals as high-temperature superconductors are treated here. The remaining two articles of this section involve liquid epitaxy of semiconductors.

The volume concludes with a section on growth of crystals from the melt. Special attention (in three of the four articles) is paid to inhomogeneities in the crystals such as point defects, inclusions, and dislocations. The final article is a review of methods for growing single crystals of fluorides.

As in previous volumes of this series, all aspects of the growth of crystals are not covered. However, a sufficiently complete picture is given of several very important aspects of the mechanisms and kinetics of crystallization and of the trends in practical growth of single crystals and crystalline films.

E. I. Givargizov
S. A. Grinberg

CONTENTS

	PAGE	RUSS. PAGE

I. PROCESSES ON GROWTH SURFACES

Concentrational Instability of the Interface
A. G. Ambrok and E. V. Kalashnikov 3   5

Quantum Chemical Investigation of Adsorption and Surface Migration
of Atoms and Molecules on Si(111) and Si(100) Surfaces
M. P. Ruzaikin and A. B. Svechnikov 15   18

Step Kinetics on Crystal Surfaces in the Presence of Anisotropy and Impurities
S. Yu. Potapenko 27   31

High-Resolution Transmission Electron Microscopic Study of Epitaxial Layers
N. A. Kiselev, V. Yu. Karasev, and A. L. Vasil'ev 37   43

Structural Reconstruction of Atomically-Clean Silicon Surface during
Sublimation and Epitaxy
A. V. Latyshev, A. L. Aseev, A. B. Krasil'nikov, and S. I. Stenin 51   61

II. MOLECULAR-BEAM EPITAXY

Molecular-Beam Epitaxy of Silicon
S. I. Stenin, B. Z. Kanter, and A. I. Nikiforov 69   81

Molecular Epitaxy of A_3B_5 Compounds
Yu. O. Kanter and A. I. Toropov 77   91

Epitaxy of Solid Solutions and Multilayered Structures in the System Cd—Hg—Te
Yu. G. Sidorov and S. I. Chikichev 87   104

δ-Structures in Gallium Arsenide
D. I. Lubyshev, V. P. Migal', V. N. Ovsyuk, B. R. Semyagin, and S. I. Stenin 99   117

III. GROWTH OF CRYSTALS AND FILMS FROM SOLUTIONS AND FLUXES

Influence of Impurities on Growth Kinetics and Morphology of Prismatic Faces
of ADP and KDP Crystals
L. N. Rashkovich and B. Yu. Shekunov 107   124

vii

CONTENTS

	PAGE	RUSS. PAGE

Controlled Flux Growth of Complex Oxide Single Crystals
G. A. Emel'chenko, V. M. Masalov, and V. A. Tatarchenko   121   139

Mechanism of Relaxation of the Nonequilibrium Liquid—Solid Interface before
Liquid-Phase Heteroepitaxy of III—V Compounds
Yu. B. Bolkhovityanov   135   158

Modelling and Control of Heat and Mass Transfer during Liquid Epitaxy
N. A. Verezub and V. I. Polezhaev   147   173

IV. GROWTH OF CRYSTALS FROM THE MELT

Aggregation of Point Defects in Silicon Crystals Growing from the Melt
V. V. Voronkov   157   183

Interaction of Crystals Growing in the Melt with Inclusions
and Concentration Inhomogeneities
O. P. Fedorov   169   197

Role of Growth Dislocations in Forming Inhomogeneous Properties in
Gallium Arsenide Single Crystals
A. V. Markov, M. G. Mil'vidskii, and V. G. Osvenskii   183   214

Multicomponent Fluoride Single Crystals
(Current Status of Their Synthesis and Prospects)
B. P. Sobolev   197   233

Part I

PROCESSES ON
GROWTH SURFACES

CONCENTRATIONAL INSTABILITY OF THE INTERFACE

A. G. Ambrok and E. V. Kalashnikov

The starting composition of a substance, as a rule the average relative concentration of the components, is one of the parameters determining the surface state during growth of single crystals from fluxes or the vapor phase onto a substrate. However, the average does not always coincide with the local concentration not only in the bulk but also at the surface. Moreover, the interfacial concentration is known [1-4] to depend nonlinearly on the starting (bulk) concentration. The local composition, numerically equal to the average concentration, can under certain conditions correspond to an unstable state of the developing surface. Such instability can cause local disruption of the starting stoichiometry both along the surface and at a distance from the surface. It has not yet been investigated at which compositions and temperatures and in which systems such disruptions are possible. The present work attempts to examine this problem.

One of the most common approaches to analyzing the behavior of the interface is based on lattice models [1-4]. If a gas—liquid, gas—solid, or liquid—solid system is examined, then, generally speaking, the coexistence of two aggregate states must be considered, as for example in [4]. In this case, a certain ratio of energy and temperature is the only guarantee that one aggregate state differs from the other. However, the interaction energy between components is admixed with the energies of the pure components even for binary systems. The relation of it to temperature can differ significantly from this of the pure components to temperature. In such a situation, there is the danger within the framework of the lattice model that a stratification, for example, in the liquid away from a liquid—solid transition, will not be distinguished. The same can be said relative to gas—solid and gas—liquid transitions. However, this problem can be approached from another angle, namely, by examining potentialities of only one of the aggregate states. For example, the liquid—solid interface can be viewed as a liquid—wall, where the wall does not interact with the liquid but limits its volume. This will reveal features of the behavior and tendency to change only in the liquid. The last model is analogous to the liquid—vacuum model developed by Ono and Kondo [1]. Finally, the same model can be used to study a condensate on a substrate. The condensate is regarded as if it existed at all compositions and over a wide temperature range. A region of compositions and temperatures at which the condensate is thermodynamically unstable can be revealed by studying the stability of this system.

In the present work, we will follow the approach of Ono and Kondo [1]. We will examine the interface (binary liquid)—vacuum or (binary liquid)—wall. The eutectic systems Sn—Pb and In—Sn are chosen as examples of binary systems. A large number of measurements has been made on these systems. Theoretical and experimental values can be compared. With respect to experimental investigations of the condensate—substrate interface, literature data on the concentration and temperature dependences of composition for this interface are fragmented. The same can be said about crystals grown from fluxes, except for the difference of compositions at the surface

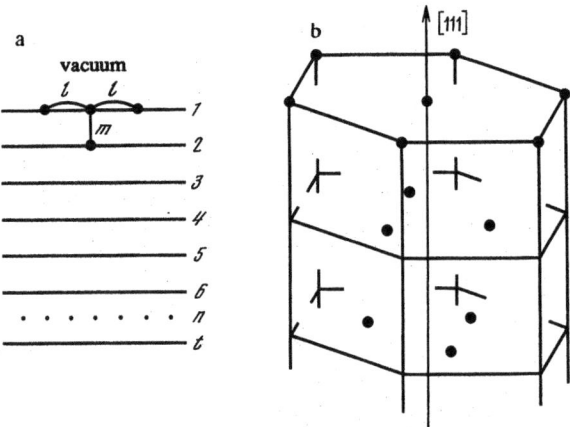

Fig. 1. Interface model: bonds to nearest neighbors (l, in the layer; m, between layers; t, layer number) (a), atomic packing at the lattice surface and in adjoining monolayers (b).

and within the crystal and the distortions of the crystal shapes as a function of the starting melt concentration that have been noted [5].

BINARY SOLUTION MODEL AT THE LIQUID–VACUUM INTERFACE

The surface limiting the liquid volume is viewed in a quasicrystalline approximation with face-centered cubic atomic packing, the [111] direction, and coordination number c. The dividing surface is placed over the first layer facing the vacuum. All layers t are parallel to the surface. The solution below the nth layer is considered homogeneous (Fig. 1). The gas density over the surface is neglected.

Let us select a homogeneous stochastic distribution of component particles among the sites as the initial state of the solution in the bulk and on the surface. (This has an advantage over models applicable to clusters [6] since continuous state changes can be followed over the whole temperature range and regions of different thermodynamic stability with a different tendency to form inhomogeneities, clusters, can be found in the homogeneous distribution and do not have to be added artificially.)

Let the system consist of $N = N_A + N_B$ particles, where N_i is the number of particles of the ith type, $i = A, B$. The fraction of bonds of a selected particle with the nearest neighbors is l in the layer and m to particles of a neighboring layer. Then the total number of nearest neighbors is $lc + 2mc$. The atomic fraction $x^{(t)}$ of component A in layer t is

$$x^{(t)} = N_A^{(t)} / (\nu\sigma),$$

where $N_A^{(t)}$ is the number of A atoms in the tth layer, σ is the interface area, and ν is the number of particles per unit area. Analogously for component B

$$1 - x^{(t)} = N_B^{(t)} / (\nu\sigma).$$

The Helmholtz free energy for a binary solution with a planar interface is written

$$F = \tfrac{1}{2} c(N_A \chi_{AA} + N_B \chi_{BB}) - \tfrac{1}{2} \sigma\nu mc \left[x^{(1)}\chi_{AA} + (1 - x^{(1)})\chi_{BB} \right] -$$

$$- N_A kT\ln\left(\lambda_A^{-3} j_A\, ev_A\right) - N_B kT\ln\left(\lambda_B^{-3} j_B\, ev_B\right) - \sigma\nu x^{(1)}kT\ln(v_A'/v_A) -$$

$$- \sigma v(1 - x^{(1)})kT\ln(v_B'/v_B) + \sigma v kT \overset{\omega}{\underset{t=1}{\Sigma}} [x^{(t)}\ln x^{(t)} + (1 - x^{(t)})\ln(1 - x^{(t)})] +$$

$$+ {}^1/_2\sigma v c \chi \{ \overset{\omega}{\underset{t=2}{\Sigma}} [2lx^{(t)}(1 - x^{(t)}) + mx^{(t)}(1 - x^{(t+1)}) + mx^{(t+1)}(1 - x^{(t)}) + \quad (1)$$

$$+ mx^{(t)}(1 - x^{(t-1)}) + mx^{(t-1)}(1 - x^{(t)})] + 2lx^{(1)}(1 - x^{(1)}) +$$

$$+ mx^{(1)}(1 - x^{(2)}) + mx^{(2)}(1 - x^{(1)})\},$$

where k is Boltzmann's constant, T is the temperature, λ_i is the deBroglie length, j_i are the internal degrees of freedom, e is the natural log base, v_i and v_i' are the free volumes of the particle in solution and the surface layer, respectively, χ_{ij} are the potential energies of particle pair interaction, and $\chi = \chi_{AB} - (\chi_{AA} + \chi_{BB})/2$ is the interaction energy.

Let us minimize the free energy (1) at constant N_A and N_B [1]. The conditions are imposed that the number of particles in the system is conserved

$$\sigma v \overset{\omega}{\underset{t=1}{\Sigma}} x^{(t)} = N_A \quad (2)$$

and that the concentration in the tth layer coincides with the bulk solution concentration $x^{(\alpha)}$:

$$\lim_{t \to \infty} x^{(t)} = x^{(\alpha)}. \quad (3)$$

Using the minimization conditions and boundary conditions of Eqs. (2) and (3), we obtain, following [1], equations for determining the local concentration over the layers:

$$\Delta x^{(t)} = x^{(t+1)} - x^{(t)}, \ \Delta^2 x^{(t)} = \Delta x^{(t+1)} - \Delta x^{(t)}, \quad (4)$$

where

$$\Delta x^{(1)} = (2v \, mc\chi)^{-1}(\gamma_A - \gamma_B) + (2m)^{-1} [l(1 - 2x^{(1)}) + m(1 - 2x^{(1)}) -$$

$$- (1 - 2x^{(\alpha)})] + kT(2mc\chi)^{-1}\{\ln[x^{(1)}/(1 - x^{(1)})] - \ln[x^{(\alpha)}/(1 - x^{(\alpha)})]\} \quad (5)$$

at $t = 1$

$$\Delta^2 x^{(t-1)} = m^{-1}(x^{(\alpha)} - x^{(t)}) + kT(2mc\chi)^{-1}[\ln(x^{(t)}/(1 - x^{(t)})) -$$

$$- \ln(x^{(\alpha)}/(1 - x^{(\alpha)}))] \quad \text{at} \ t \geqslant 2, \quad (6)$$

$\gamma_i = (1/2)vmc\chi_{ii} - vkT\ln(v_i'/v_i)$ is the surface tension of the ith liquid. The Gibbs adsorption Γ_A and Γ_B of components A and B is defined by

$$\Gamma_A = v \overset{\omega}{\underset{t=1}{\Sigma}} (x^{(t)} - x^{(\alpha)}), \ \Gamma_B = v \overset{\omega}{\underset{t=1}{\Sigma}} [(1 - x^{(t)}) - (1 - x^{(\alpha)})]. \quad (7)$$

Now the dependence of surface tension of the binary solution on concentration and temperature is written:

$$\gamma = \sigma^{-1}(F - F^{(\alpha)}) - (\mu_A - \mu_B)\Gamma_A, \quad (8)$$

where F is the free energy of a solution with a planar liquid—vacuum interface taking into account the multilayered transition zone and $F^{(\alpha)}$ is the free energy of the bulk solution that is homogeneous up to the dividing surface:

$$F^{(\alpha)} = \tfrac{1}{2} c(N_A^{(\alpha)}\chi_{AA} + N_B^{(\alpha)}\chi_{BB}) - N_A^{(\alpha)}kT\ln(\lambda_A^{-3}j_A ev_A) -$$

$$- N_B^{(\alpha)}kT\ln(\lambda_B^{-3}j_B ev_B) + NkT[x^{(\alpha)}\ln x^{(\alpha)} + (1 - x^{(\alpha)})\ln(1 - x^{(\alpha)})] + Ncx^{(\alpha)}(1 - x^{(\alpha)})\chi, \quad (9)$$

μ_A and μ_B are the chemical potentials of components in the bulk solution. Combining Eqs. (1), (4), (5)-(7), and (9) and substituting them into Eq. (8), we obtain the complete dependence of the solution surface tension as a function of the concentration distribution over the layers:

$$\gamma = x^{(1)}\gamma_A + (1 - x^{(1)})\gamma_B + \nu kT \sum_{t=1}^{\omega} \{x^{(t)}\ln(x^{(t)}/x^{(\alpha)}) - (1 - x^{(t)})\ln[(1 - $$
$$- x^{(t)})/(1 - x^{(\alpha)})]\} - \tfrac{1}{2}\,\nu c\chi[2l(x^{(1)} - x^{(\alpha)})^2 + m\{4(x^{(1)} - x^{(\alpha)})^2 - $$
$$- (x^{(2)} - x^{(1)})(1 - 2x^{(1)}) + 2x^{(1)}(1 - x^{(1)})\} + \sum_{t=2}^{\omega} \{2l\,(x^{(t)} - x^{(\alpha)})^2 + $$
$$+ 4m(x^{(t)} - x^{(\alpha)})^2 - m(x^{(t-1)} + x^{(t+1)} - 2x^{(t)})(1 - 2x^{(t)})\}. \tag{10}$$

A system of n simultaneous nonlinear equations must be solved to determine the local concentration over the layers and to calculate subsequently the surface tension using Eq. (10). This can be done by two methods.

The first consists of *a priori* assignment of the difference of concentrations between the $(t + 1)$th and tth layers of Eq. (4) and successive approximation of the local concentrations so that the boundary conditions of Eq. (3) are satisfied.

The second approach consists of the following. The interfacial region is bounded by a certain number of monolayers. Calculations carried out for analogous systems [1-4] indicate that this region is bounded by 3-4 layers whereas the concentration in subsequent layers is the same as in the bulk [Eq. (3)]. We set the thickness of this region, the transition zone, equal to six monolayers in order certainly to cover the interfacial region:

$$x^{(t)} = x^{(\alpha)} \quad \text{at} \quad t = 7. \tag{11}$$

We assumed that the concentration in the seventh layer was equal to the bulk and thus broke the chain of simultaneous equations (4). As a result, we obtained a closed self-consistent system of six equations in six unknowns, a system of nonlinear equations solved only by numerical methods. We chose the Newton method [7]. The solution was terminated if the difference of the previous iteration from the next one was 10^{-5}. The initial approximation for all layers was the bulk concentration. Calculations were carried out for the concentration range $(6-98) \cdot 10^{-2}$ at. fract. in steps of $1 \cdot 10^{-2}$ at. fract. and in the temperature range 620-1100 K in steps of 10 K. In the temperature range 620-780 K the Newton process diverged on reaching a certain bulk concentration and the program was stopped. The initial conditions were then set at $x_0 = 0.98$ at. fract. for all layers and the calculation was continued.

The hidden nature of the stratification in the selected eutectic systems had to be considered to evaluate the behavior of the interface and layers adjoining it. Traditional eutectic diagrams do not portray this feature. Nevertheless, it was demonstrated both in experimental [8] and in theoretical works [9, 10] that the liquid state is stratified in the eutectic systems. The decomposition of the liquid solution in these systems is more sluggish due to the lower energy of mixing compared with typically stratifying systems.

The critical mixing (decomposition) temperature for a Sn—Pb solution, $T_c = 775$ K, was found experimentally [11]. Theoretical estimates gave a slightly elevated value, $T_c = 867$ K [12]. The exchange energy was determined in [11] using

$$\chi = 2k\,T_c\,/c \quad \text{at} \quad T_c = 775 \text{ K}. \tag{12}$$

The values $\gamma_i(T)$ and density $\rho_i(T)$ for pure ith components [where $i = A, B = $ (Sn—Pb) and (In—Sn)] were taken from [13] for calculating the concentration distribution over layers [Eq. (4)] and the surface tension [Eq. (10)]. Moreover, the number of sites according to the model used was defined as

$$\nu = (\nu_A + \nu_B)/2,$$

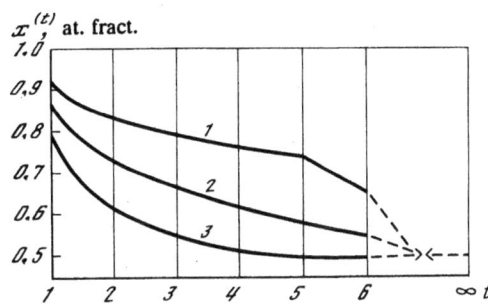

Fig. 2. Concentration distribution $x^{(t)}$ over layers t along the normal to the interface at temperatures $T = 673$ (1), 775 (2), 900 K (3).

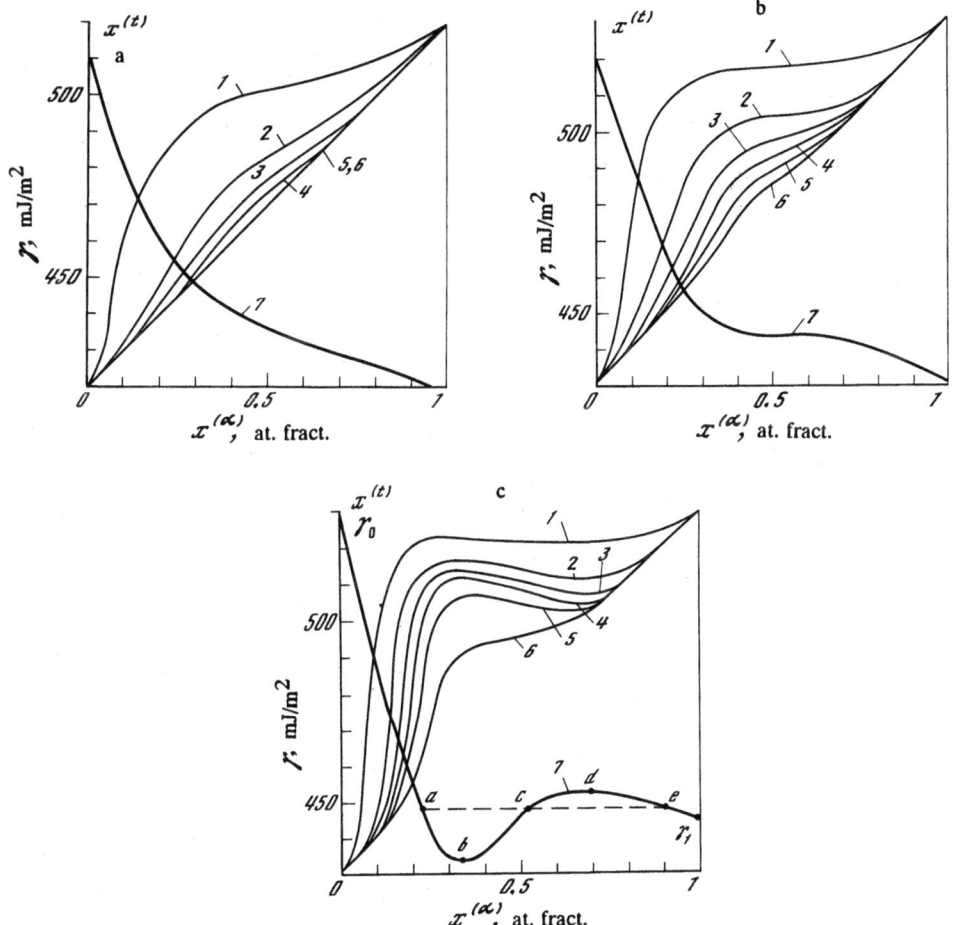

Fig. 3. Surface concentration $x^{(t)}$ in various layers t (1-6) and surface tension γ (7) as functions of bulk concentration $x^{(\alpha)}$ of surface active Pb. $T = 900$ (a), 775 (b), 673 K (c). Layer 1 was grown in vacuum; successive monatomic layers $t = 2, 3, 4, 5, 6$, in solution. The dashed line is a portion of the equilibrium isotherm $\gamma_0 ace\gamma_1$ satisfying Maxwell's rule.

where $v_i = 3^{-1/2} \cdot 4^{1/3}(v_i/N_0)^{-2/3}$, v_i is the atomic volume of the ith component, and N_0 is Avogadro's number.

RESULTS

The concentration distribution over layers is well studied [1, 3, 4]. An example of such a distribution near the surface at three different temperatures is given in Fig. 2. The bulk concentration in all cases was the same

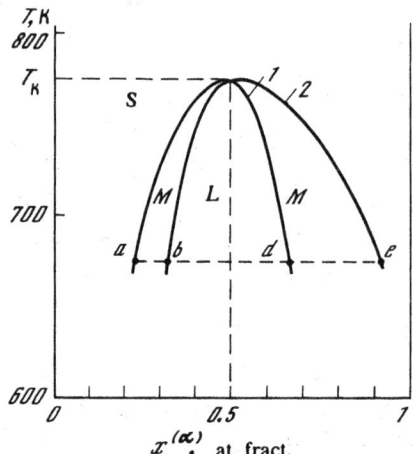

Fig. 4. Regions of different thermodynamic stability of a homogeneous stochastic atomic distribution over layers. S, stable; M, metastable; L, labile states. Curve 1 is a spinodal; curve 2, binodal.

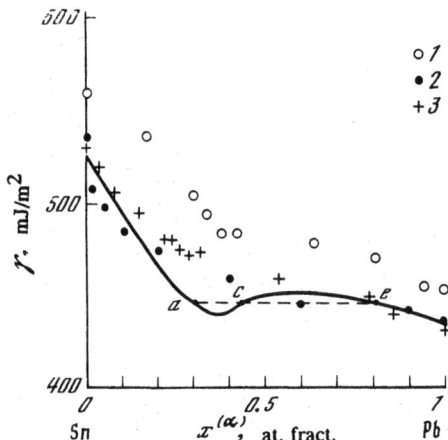

Fig. 5. Comparison of the calculated dependence (solid line) of surface tension γ on bulk concentration $x^{(\alpha)}$ with experimental data for the system Sn–Pb at $T = 723$ K. Data from [15] (1), [16] (2), [17] (3). The dashed line (ace) is a portion of the equilibrium isotherm.

$x^{(\alpha)} = 0.5$ at. fract. Pb. As noted above, not every average and bulk concentration $x^{(\alpha)}$ has a stable state. It is difficult from Fig. 2 to make a conclusion about the stability and, therefore, about the existence of the system state at the selected composition $x^{(\alpha)}$.

The system must be examined at all concentrations for a more complete understanding of its behavior since only by comparing the succession of free energies, the concentration distribution in layers $x^{(t)}$, and the surface tension at various compositions and temperatures can it be seen how stable or achievable is one state or another.

Dependences of γ and $x^{(t)}$ on bulk concentration $x^{(\alpha)}$ at $0 \leq x^{(\alpha)} \leq 1$ for various temperatures are presented in Fig. 3a, b, and c. Three characteristic cases can be arbitrarily identified: 1) $T > T_c$ (Fig. 3a), the interfacial layer is represented by four monolayers whereas the concentration $x^{(t)}$ for the fifth and sixth monolayers are the same as the bulk and there are no features on the surface tension isotherm; 2) $T = T_c$ (Fig. 3b), the interfacial layer consists of six monolayers and inflection points appear in the functions $x^{(t)}[x^{(\alpha)}]$ and $\gamma[x^{(\alpha)}]$; 3) $T < T_c$, the path of the curve for the function $x^{(t)}[x^{(\alpha)}]$ has a clearly anomalous nature in a certain concentration region. The surface tension isotherm $\gamma[x^{(\alpha)}]$ has a van der Waals shape indicating that the stochastic homogeneous atomic distribution is unstable. The set of points b and d (Fig. 3c) on these at various temperatures give a spinodal (Fig. 4). The set of points a and e (Fig. 3c) are positioned using Maxwell's rule [14] and give a binodal (Fig. 4). Thus, the whole temperature—concentration region of existence of the transition layer with a homogeneous stochastic atomic distribution over the sites in each layer is divided by these two curves into three regions of different solution thermodynamic stability: a) stable, all states lie above the binodal; b) metastable, between the binodal and spinodal; c) labile, inside the spinodal.

Experimental values from [15-17] and those calculated for $\gamma[x^{(\alpha)}]$ are compared in Fig. 5. The general path of the theoretical curve clearly agrees well with the general path of the experimental isotherms. Of course if the metastable states ab and de cannot be achieved in the experiment, then states bcd corresponding to the labile region cannot be found under usual conditions. Equilibrium isotherms with a planar portion $\gamma_0 ace\gamma_1$ that satisfy Maxwell's rule are most frequently found from measurements.

DISCUSSION

One of the initial objectives of the work was to calculate the tendency of the liquid in eutectic systems to form inhomogeneities and to stratify. This enabled the exchange energy of Eq. (12) to be specified by the critical temperature. Nevertheless, we did not include inhomogeneity in calculations of the liquid model but did take into consideration possible instability of the homogeneous state. The sensitivity of the surface to a change in the behavior of the system as a whole could be found by this approach. In fact, no less than three monolayers are drawn into the transition zone at temperatures significantly greater than T_c where the liquid acts ideally. Thus, even if the components are completely miscible the transition zone cannot be limited by only one monolayer. An interface of one layer is found only at $T \gg T_c$ and at low concentrations, i.e., in an approximately ideal solution.

It is noteworthy that the concentration of the active component, Pb, at high temperatures in the surface increases smoothly with increasing bulk concentration. A sharp increase in the number of monolayers is seen at $T = T_c$. The transition zone includes all monolayers from the model. Clearly defined anomalies appear at $T < T_c$ (Fig. 3c). These are consistent with a decrease of Pb concentration in the surface layer, although it is surface-active. The anomaly indicates that the surface layers are unstable relative to the homogeneous concentration distribution in them and that the bulk becomes unstable due to this. In fact, descending to the critical temperature T_c and below, it can be seen from Fig. 3a-c that the lower the temperature the greater the number of monolayers in the transition zone. Equations (5), (6), and (10) were solved again (by the first method) to check this fact using a common iteration procedure, as was done in [1]. As it turned out, anomalies arise in the first two layers at $T = T_c$. But rather than distort deeper layers, the system prefers to incorporate a larger number of layers from the bulk into the transition zone. The lower the temperature, the greater the number of these layers added to the transition zone. This is illustrated in Fig. 6 and completely confirms calculations for the six-layered model. In fact, for compositions within the spinodal, no matter how many layers are included, they will be insufficient since the transition encompasses the whole system at once and is initiated by the surface. The latter is also illustrated by the analysis given above of the behavior of the layer concentration as a function of the bulk concentration with decreasing temperature (cf. Fig. 3a-c and Fig. 6). Therefore, T_c obtained for the transition zone (cf. Fig. 4) coincides with that for the bulk at a critical concentration $x_c = 0.5$ at. fract. Stratification in the spinodal decomposition serves only as a qualitative illustration of the incorporation of the bulk into the transition. It should be expected that x_c in the surface will shift due to a nonlinear relation between the bulk and layer concentrations of Eqs. (4) and (6). However, although the binodal is asymmetric, x_c for the transition zone is also $x_c = 0.5$ at. fract. This is caused by the statistical independence of fluctuations of the particle and energy densities within the frame-

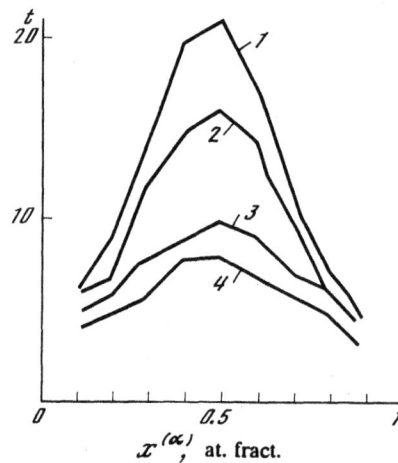

Fig. 6. Increase of number of layers t incorporated into the transition zone where the limitation of Eq. (11) is lifted for various temperatures and concentrations. $T = 673$ (1), 723 (2), 775 (3), 900 K (4).

work of the lattice model selected [18]. Nevertheless, the theoretical dependences within the framework of this model agree satisfactorily with the experimental ones (Fig. 5) and reveal features of the behavior of $x^{(t)}[x^{(\alpha)}]$ and $\gamma[x^{(\alpha)}]$.

For the In–Sn system, analogous calculations gave the functions $x^{(t)}[x^{(\alpha)}]$ and $\gamma[x^{(\alpha)}]$ in Figs. 7 and 8. A unique inversion of the component concentrations in transition layers is seen in this system, i.e., the Sn concentration in the surface increases as the bulk concentration is increased to $x^{(\alpha)} \approx 0.3$ at. fract. As the Sn bulk concentration is increased further, its concentration in the surface layer decreases sharply and the surface is enriched in In. Although the general shape of the isotherms $\gamma[x^{(\alpha)}]$ for both systems is different (cf. Fig. 3a-c and Figs. 7 and 8), the isotherms at $T < T_c$ to a certain extent exhibit van der Waals character, indicating that the stochastic homogeneous distribution over the transition layers is unstable. This is evident in the experiment as a plateau in the functions $\gamma[x^{(\alpha)}]$, in agreement with Maxwell's rule as noted above. Isotherms γ for In–Sn agree well with the experimental function $\gamma[x^{(\alpha)}]$ [19] (cf. Fig. 8). Finally, the isotherms of concentration $x^{(t)}$ as a function of the initial [bulk $x^{(\alpha)}$] near the interface are consistent with a nonlinear relation and a significant difference between them. In other words, the bulk composition cannot be achieved at the interface except at low concentrations. The calculated behavior of the transition zone resulted from a consideration of the tendency of the liquid eutectic systems to stratify. Such a tendency to stratify is a result of a change in the short-range order. This in turn should be reflected in structurally sensitive values, in particular, in the long-wavelength structure factors included by Bhatia and Thornton [20, 21]. Moreover, generation of changes in the short-range order can cause the appearance of clusters consisting primarily of one type of atoms (ions). However, charge neutrality should be maintained in the systems [21] since if such clusters arise due to spontaneous decomposition of a homogeneous solution at T_c then the whole system should redistribute the valence electrons so that possible excess charge is compensated for (shielded). The charge neutrality condition reflecting a change in the short-range order is known to be [21]:

$$(x_B^{(\alpha)} - x_A^{(\alpha)}) + x_B^{(\alpha)} x_A^{(\alpha)} (a_{AA} - a_{BB}) = 0, \tag{13}$$

where a_{ij} are Faber–Ziman structure factors [21] at the long-wavelength limit. Let us express a_{ij} in terms of the Bhatia–Thornton structure factors S_{xx} [20, 21]:

$$a_{AA} = \Phi - x_B^{(\alpha)}/x_A^{(\alpha)} + S_{xx}(\delta - 1/x_A^{(\alpha)})^2,$$

$$a_{BB} = \Phi - x_A^{(\alpha)}/x_B^{(\alpha)} + S_{xx}(\delta - 1/x_B^{(\alpha)})^2, \tag{14}$$

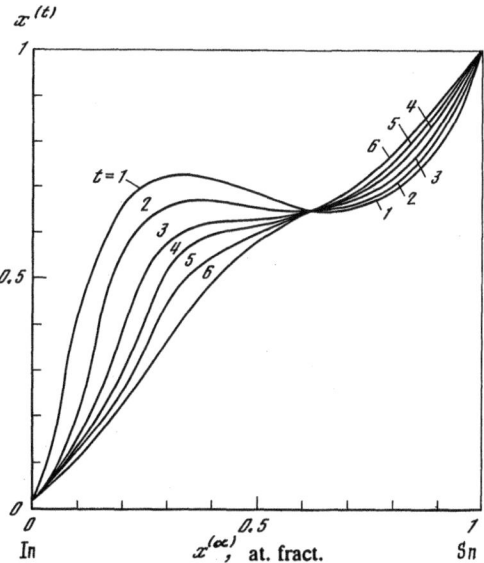

Fig. 7. Surface concentration of Sn $x^{(t)}$ in various layers t (1-6) as a function of bulk concentration $x^{(\alpha)}$ for the system In–Sn. $T = 683$ K.

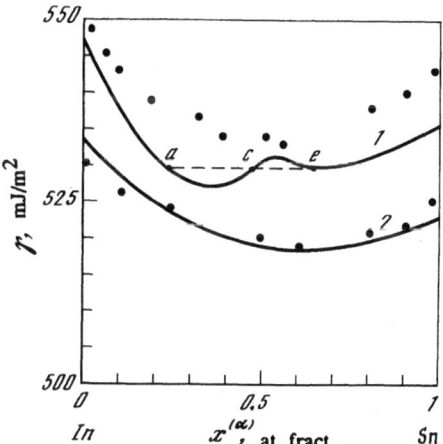

Fig. 8. Comparison of calculated surface tension γ (solid line) with experimental data for the system In–Sn [19]. $T = 673$ (1), 773 K (2). The dashed line (*ace*) is a portion of the equilibrium isotherm.

where $\Phi = N_0 k T \chi_T / V_M$, χ_T is the isothermal compression, $\delta = (v_A - v_B)/(x_A^{(\alpha)} v_A + x_B^{(\alpha)} v_B)$, V_M is the volume of mixing, and v_A and v_B are the atomic volumes of the pure components. Substituting Eq. (14) into Eq. (13), we obtain (for an infinitely large system) a different expression reflecting the change in the short-range order:

$$S_{xx} = [\delta/(1 - 2x^{(\alpha)}) + \tfrac{1}{2} x^{(\alpha)}(1 - x^{(\alpha)})]^{-1} = S_{loc}. \tag{15}$$

The structure factor obtained indicates the presence or absence of rearrangement in the short-range order. Shapes of this S_{loc} for various systems are given as an example in Fig. 9. For example, S_{loc} coincides with the particular behavior of the surface concentration for Ag–Cu (Fig. 10) observed in [22]. The shape of S_{loc} for the eutectic systems is clearly the same as for typically stratified ones but different from the system in which there are no

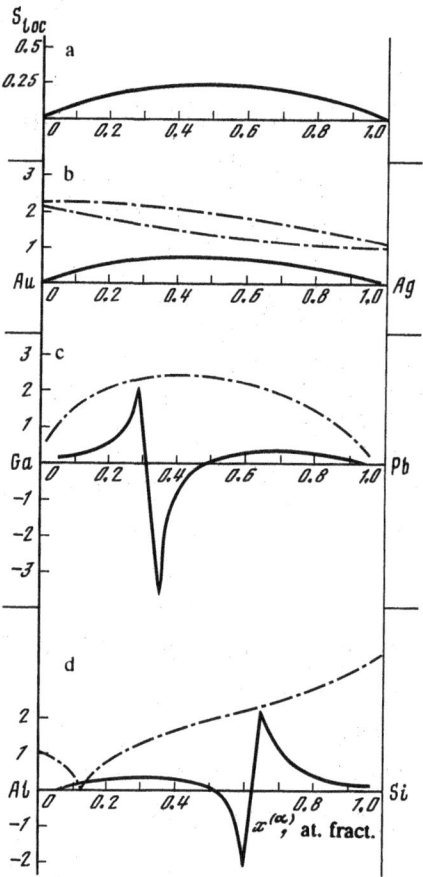

Fig. 9. Shape of the structure factor S_{loc} for various systems: ideal solution (*a*), completely miscible solution, $v_{Au} = 12.5$, $v_{Ag} = 12.5$ (*b*), stratifying system, $v_{Ga} = 11.25$, $v_{Pb} = 22.45$ (*c*), eutectic system, $v_{Al} = 11.25$, $v_{Si} = 6.88$ (*d*). The quantities $v_i \cdot 10^{-3}$ m^3/(kg·at) are atomic volumes of pure components. The solid line is S_{loc}; the dashed line, a diagram of aggregate equilibria.

changes in the short-range order (no stratification) (Fig. 9*a* and *b*). Thus, it can be assumed that the tendency for eutectic systems to stratify is rather general and the same as that for typically stratifying systems. However, since the bulk states of the system are inextricably related to the surface states that limit the volume, it can be expected that the van der Waals shape of the theoretical surface tension isotherms is characteristic for any eutectic system.

The demonstrated relation of bulk and surface states suggests the following conclusions:

1. Within the framework of the model used, the interface in a system with a stochastic distribution of atomic components is composed of 3-4 layers. This is applicable to seemingly ideal solutions at temperatures greater than the critical temperature for solution stratification and to solutions in which the short-range order does not rearrange.

2. Calculation of the rearrangement of the short-range order in the liquid bulk leads to regions of different thermodynamic stability for the homogeneous stochastic distribution of atomic components over sites of the quasilattice of the transition zone. The regions are limited by a binodal and spinodal.

3. A homogeneous surface state retaining the initial stoichiometry can be achieved at infinitely low concentrations. For metastable states, a homogeneous state can be expected to be retained at the surface but with a composition different from the initial one (bulk). In labile states (compositions within the spinodal), the system

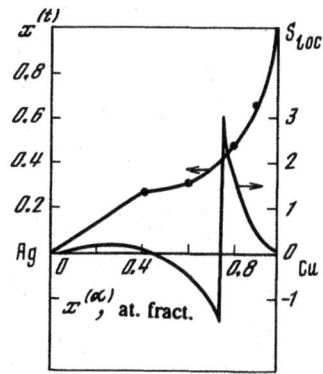

Fig. 10. Comparison of S_{loc} with the dependence of surface $x^{(t)}$ on bulk $x^{(\alpha)}$ concentration for the system Ag—Cu. Points are experimental values from [22].

will strive to form clusters. The concentration inhomogeneities, clusters, arising will emerge at the interface and determine the local composition and properties.

REFERENCES

1.   S. Ono and S. Kondo, *Molecular Theory of Surface Tension in Liquids* [Russian translation], Izd. Inostr. Lit, Moscow (1963).
2.   J. E. Lane, "A multilayer model of solid-regular solution interface," *Aust. J. Chem.*, **21**, No. 4, 827-851 (1968).
3.   N. A. Smirnova and E. M. Piotrovskaya, "Investigation of surface properties of ternary alloys within the framework of a multilayered lattice model," in: *Surface Properties of Alloys*, Naukova Dumka, Kiev (1982), pp. 11-22.
4.   W.-H. Shih and D. Stroud, "Two-component lattice-gas model for surface segregation in liquid alloys," *Phys. Rev. B: Condens. Matter*, **33**, No. 12, 8048-8052 (1986).
5.   A. A. Chernov, E. I. Givargizov, Kh. S. Bagdasarov, et al., *Modern Crystallography*, Vol. 3 [in Russian], Nauka, Moscow (1980).
6.   Yu. K. Tovbin, "Concentrational boundary profile of a regular solution for a number of molecular-interaction models," *Poverkhnost*, No. 6, 33-45 (1985).
7.   B. P. Demidovich and I. A. Maron, *Principles of Calculational Mathematics* [in Russian], Nauka, Moscow (1970).
8.   V. M. Zalkin, *Nature of Electrical Alloys and the Contact Melting Effect* [in Russian], Metallurgiya, Moscow (1987).
9.   J. L. Murray, "Calculations of stable and metastable equilibrium diagrams of the Ag—Cu and Cd—Zn systems," *Metall. Trans. A*, **15**, No. 1, 261-268 (1984).
10.   E. V. Kalashnikov, "On the state of a binary liquid system with a eutectic point," *Zh. Fiz. Khim.*, **55**, No. 6, 1416-1424 (1981).
11.   E. L. Demina, V. I. Sakovich, Yu. P. Mukhachev, and P. S. Popel', "Microstratification region of liquid binary alloys Sn—Pb," in: *Features of Structure Formation of Eutectic Alloys* [in Russian], Proceedings of the Third All-Union Conf., Part 1, Dnepropetrovsk (1986), pp. 151-153.
12.   E. V. Kalashnikov, "Regions of different thermodynamic stability in a liquid in relation to crystal growth and regular structures," in: *Relationship of Liquid and Solid Metallic States* [in Russian], Abstracts of Papers, Sverdlovsk (1987), p. 90.
13.   V. I. Nizhenko and L. I. Floka, *Surface Tension of Liquid Metals and Alloys* [in Russian], Metallurgiya, Moscow (1981).
14.   L. D. Landau and E. M. Lifshits, *Statistical Physics*, Part 1 [in Russian], Nauka, Moscow (1976).
15.   V. Somol and M. Boranec, "Surface tension of molten Pb—Sb and Pb—Bi alloys," *Sb. Vys. Sk. Chem.-Technol. Praze, Anorg. Chem. Technol. B*, **30**, 199-206 (1984).
16.   S. I. Popel', V. N. Kozhurkov, and G. V. Zakharova, "Density and surface tension of Pb—Sn alloys and their adhesion to iron," *Zashch. Met.*, **7**, No. 4, 421-426 (1971).
17.   Kh. I. Ibragimov, N. L. Pokrovskii, P. P. Pugachevich, and V. K. Semenchenko, "Study of the surface tension of the systems Sn—Bi and Sn—Pb," in: *Surface Effects in Alloys and Solid Phases Arising in Them* [in Russian], Kabard.-Balkar. Kn. Izd., Nal'chik (1965), pp. 269-276.
18.   A. Z. Patashinskii and V. L. Pokrovskii, *Fluctuation Theory of Phase Transitions*, 2nd Edn. [in Russian], Nauka, Moscow (1982).
19.   R. Kh. Dadashev, Kh. I. Ibragimov, and S. É. Yushaev, "Investigation of surface properties of alloys of the ternary system In—Sn—Pb," in: *Physics of Surface Effects in Alloys* [in Russian], Chech.-Ing. Univ., Groznyi (1977), pp. 129-135.
20.   A. B. Bhatia and D. E. Thornton, "Structural aspects of electrical resistivity of binary alloys," *Phys. Rev. B: Solid State*, **2**, No. 8, 3004-3012 (1970).
21.   S. Tamaki, "Charge distribution in liquid metals and alloys," *Can. J. Phys.*, **65**, No. 3, 286-308 (1987).
22.   C. Norris, "Photoelectron spectroscopy of liquid metals and alloys," in: *Liquid Metals*, Third International Conf. on Liquid Metals, Inst. of Phys. Bristol, London (1976), pp. 171-180.

QUANTUM-CHEMICAL INVESTIGATION OF ADSORPTION
AND SURFACE MIGRATION OF ATOMS AND MOLECULES ON
Si(111) AND Si(100) SURFACES

M. P. Ruzaikin and A. B. Svechnikov

Characteristics of interatomic interactions and structural features in the phase contact region are of great interest in studying growth processes of crystals and epitaxial layers of semiconductors. Two phases come into contact through an adsorption layer in gas-transport systems and during molecular-beam growth. Processes occurring in the adsorption layers (adsorption, desorption, surface diffusion, possible chemical reactions) are known to control the crystallization rate and the quality of the crystals and epitaxial layers grown. Investigations of thermodynamic properties and kinetic processes occurring in adsorption layers require knowledge of the adsorption energy, the surface potential relief relative to adparticles, and changes in the geometric and energetic characteristics of molecules on adsorption [1-3].

The energetic and geometric (bond lengths and angles) characteristics of atoms and molecules adsorbed on crystal faces can be found by accurately using a solid cluster model and quantum-chemical methods for calculating electronic structure [4]. However, the structural and energetic characteristics of particles adsorbed on surfaces of widely used semiconductors have not been described systematically or analyzed in the literature.

In the present work, results are presented of quantum-chemical investigations of adsorption of H, Si, and Cl atoms and molecules consisting of these atoms on Si(111) and Si(100) surfaces.

Cluster Approximation. Calculation Method. The cluster approximation has been most widely applied to theoretical study of local crystal properties [4-6]. The essence of the method is as follows. A collection of atoms (cluster) adjoining the local defect studied (an impurity atom, vacancy, or adsorbed atom or molecule) is isolated from the crystal. It is assumed that the boundary atoms in this cluster are located in positions corresponding to those in the perfect crystal whereas the inner atoms can be shifted relative to these positions. Features of the local defect studied control the atomic shifts. The rate of decrease of the effect of the local defect on the crystal atoms as the defect is removed from the crystal and the requirement that this effect on the cluster boundary atoms be small are considered in choosing the cluster size.

The cluster boundary atoms have broken bonds. A procedure that artificially saturates the broken bonds with pseudoatoms is used to minimize their effect on the calculations for ionic—covalent crystals [7]. The parameters of the pseudoatoms are selected such that a uniform electron density distribution is achieved in the cluster and at its boundary. A special case of such an approach is the use of H atoms as pseudoatoms [4, 5].

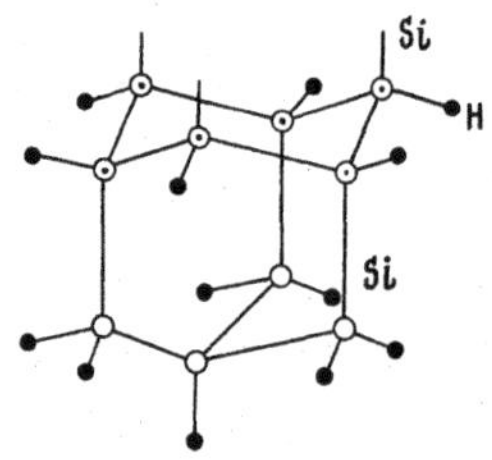

Fig. 1. Structure of the $Si_{10}H_{13}$ cluster.

In the present work, clusters containing up to 15 Si atoms are examined to study chemisorption of atoms and molecules on the Si(111) and Si(100) surfaces. As an example, the cluster $Si_{10}H_{13}$ is shown in Fig. 1. In this cluster 6 Si atoms (denoted by points in Fig. 1) model the Si(111) surface, 4 model the crystal bulk, and 13 H atoms are included to saturate broken bonds of boundary atoms.

The adsorption energies E_{ad} of atoms and molecules were determined by the formula

$$-E_{ad} = E_2 - (E_1 + E_{a,M}),$$

where E_1 is the total cluster energy without adparticles (atoms or molecules), E_2 is the total cluster energy with adparticles, and $E_{a,m}$ is the total energy of isolated atoms and molecules. The total energies of clusters and molecules were calculated by the quantum-chemical method MINDO/3 [7]. Geometric characteristics of adsorbed atoms and molecules were optimized using the Davidon—Fletcher—Powell method [8, 9]. Calculations were performed using the program "Cluster 1" developed by Zakharov and Litinskii [10].

Adsorption of H, Si, and Cl Atoms on Si(111). The symmetric points A, B, C, D, and E (Fig. 2) are possible sites of atomic adsorption. The A sites are located over the surface atoms (single-center adsorption sites); B and C, over centers of triangles with surface atoms at their apices (three-center adsorption sites); E, over the midpoints of lines joining surface atoms (two-center adsorption sites). The B and C adsorption sites differ in the placement of the inner surface layer atom.

The atomic energies were calculated in different positions relative to the surface along the $ABCA$, AEA, and BDB directions. Analysis of the potential surface cross sections obtained this way indicates that the sites most suitable for H atom adsorption are the single-center sites (A positions). Adsorption at the three-center sites (B and C positions) is unfavorable. A covalent Si—H σ-bond is formed on adsorption at the A position. Overlap of the $P_z(Si)$—$S(H)$ and $S(Si)$—$S(H)$ orbitals is mostly responsible for the bond. The bond energy of the H atom to the surface is 291.1 kJ/mole. The bond length is 0.149 nm. The effective atomic charge is $-0.029e$.

The two-center position E corresponds to a saddle point on the potential surface ($E_{ad} = 1.45$ kJ/mole) with decreases in the direction of the A points. This indicates that migration of an atom along the surface from the stable A position into an analogous neighboring position is possible only through the E position. An energy barrier of 293.6 kJ/mole must be overcome to accomplish this.

Analysis of the two-center interaction energies in the cluster indicates that adsorption of H at the A position shortens the Si—Si bond in the cluster. For adsorption at the B and C positions, the Si—Si bond in the cluster is weakened. This is one reason for the decrease in total cluster energy and eventually the decrease of adsorption energy.

The calculated bond lengths and adsorption energies for single-center adsorption agree well with the experimental [11, 12] and theoretical [12, 13] values. The good agreement of the Si—H bond lengths and energies obtained with the corresponding values in silane [14] should be noted.

Calculated adsorption energies E_{ad}, equilibrium distances r_0, and effective charges q of adsorbed Si atoms on Si(111) in the A, B, C, and E positions are presented in Table 1 [15]. As seen from Table 1, the sites most favorable for Si adsorption are the three-center B and C positions. A Si adatom at the two-center position lies at

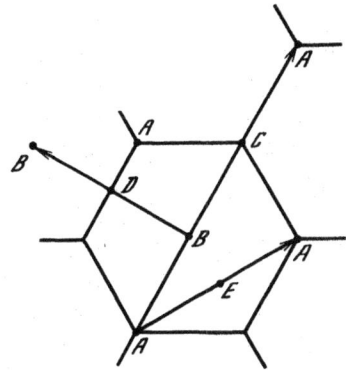

Fig. 2. Atomic adsorption sites on the Si(111) surface.

a saddle point of the potential surface. An adatom can shift from the B position to the C position through the E position by overcoming an energy barrier of 19.4 kJ/mole. For shifting from the C into the B through the E position, an activation barrier of 17.2 kJ/mole must be overcome.

The atomic orbitals P_x and P_y of the adatom (a) and S and P_z of surface atoms (s) with the overlap $S(Si_s)-P_{x,y}(Si_a)$ and $P_z(Si_s)-P_{x,y}(Si_a)$ are involved in bond formation on adsorption of Si atoms at the three-center sites. The significant effective charges of Si adatoms are explained by a shift of electrons into the surface layer, toward the adatom. The interaction between surface atoms decreases whereas that with neighboring atoms of the cluster bulk decreases.

A Si adatom in the A position is located at a local minimum of the potential surface. The bond of the adatom with the Si surface atom is similar in nature to a Si—Si bond in the crystal bulk. However, the adsorption energy agrees well with the corresponding value obtained by a crystal chemical approximation [1, 2].

Results calculated for adsorption of Cl atoms on Si(111) [16] are presented in Table 2. Analysis of the dependence of E_{ad} on the adsorption position suggests that the sites of primary Cl atom adsorption are the single-center A positions. In the E position, the adatom is located at a saddle point in the potential surface. Therefore, a Cl adatom shifts from one A position to the next by going through the E position and overcoming an energy barrier of 53.8 kJ/mole. The Si—Cl bond in the single-center position is due to the atomic overlaps $S(Si)-P_z(Cl)$ and $P_z(Si)-P_z(Cl)$. The interatomic interaction energy in the crystal bulk is slightly increased.

The value E_{ad} given above for Cl atoms agrees with the estimates in [1, 2]. The calculated equilibrium bond length r_0 is slightly greater than the values obtained experimentally [17] and by nonempirical calculations [18].

Adsorption of H, Si, and Cl Atoms on Si(100). The atomic adsorption sites possible on the intact Si(100) surface are the points A, B, C, and D. The point A is located over a surface atom (single-center adsorption position); B and C, over midpoints of lines joining surface atoms; D, over the center of squares of surface atoms (Fig. 3). The point B lies over a Si atom of the first inner layer. The potential surface cross section for adatom interactions with the surface were calculated along the AC, AB, and AD directions.

The potential surface cross section for H adatoms was analyzed in [19]. The most favorable site is the single-center A position. The adsorption energy here is 266.3 kJ/mole. The bond length is 0.147 nm. The effective charge of the adatom is 0.022e. The H adsorption energy in the A position is less than the Si—H bond energy in silane (304 kJ/mole [14]). This is consistent with incomplete bond saturation. There is a weak dependence of the H adsorption energy on position along AC in the vicinity of the point A. This is due to the fact that increasing the deviation from A increases the contribution to the bond energy from overlap of the $S(H)$ and $P_x(Si)$ orbitals such that the sum of the off-diagonal bond-order matrix elements for the $S(H)-S(Si)$, $S(H)-P_x(Si)$, and $S(H)-P_z(Si)$

Table 1

Adsorption position	E_{ad}, kJ/mole	r_0, nm	q, e
A	212.1	0.229	0.094
B	376.6	0.130	0.497
C	378.4	0.132	0.512
D	357.2	0.155	0.474

Table 2

Adsorption position	E_{ad}, kJ/mole	r_0, nm	q, e
A	384.8	0.209	−0.246
B	193.1	0.095	0.227
C	145.7	0.100	0.178
E	331.0	0.120	0.049

bonds changes little. Analysis of the two-center energies in the cluster showed that adsorption of H atoms at the A position strengthens the Si—Si bond in the cluster. The Si—Si bond in the cluster weakens on adsorption at the B or C position. This is a reason for the unfavorable energetics of adsorption at these positions.

The inability of H adatoms to shift from one A position into another through B indicates that surface H migrates only along atomic surface rows ...ACA... The activation energy for the shift from A to A through C is 73.5 kJ/mole.

Characteristics calculated for a Si adatom on Si(100) at points A, B, C, and D are presented in Table 3 [19]. It can be seen from Table 3 that the primary positions for Si adsorption are the two-center C positions. Here the Si adatom is bonded by two bonds to surface atoms so that the two surface valences are saturated. The bond energy of each of these is 202.7 kJ/mole. This is 28.3 kJ/mole less than the Si—Si bond energy in the crystal bulk (231 kJ/mole [20]). The Si—Si bonds in the cluster bulk weaken slightly on adsorption of atoms at all positions. The A position, corresponding to single-center Si adsorption, is a saddle point on the potential surface with decreases in the direction of the neighboring points C. Therefore, adatoms can diffuse along the line ...CAC... An energy barrier of 58.9 kJ/mole must be overcome. A path through point D (...CDC...) is also possible with activation energy 78 kJ/mole.

Results calculated for chemisorption of Cl atoms on Si(100) are presented in Table 4. Like Si adatoms, Cl atoms on the surface slightly weaken the interaction between Si atoms in the cluster. The Cl adatoms can shift from the preferred position to a neighboring one through the single-center A position by overcoming an energy barrier of 159.6 kJ/mole. Migration of Cl adatoms in directions different from ...CAC... is less probable due to large activation energies.

Adsorption of H, Si, and Cl Atoms on the Si(100)-2 × 1 Surface. Three conditions must be considered in choosing a cluster model for the Si(100)-2×1 surface. First, the structure of the reconstructed surface has bonds in several atomic layers adjoining the crystal surface that are significantly distorted from the ideal ones. Second, Si—Si dimers with high stability relative to atomic adsorption are found on the surface. Third, the geometry of practically the whole cluster should be optimized when calculating chemisorption of atoms on this surface. In the present work, the cluster Si_9H_{12} was used to study the Si(100)-2×1 surface and chemisorption of atoms on it. In this cluster, two Si atoms model the surface and can form a dimer whereas seven atoms belong to three atomic planes in the crystal bulk.

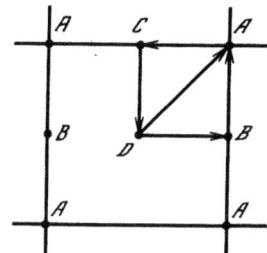

Fig. 3. Atomic adsorption sites on the Si(100) surface.

Table 3

Adsorption position	E_{ad}, kJ/mole	r_0, nm	q, e
A	346.6	0.204	0.161
B	252.1	0.110	0.364
C	405.5	0.131	0.103
D	327.5	0.025	0.324

Table 4

Adsorption position	E_{ad}, kJ/mole	r_0, nm	q, e
A	347.4	0.205	−0.133
B	108.8	0.119	−0.312
C	507.0	0.106	−0.077
D	245.3	0.058	−0.270

The optimized geometry of the Si_9H_{12} cluster with initial atomic positions corresponding to the ideal Si(100) surface demonstrated that neighboring surface atoms are repelled near their crystallographic positions. They must be drawn together to form a bond. The bond deviation angle at which dimerization begins is three degrees. At this angle, an energy barrier of 0.9 kJ/mole is overcome. Further optimization of the geometric position of surface cluster atoms leads to formation of a dimer corresponding to the asymmetric dimer model [22]. The bond length in the dimer is 0.232 nm. The cluster energy gain on forming the dimer is 96.5 kJ/mole. The two-center interaction energy between dimer atoms is 901.5 kJ/mole. This indicates formation of a strong covalent bond. The atoms forming the dimer have different effective charges. One atom is negatively charged ($-0.154e$); the other, positively ($0.274e$). The calculated bond length in the dimer agrees well with that from [23]. The dimer formation energy agrees with the data of [24, 25].

The positions of adatoms were determined by optimizing all bonds in the cluster and the distance from the adatom to the dimer. The adsorption energy E_{ad} for successive adsorption of atoms on the dimer was determined using the formula

$$-E_{ad} = E_{k+1} - (E_k + E_a),$$

where E_k and E_{k+1} are the energies of clusters containing k and $k + 1$ adatoms, respectively, and E_a is the energy of an isolated atom.

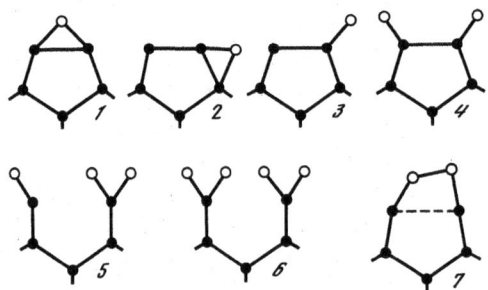

Fig. 4. Atomic adsorption positions on the Si(100)-2×1 surface.

Table 5

Adsorption position	3	4	5	6
E_{ad}, kJ/mole	297.2	297.0	176.0	256.7

The H adsorption energies on the dimer of the Si(100)-2×1 surface for the positions shown in Fig. 4 are presented in Table 5.

Adsorption of one H atom on the dimer (Fig. 4, *3*) leads to formation of a symmetric dimer with a Si—Si bond length 0.243 nm. Adsorption of a H atom in the center of the dimer is unfavorable since this requires dissociation of the dimer. The optimization procedure in this case shifts the adatom to the edge of the dimer. The effective charges of the dimer atoms become 0.04e. The energy of the two-center interaction between dimer atoms decreases to 137.9 kJ/mole.

A Si—H bond equivalent to the first is formed by adsorption of a second H atom on the opposite edge of the dimer (Fig. 4, *4*). The bond strength in the dimer does not change. An attempt to adsorb two H atoms simultaneously as an H_2 molecule leads to dissociation of the molecule and formation of two Si—H bonds with a configuration equivalent to successive adsorption of two H atoms.

The Si—Si bond of the surface dimer is broken on adsorption of a third H atom (Fig. 4, *5*). The adsorption energy of the third atom is much less than that of the first and second atoms. This is mainly due to the necessity of dissociating the dimer on adsorption. Adsorption of a fourth H atom forms a Si—H bond very similar to the previous one. Adsorption of an H_2 molecule on a dimer with two adsorbed H atoms was calculated. The adsorption is accompanied by dissociation of H_2 and the dimer. A significant energy barrier (of the order of 540 kJ/mole) is overcome. Associative desorption of two H atoms to form H_2 requires that an energy barrier of about the same magnitude be overcome.

The data presented above indicate that the principal forms of H adsorption on contact of the Si(100)-2×1 surface with gaseous H_2 at various pressures and temperatures will be different. At low pressures and (or) sufficiently low temperatures, H_2 should be adsorbed as in Fig. 4, *4*. At high pressures and (or) temperatures, it takes the form shown in Fig. 4, *6*. In the literature, such forms of H_2 adsorption are known as the monohydride and dihydride phases. Their existence was verified experimentally in [26-28]. The high activation energy of adsorption on the dimer with two H atoms indicates that the monohydride phase is stable. This agrees with the experimental data [28].

Possible sites of Si adsorption on the dimer of the Si(100)-2×1 surface at which two (Fig. 4, *1*) and three (Fig. 4, *2*) bonds are formed are shown in Fig. 4. For Si adsorption over the center of the dimer (cf. Fig. 4, *1*), two stable bonds with the dimer atoms are formed (E_{ad} = 373.2 kJ/mole). The bond in the dimer is substantially weakened (the energy of the two-center interaction decreases to 212 kJ/mole). The bond length increases to 0.309 nm. On adsorption of Si on the dimer edge, the adatom forms one bond to a Si atom of the dimer and two bonds to atoms of the subsurface layer. The adsorption energy at this position is 317.9 kJ/mole. The dimer bond is not substantially weakened by formation of only one bond of the Si adatom to the dimer (the energy of the two-center interaction decreases to 885.1 kJ/mole, the bond length decreases by 0.0044 nm). The bonds between the dimer atom and atoms of the subsurface layer are strongly deformed by formation of two bonds between the Si adatom and atoms of the subsurface layer. The bond lengths increase to 0.275 nm. The energy of the two-center interaction decreases to 265.2 kJ/mole.

Of the two possible positions for Si adsorption on the dimer, that over the center of the dimer is energetically favorable. However, the adsorption position on the edge of the dimer is stable due to the significant energy barrier separating these two positions. The substantial magnitude of this barrier is due to the necessity of breaking two bonds of the adatom to the subsurface layer and the weakening of the bond within the dimer on going from the edge of the dimer to the position over the center.

Use of the optimization procedure to bond a second Si atom to the previous structures leads to deactivation through formation of a Si_2 molecule adsorbed on the dimer (Fig. 4, 7). The two-center energy in this molecule is 1204.7 kJ/mole. The interatomic distance is 0.207 nm. The atoms of the molecule formed in this manner are strongly bonded to the dimer atoms (the energies of the two-center interaction are 978.4 and 704.0 kJ/mole). The bond in the dimer in this case is considerably weakened.

Positions for Cl adsorption on the dimer of the Si(100)-2×1 surface obtained by optimizing the cluster geometry are shown in Fig. 4. The adsorption energy is 312.0 kJ/mole for adsorption over the center of the dimer (cf. Fig. 4, 1). The Cl atom forms two nonequivalent bonds to the dimer (the energies of the two-center interaction are 284 and 690 kJ/mole, the bond lengths are 0.246 and 0.221 nm). The bond in the dimer is weakened to the value of the two-center interaction energy 769 kJ/mole. The position on the edge of the dimer is energetically more favorable for adsorption. Here the adsorption energy is 356.7 kJ/mole. The bond is similar in characteristics to the Si—Cl bond formed by adsorption of Cl on Si(111). Adsorption of a second Cl atom (E_{ad} = 353.2 kJ/mole) results in the configuration shown in Fig. 4, 4. A new Si—Cl bond equivalent to the first is formed. The bond characteristics in the dimer do not change.

Adsorption of four Cl atoms on the surface dimer breaks the bond in the dimer. Four equivalent Si—Cl bonds are formed (Fig. 4, 6). The change from two adsorbed Cl atoms on the dimer to four has an activation energy close to the formation energy of the dimer.

From the data presented it follows that two stable configurations are formed by adsorption of Cl atoms on Si(100)-2×1 (Cl—Si—Si—Cl and Cl—Si—Cl...Cl—Si—Cl). This agrees with the hypothesis about Cl adsorption positions made on the basis of photoemission spectra [29].

Adsorption of Molecules on the Si(111) Surface. A Quantum-Chemical Calculation for Adsorption of the Molecules HCl, SiH, SiCl, $SiCl_2$, $SiCl_3$, $SiHCl_3$, and $SiCl_4$ was carried out. The height of the molecular center of mass over the surface and all of its geometric characteristics (bond lengths and angles and dihedral angles) were optimized in order to determine the sites and energies of adsorption on the Si(111) surface. Possible configurations of the admolecules on the Si(111) surface are shown in Fig. 5. The adsorption energies (kJ/mole) in various configurations are presented in Table 6. More complete data for the geometric characteristics of admolecules on Si(111) and for the energies of the two-center interaction between atoms of the admolecules and surface can be found in [16].

Two possible cases arise from a calculation of the adsorption energy of HCl with optimized geometric characteristics. In the first case (Fig. 5, 1) the molecule strives to be adsorbed on two centers. However, this results in complete dissociation with formation of Si—H and Si—Cl bonds characteristic of single-center adsorption of H and Cl atoms. In the second case the molecule is adsorbed as depicted in Fig. 5, 2. The bond energy of the

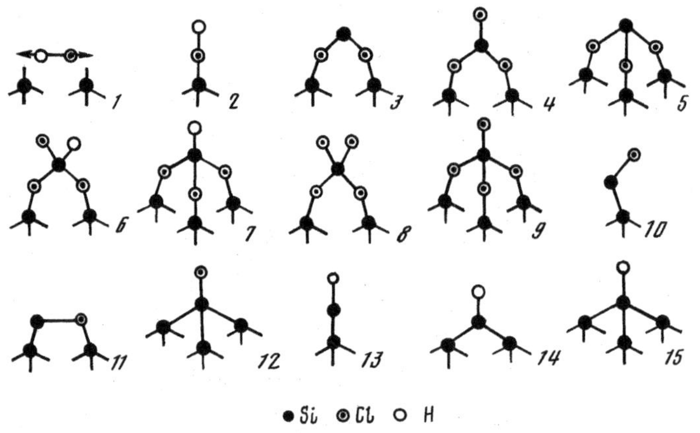

Fig. 5. Configurations of molecules adsorbed on the Si(111) surface: 1, 2) HCl; 3) $SiCl_2$; 4, 5) $SiCl_3$; 6, 7) $SiHCl_3$; 8, 9) $SiCl_4$; 10-12) SiCl; 13-15) SiH.

Table 6

Adsorbed molecules	$K = 1$	$K = 2$	$K = 3$
HCl	36.4 (2)	–	–
$SiCl_2$	–	205.8 (3)	–
$SiCl_3$	–	384.9 (4)	435.1 (5)
$SiHCl_3$	–	403.1 (6)	396.8 (7)
$SiCl_4$	–	451.9 (8)	487.2 (9)
SiCl	265.5 (10)	341.0 (11)	295.4 (12)
SiH	147.3 (13)	185.1 (14)	243.4 (15)

Note. The quantity K is the number of surface atoms involved in bond formation to admolecules. The numbers of the configurations in Fig. 5 are shown in parentheses.

molecule to the surface is small. The H—Cl bond length and the distance between Si and Cl atoms are 0.126 and 0.240 nm, respectively. The difference between the effective charges of the atoms in the molecules decreases sharply. The charge on the Cl atom falls from −0.226 to −0.171e; on the H atom, from 0.226 to 0.082e. This fact indicates that adsorption of HCl is accompanied by electron transfer from the surface atom to the Cl atom and by bond formation.

The molecules $SiCl_2$, $SiCl_3$, $SiHCl_3$, and $SiCl_4$ are adsorbed by forming bonds with two and three surface atoms (Fig. 5, 4-9). The large adsorption energies of $SiHCl_3$ and $SiCl_4$, which do not have free valence orbitals, should be noted. It follows from an analysis of the bond-order matrices and overlap integrals that adsorption is effected by forming σ-bonds between P_z atomic orbitals of the Cl atom and the Si surface atom. The σ-bond in the admolecule weakens considerably whereas the π-bonds are somewhat stabilized. As a result, the Si_s—Cl bond lengths (Si_s is a Si surface atom) are shorter than the Si_m—Cl bond lengths (Si_m is a Si molecular atom). The energies of the two-center interaction in the Si_s—Cl bonds are greater than those in the Si_m—Cl bonds. This means that the Cl atoms are not localized in the admolecules but are intermediate between adatoms and component parts of the molecule. Such adsorbed molecules should be viewed as surface complexes containing Cl atoms bonded strongly to the surface. However, bonds in the remaining parts of the molecule are not broken. The complexes can

Table 7

Adsorbed molecules	$K = 1$	$K = 2$	$K = 4$
HCl	69.3 (2)	163.2 (3)	–
SiH	236.1 (4)	357.7 (5)	–
SiCl	209.3 (6)	288.4 (7)	–
		269.3 (8)	
		245.8 (9)	
SiCl$_2$	129.3 (10)	56.5 (11)	–
		47.5 (12)	
SiCl$_3$	224.5 (13)	168.5 (14)	351.7 (16)
		163.0 (15)	363.1 (17)
SiHCl$_3$	192.7 (18)	158.5 (19)	363.8 (21)
		178.4 (20)	
SiCl$_4$	258.2 (22)	165.4 (23)	347.3 (25)
		170.5 (24)	314.6 (27)
		186.3 (26)	430.6 (28)

Note. The quantity K is the number of surface atoms involved in bond formation to admolecules. The numbers of the configurations in Fig. 6 are given in parentheses.

dissociate to form adsorbed Cl atoms. On the other hand, the complexes can be viewed as resulting from adsorption of unsaturated molecules containing Si onto adsorbed Cl atoms. The situation is analogous for adsorption of SiCl$_2$ and SiCl$_3$ but with a weaker Si$_s$—Cl bond.

The SiCl molecule during adsorption is strongly bonded to two surface atoms (Fig. 5, 11). The intramolecular Si—Cl bond is noticeably weakened. Adsorption of SiCl with formation of only Si$_s$—Si$_m$ bonds is less favorable. The SiH molecule is adsorbed mainly at the three-center position (Fig. 5, 15). The two-center position (Fig. 5, 14) can be viewed as an intermediate position during a jump from one three-center position into another.

Adsorption of Molecules on the Si(100) Surface. The adsorption energy was calculated for the molecules H$_2$, HCl, SiH, SiCl, SiCl$_2$, SiCl$_3$, SiHCl$_3$, and SiCl$_4$. The sites and adsorption energies on the Si(100) surface were determined by optimizing the height of the center of mass of the admolecule and all of its geometric characteristics. The calculated adsorption energies (kJ/mole) of molecules in various configurations on the Si(100) surface are presented in Table 7. The surface configurations corresponding to the energies in Table 7 are shown in Fig. 6. The calculation demonstrated that H$_2$ is adsorbed in a single-center position (Fig. 6, 1) and that it dissociates into two H adatoms. The adsorption energy is 175.6 kJ/mole. The sites of HCl adsorption are the single- and two-center positions. The Cl atom bonds to the surface. The intramolecular HCl bond changes slightly during adsorption. However, the molecule as a whole acquires a charge of −0.119e.

The most favorable adsorption sites for SiH, SiCl, SiCl$_2$, SiCl$_3$, SiHCl$_3$, and SiCl$_4$ are sites at which the molecules form bonds with two (SiH, SiCl, and SiCl$_2$) and four (SiCl$_3$, SiHCl$_3$, and SiCl$_4$) surface atoms. The molecules containing Cl are adsorbed with formation of bonds involving Cl atoms of the molecule and Si surface atoms. The intramolecular Si—Cl bonds are considerably weakened. This indicates that the Cl atoms forming bonds to the Si surface atoms are not localized in the admolecules but are intermediate between adatoms and molecular atoms. Such adsorption of molecules containing Cl is similar to adsorption of molecules on the Si(111) surface. It differs in that the molecules are much more deformed on adsorption on Si(100) than on Si(111). This is due to a substantial misfit between the surface geometry (especially for the four-center adsorption) and the molecules.

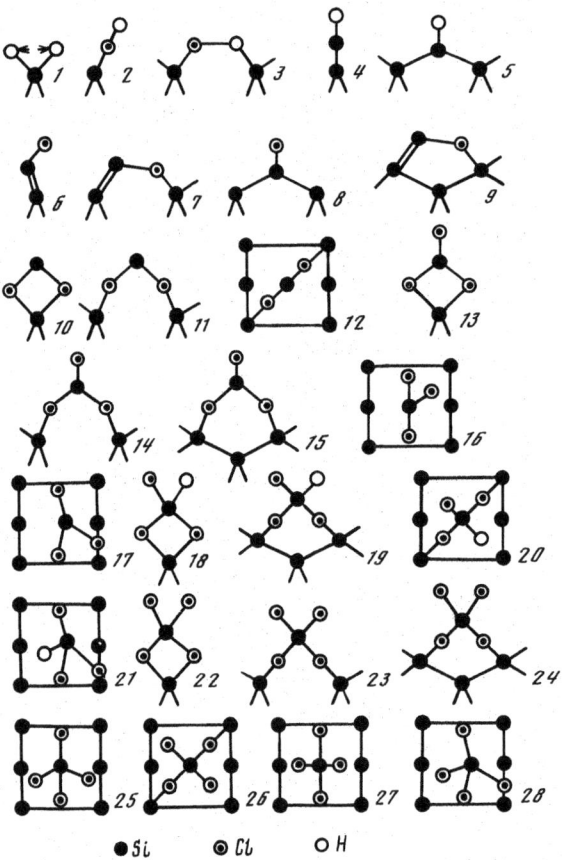

Fig. 6. Configurations of molecules adsorbed on the Si(100) surface: 1) H_2; 2, 3) HCl; 4, 5) SiH; 6-9) SiCl; 10-12) $SiCl_2$; 13-17) $SiCl_3$; 18-21) $SiHCl_3$; 22-28) $SiCl_4$.

REFERENCES

1. A. A. Chernov and N. S. Papkov, "Adsorption layer and formation of nuclei during crystallization in the system Si—H—Cl," *Dokl. Akad. Nauk SSSR*, **228**, No. 5, 1083-1086 (1976).

2. A. A. Chernov, M. P. Ruzaikin, and N. S. Papkov, "Surface processes of adsorption and gas-phase epitaxy of semiconductors (GaAs, InAs, Si)," *Poverkhnost*, No. 2, 94-108 (1982).

3. A. A. Chernov and M. P. Ruzaikin, "Equilibrium adsorption layers on GaAs(111) and Si(111) surfaces during chemical crystallization from the gas phase," in: *Growth of Crystals*, Vol. 13, Consultants Bureau, New York (1985).

4. G. M. Zhidomirov, A. L. Shlyuger, and L. N. Kantorovich, "Current chemisorption models," in: *Current Problems of Quantum Chemistry* [in Russian], Nauka, Leningrad (1987), pp. 224-281.

5. R. A. Évarestov, *Quantum-chemical Methods in Solid-state Theory* [in Russian], Izd. Leningr. Gos. Univ., Leningrad (1982).

6. A. A. Levin, *Introduction to Solid-state Quantum Chemistry* [in Russian], Khimiya, Moscow (1974).

7. G. M. Zhidomirov and I. D. Mikheikin, *Cluster Approximation in Quantum-chemical Studies of Chemisorption and Surface Structures* [in Russian], All-Union Inst. of Scientific and Technical Information (VINITI), Akad. Nauk SSSR, Moscow (1984) (Progress in Science and Technology. Structure of Molecules and the Chemical Bond, No. 9).

8. W. C. Davidon, "Variance algorithm for minimization," *Comput. J.*, **10**, No. 4, 406-410 (1968).

9. R. Fletcher and M. J. D. Powell, "A rapidly convergent descent method for minimization," *Comput. J.*, **6**, No. 2, 163-168 (1963).

10. I. P. Zakharov and A. O. Litinskii, "Program for calculating cyclic models of solids and surface structures using the MINDO/3 approximation," *Zh. Strukt. Khim.*, **24**, No. 6, 111-112 (1983).

11. T. N. Kompaniets, "Interaction of hydrogen, oxygen, and certain other gases with the surface of Ge and Si single crystals," *Zh. Tekh. Fiz.*, **46**, No. 7, 1361-1372 (1976).

12. V. Barone, F. Leli, N. Russo, et al., "Nonempirical cluster model study of the chemisorption of atomic hydrogen on the (111) surface of diamondlike crystals," *Phys. Rev. B: Condens. Matter*, **34**, No. 10, 7203-7208 (1986).

13. M. Seel and P. S. Bagus, "Adsorption and surface penetration of atomic hydrogen at the open site of Si(111)," *Phys. Rev. B: Condens. Matter*, **23**, No. 10, 5464-5471 (1981).

14. K. S. Krasnov (ed.), *Molecular Constants of Inorganic Compounds: Handbook* [in Russian], Khimiya, Leningrad (1979).

15. M. P. Ruzaikin and A. B. Svechnikov, "Quantum-chemical calculation of the adsorption energies of H and Si on the Si(111) surface," *Poverkhnost*, No. 8, 17-21 (1987).

16. M. P. Ruzaikin and A. B. Svechnikov, "Calculation of the adsorption energies of Cl atoms and molecules containing Cl on the Si(111) surface," *Poverkhnost*, No. 3, 59-63 (1988).

17. P. H. Citrin, J. E. Rowe, and P. Eisenberger, "Direct structural study of Cl on Si(111) and Ge(111) surfaces: new conclusion," *Phys. Rev. B: Condens. Matter*, **28**, No. 4, 2299-2301 (1983).

18. B. N. Dev, K. C. Mishra, W. M. Gibson, et al., "First-principles investigation of location and electronic structure of adsorbed halogen atoms on seimconductor surfaces," *Phys. Rev. B: Condens. Matter*, **29**, No. 2, 1101-1104 (1984).

19. M. P. Ruzaikin, A. B. Svechnikov, and O. G. Libedenets, "Quantum-chemical analysis of adsorption of H and Si atoms on Si(100)," *Poverkhnost*, No. 2, 65-68 (1988).

20. V. P. Glushko (ed.), *Thermodynamic Constants of Substances: Handbook*, Vol. 2, Book 2 [in Russian], Nauka, Moscow (1979).

21. M. P. Ruzaikin and A. B. Svechnikov, "Adsorption energy of Cl atoms on Si(100)," *Poverkhnost*, No. 12, 114-118 (1988).

22. D. J. Chadi, "Atomic and electronic structures of reconstructed Si(100) surfaces," *Phys. Rev. Lett.*, **43**, No. 1, 43-47 (1979).

23. J. D. Levin, "Structural and electronic model of negative electron affinity on the Si/Cs/O surface," *Surf. Sci.*, **34**, No. 1, 90-107 (1973).

24. R. E. Schlier and H. E. Farnsworth, "Structure and adsorption characteristics of clean surface of germanium and silicon," *J. Chem. Phys.*, **30**, No. 4, 917-926 (1959).

25. W. S. Verwoerd, "Cluster calculations of the surface dimer structure on Si(100) surfaces," *Surf. Sci.*, **90**, No. 3, 581-597 (1980).

26. H. Ibach and J. E. Rowe, "Hydrogen absorption and surface structures of silicon," *Surf. Sci.*, **43**, No. 2, 481-492 (1974).

27. T. Sakurai and H. D. Hagstrum, "Interplay of the monohydride phase and a newly discovered dihydride phase in chemisorption of H on Si(100)-2×1," *Phys. Rev. B: Solid State*, **14**, No. 4, 1593-1596 (1976).

28. C. M. Garner, L. Lindau, C. G. Su, et al., "Electron-spectroscopic studies of early stages of the oxidation of Si," *Phys. Rev. B: Condens. Matter*, **19**, No. 8, 3944-3956 (1979).

29. J. E. Rowe, G. Margaritondo, and S. B. Christman, "Chlorine chemisorption on silicon and germanium surfaces," *Phys. Rev. B: Solid State*, **16**, No. 4, 1581-1589 (1977).

STEP KINETICS ON CRYSTAL SURFACES IN THE PRESENCE
OF ANISOTROPY AND IMPURITIES

S. Yu. Potapenko

Atomically smooth crystal faces are known to grow by step advancement. Screw dislocations often serve as step sources at relatively small supersaturations. An experimental technique has recently appeared that enables the dislocation structure [3] or surface relief to be observed at the level of monatomic steps [4]. Thus, extremely fine details can be studied.

A quantitative description of the kinetics of actual crystal growth requires consideration of several factors. Several of these are examined in the present work. The first section involves a study of the kinematics of step and dislocation growth on anisotropic surfaces. In the second section, the influence of impurity adsorption on spiral step migration is investigated.

We will assume that the crystal face is smooth, i.e., the temperature is lower than the transition temperature to a coarsened phase [5]. Let the relative supersaturation be small so that the probability of forming two-dimensional nuclei is negligible. In the present work a kinetic growth regime is examined where step migration is controlled by the rate of incorporation processes and not by the availability of crystallizing components.

1. STEP KINETICS ON AN ANISOTROPIC CRYSTAL FACE

Dislocational growth of an isotropic crystal face was studied in [1]; vaporization, in [6]. A polygonal square spiral model was examined in [7]. In the present section, step migration is studied. The rate of step advancement and the radius of a two-dimensional crystalline nucleus are continuous functions of the angle φ between the tangent to the step and the crystallographic axis.

Step Migration Equation. Microscopic anisotropy causes the step migration rate in the direction normal to it to have the form [8]

$$v = v_\infty(\varphi)\,(1 + k\rho_c(\varphi)),\tag{1}$$

where $v_\infty(\varphi)$ is the migration rate of a rectilinear step, $\rho_c(\varphi)$ is the critical-nucleus radius on the surface, and k is the step curvature (for a convex step $k < 0$). If the surface is isotropic, then v_∞ and ρ_c do not depend on φ.

Let $\{x(s, t), y(s, t)\} = \mathbf{r}(s, t)$ be the Cartesian coordinates of points on the step at time t, where s is the arclength. Then $x' = \partial x/\partial s = \cos\varphi$, $y' = \sin\varphi$, where φ is the angle between the tangent and the x axis. It is easy to demonstrate that $\mathbf{r}(s, t)$ satisfies the equation

$$\begin{aligned} \dot{\mathbf{r}} &= \hat{\Omega}\mathbf{r} \\ u' &= kv \end{aligned}\,; \quad \hat{\Omega} = \begin{pmatrix} u & -v \\ v & u \end{pmatrix},\ \dot{\mathbf{r}} = \partial\mathbf{r}/\partial t.\tag{2}$$

Here v is determined by Eq. (1), in which the curvature $k = -(x''^2 + y''^2)^{1/2}$ and $u(s, t)$ is the rate of migration of point s along the tangent. Since $x'\dot{y} - \dot{x}y' = v$ and $x'^2 + y'^2 = 1$, it is convenient to switch from Eq. (2) to that for $\varphi(s, t)$:

$$\begin{cases} \dot{\varphi} = u\varphi' + v' \\ u' = v\varphi' \end{cases}; \quad v = v_\infty + q\varphi', q = v_\infty \rho_c. \tag{3}$$

Analogous kinematic equations for the isotropic steady-state ($\dot{\varphi} = \text{const}$) case were studied in [9].

We note that $u(s, t)$ is not a dynamic variable. The second Eq. (3) represents an anholonomic bond. A dynamic variable does not appear in Eq. (3) since v is the step advancement rate along the normal. If the step is not rectilinear, then the arclength between points on the step that are being displaced along its normal is changed and the derivative with respect to time at constant arclength occurs in the migration equation.

Let us examine a step formed by a screw dislocation terminating at point $r = 0$. In this case $r(0, t) = 0$ and for Eq. (3) we have the boundary conditions

$$1 + \rho_c(\varphi(0, t))\varphi'(0, t) = 0, \quad u(0, t) = 0. \tag{4}$$

Symmetry of the Kinematic Equation. The group symmetry of the step migration equation includes transformations preserving the form of the equations. This means that the equations after the transformations are described as Eq. (2) or Eq. (3) where only the functions $v_\infty(\varphi)$ and $\rho_c(\varphi)$ can change. The existence of nontrivial transformations provides the possibility to convert the known equation into an equation with other values of anisotropy. We now examine spatial transformations where the distance r from a certain point to the origin changes by $\varkappa$ times depending on the direction θ. The new polar coordinates of the point are

$$R = \varkappa(\Theta)r, \quad \Theta = \Theta(\theta). \tag{5}$$

It is convenient to express the kinematic equation in polar coordinates:

$$\dot{r} = (1 + \text{tg}^2\gamma)^{1/2}(1 + k\rho_c(\varphi))v_\infty(\varphi), \quad \varphi = \theta - \gamma,$$
$$\text{tg}\,\gamma = \frac{1}{r}\frac{\partial r}{\partial \theta}, \quad k = -(1 + \text{tg}^2\gamma)^{-3/2}\left(\frac{1}{r} - \frac{\partial^2}{\partial\theta^2}\frac{1}{r}\right). \tag{6}$$

Equation (6) after transformation according to Eq. (5) is

$$\dot{R} = (1 + \text{tg}^2\Gamma)^{1/2}V_\infty(\Gamma, \Theta)(1 + KP_c(\Gamma, \Theta)),$$
$$\text{tg}\,\Gamma = \frac{1}{R}\frac{\partial R}{\partial\Theta}, \quad K = -(1 + \text{tg}^2\Gamma)^{-3/2}\left[1 - \frac{\partial^2}{\partial\Theta^2}\right]\frac{1}{R}, \tag{7}$$

where

$$V_\infty(\Gamma, \Theta) = (\varkappa^2 + \omega^2(\varkappa' - \varkappa\,\text{tg}\Gamma)^2)^{1/2}(1 + \text{tg}^2\Gamma)^{-1/2}v_\infty(\varphi(\Theta, \Gamma)), \tag{8}$$

$$P_c = \frac{[1 - (\omega\partial/\partial\Theta)^2](\varkappa/R)}{[1 - \partial^2/\partial\Theta^2]/R}\left(\frac{1 + \text{tg}^2\Gamma}{1 + (\omega^2/\varkappa^2)(\varkappa' - \varkappa\,\text{tg}\Gamma)^2}\right)^{3/2}, \quad \omega = d\Theta/d\theta, \quad \varkappa' = d\varkappa/d\Theta. \tag{9}$$

Equation (7) retains the form of the kinematic equation if the functions for the rate V_∞ and the critical-nucleus radius P_c as well as the angle between the normal to the step and the crystallographic direction $\varphi = \theta(\Theta) - \gamma(\Theta, \Gamma)$ depend only on $\Phi = \Theta - \Gamma$, i.e.,

$$\hat{L}V_\infty = 0, \quad \hat{L}P_c = 0, \quad \omega^{-1} - \hat{L}\gamma = 0, \tag{10}$$

where $\hat{L} = \partial/\partial\Theta + \partial/\partial\Gamma$. Each of these relations leads to $\omega = (\varkappa^2\cos\delta)^{-1}$ and the differential equation

$$\varkappa\varkappa'' - 2\varkappa'^2 + (\varkappa^2\cos^2\delta - 1)\varkappa^2 = 0, \tag{11}$$

where δ is the integration constant. The solution to Eq. (11) is

$$\varkappa = (1 + \sin\delta \, \sin2\Theta)^{-\frac{1}{2}}\cos\delta. \tag{12}$$

The angular variable and $\varphi = \theta - \gamma$ are transformed according to the equations

$$\begin{aligned}
\cos\delta \, \mathrm{tg}\,\theta &= \mathrm{tg}\,\Theta + \sin\delta, \\
\mathrm{tg}\,\varphi &= \cos\delta \, \mathrm{tg}\,\Phi/(1 - \sin\delta\,\mathrm{tg}\,\Phi).
\end{aligned} \tag{13}$$

Using Eqs. (12) and (13), we obtain

$$\begin{aligned}
V_\infty &= (1 - \sin\delta\sin2\Phi)^{\frac{1}{2}} v_\infty\,(\varphi(\Phi)), \\
P_c &= (\cos\delta)^{3/2}(1 - \sin\delta\sin2\Phi)^{-3/2}\rho_c\,(\varphi(\Phi)).
\end{aligned} \tag{14}$$

Thus, transformation (5) with $\varkappa(\Theta)$ and $\theta(\Theta)$ determined by Eqs. (12) and (13) at an arbitrary initial anisotropy of rate and critical-nucleus radius retained the form of the equation by changing only the dependences of v_∞ and ρ_c on the angle between the normal to the step and the crystallographic direction. Analogous transformations can be made for the rectilinear polygonal step examined in [7]. The change of migration rates $V_{x\infty}$ and $V_{y\infty}$ and of the critical-nucleus radii ρ_{xc} and ρ_{yc} at $V_{x\infty}\rho_{xc}/(V_{y\infty}\rho_{yc}) = 1$ leads to the same system of equations for the segments comprising the steps.

Step Shape Far from a Dislocation. A step at a terminating screw dislocation with an arbitrary initial shape acquires the shape of a rotating spiral after a certain time. Let us write Eq. (3) in the variables $\tau = \ln t$ and $\xi = s/t$:

$$\begin{aligned}
\partial_\tau\varphi &= (u + \xi + v'_\infty)\,\partial_\xi\varphi + \frac{1}{t}\partial_\xi q\,\partial_\xi\varphi, \\
\partial_\xi u &= v_\infty\partial_\xi\varphi + \frac{1}{t}q(\partial_\xi\varphi)^2, \quad v'_\infty = dv_\infty/d\varphi.
\end{aligned} \tag{15}$$

The distance between turns of the spiral is of the order of the characteristic critical-nucleus radii. At distances from the dislocation that are large compared to ρ_c, where the radius of curvature of the step is much greater than the distance between turns of the spiral, they can be represented as closed loops and their evolution with time can be examined. Then in Eq. (15) as $t \to \infty$, terms containing $1/t$ can be neglected everywhere except for the final number of points. The solution near these will be examined below. At this limit the curvature decreases with time and the step becomes locally rectilinear. It is easy to demonstrate that this shape is stable. The step shape is described by the steady-state equations

$$\begin{aligned}
(u + v'_\infty + \xi)d\varphi/d\xi, \quad du/d\xi &= v_\infty d\varphi/d\xi, \\
\varphi|_{\xi=0} = \varphi_0, \quad u|_{\xi=0} &= 0.
\end{aligned} \tag{16}$$

Equation (16) has two solutions: $\varphi = $ const, i.e., the step consists of rectilinear segments and

$$\xi + \int_0^{\varphi(\xi)} d\varphi^*(v_\infty\,(\varphi^*) + v''_\infty\,(\varphi^*)) = 0. \tag{17}$$

If $v_\infty(\varphi) + v''_\infty(\varphi) > 0$ for all φ, then Eq. (17) describes a smooth closed curve. The curve has special points if this condition is relaxed.

As an example, we will examine a surface with group symmetry C_{nv} for $v_\infty(\varphi) = (1 + E\cos n\varphi)v_0$, where $0 \le E < 1$ characterizes the anisotropy. Using Eq. (17), we obtain the equation of a curve in parametric form:

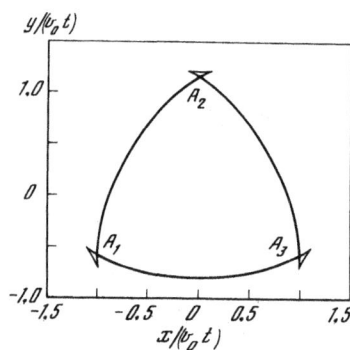

Fig. 1. Shape of a step far from a dislocation at $E = 0.2$
(closed line $A_1 A_2 A_3$).

$$x/(v_0 t) = -\sin\varphi + E(n \sin n\varphi \cos\varphi - \cos n\varphi \sin\varphi),$$

$$y/(v_0 t) = \cos\varphi + E(n \sin n\varphi \sin\varphi + \cos n\varphi \cos\varphi), \tag{18}$$

where φ ($0 \le \varphi < 2\pi$) is the angle between the tangent and the x axis. At $E > E_n = (n^2 - 1)^{-1}$, the curve has $2n$ inversion points and n self-intersections. The shape of the curve at $n = 3$ and $E = 0.2$ is shown in Fig. 1. The physical meaning is a closed line of segments $A_1 A_2$, $A_2 A_3$, and $A_3 A_1$. The existence of special points in this approximation is due to the degeneracy of the step migration directions, the rate along which is rather high. Near the special points A_i, the dependence of the migration rate on the curvature should be considered.

We will demonstrate that for every function $\rho_c(\varphi)$ the model has a single solution for Eq. (18) and no closed partially linear solutions. A line with a kink stretching in the direction $\varphi_0 = \varphi(0, t)$ is described by the steady-state solution of Eqs. (3)

$$s + \int_{\varphi_0}^{\varphi(s)} d\varphi^* q(\varphi^*)/(-Qv_\infty(\varphi_0)\cos(\varphi^* - \varphi_0) + v_\infty(\varphi^*)) = 0, \tag{19}$$

where $Q = 1 + \rho_c(\varphi_0)\varphi'(0) < 1$. In order that the step was linear at $s \to \pm\infty$, $\varphi(\pm\infty) = \varphi^\pm$, $\varphi^- < \varphi^0 < \varphi^+$, the denominator of the expression under the integral of Eq. (19) should have zeroes:

$$v_\infty(\varphi^\pm) - Qv_0(\varphi_0)\cos(\varphi^\pm - \varphi_0) = 0, \tag{20}$$

since $v_\infty(\varphi) - Q(\varphi_0)\cos(\varphi - \varphi_0) > 0$ at $\varphi^- < \varphi < \varphi^+$. The inequality below follows from Eqs. (19) and (20)

$$\frac{v_\infty(\varphi^+)}{v_\infty(\varphi_0)\cos(\varphi^+ - \varphi_0)} = \frac{v_\infty(\varphi^-)}{v_\infty(\varphi_0)\cos(\varphi^- - \varphi_0)} < 1, \tag{21}$$

and limits possible directions of kink propagation. For the case at hand, Eq. (21) acquires the form

$$(1 + E\cos n\varphi^\pm)/[(1 + E\cos n\varphi_0)\cos(\varphi^\pm - \varphi_0)] < 1. \tag{22}$$

If follows from Eq. (22) that $\varphi^+ > -\varphi_0$ at $-\pi/n < \varphi_0 < 0$ and $\varphi^- < -\varphi_0$ at $0 < \varphi_0 < \pi/n$. Taking these inequalities into account, simple geometric plots can demonstrate that closed steps composed of line segments do not exist. Equation (18) gives a single solution. The magnitude of the kinks γ propagated in the directions $\varphi_k = 2\pi k/n$ ($k = 0, 1, \ldots, n - 1$) at $|E| > E_n$ is determined by the equation

$$\frac{n\gamma}{2} = \frac{E}{E_n}\sin\frac{n\gamma}{2}.$$

The appearance of the kink in the curve is analogous to generation of an impact wave in a nonlinear medium. Equation (15) for small φ and $t \to \infty$ leads to the equation

$$\partial_\tau \varphi + (E/E_n - 1)v_0 \varphi \partial_\xi \varphi = 0,$$

that describes the appearance of a jump in $\varphi(\tau, \xi)$ if $E > E_n$.

We note that if the step shape far from a dislocation $\varphi(\xi)$ is known, for example, experimentally, then using Eq. (16) the function $v_\infty(\varphi)$ can be determined:

$$v_\infty(\varphi) = \xi(\varphi)\sin\varphi + v_\infty(0)\cos\varphi - \int_0^\varphi d\psi \xi(\psi)\cos(\varphi - \psi). \tag{23}$$

Here it is assumed that $v_\infty(\varphi)$ is an even function. Finally, in the presence of kinks Eq. (23) determines the step migration rates only for discrete directions.

In [10] experimental observations of a spiral step on the (111) surface of a $Ba(NO_3)_2$ crystal are reported. The step shape at supersaturation of the order of 2% far from the source is described to an accuracy of 15% by Eq. (18) at $n = 3$ and $E = 0.2$ (cf. Fig. 1). For $n = 2$, Eq. (18) describes the step shape on the (100) surface of a KH_2PO_4 crystal [11].

Dependence of Spiral Step Rotation Frequency on Anisotropy Parameters. The crystal face growth rate is controlled by the generation of steps by a dislocation source and is proportional to the spiral rotation frequency. The symmetry of the kinematic equation determined by us enables certain exact relations to be found for the dependence of F on the anisotropy parameters for the case where

$$\begin{aligned} v_\infty(\varphi) &= (1 + H_0\sin(2\varphi - \Delta))^{1/2}v_0, \\ \rho_c(\varphi) &= (1 + G_0\sin(2\varphi - \Delta))^{-3/2}\rho_0. \end{aligned} \tag{24}$$

In new coordinates the rate and critical-nucleus radius acquire the form

$$\begin{aligned} V_\infty &= 1 - H_0\sin\delta\cos(\delta - \Delta) + (H_0\cos(\delta - \Delta) - \cos\delta)\sin 2\Phi + H_0\cos\delta\sin(\delta - \Delta)\cos 2\Phi, \\ P_c &= (\cos\delta)^{3/2}(1 - G_0\sin\delta\cos(\delta - \Delta) + (G_0\cos(\delta - \Delta) - \cos\delta)\sin 2\Phi + G_0\cos\delta\sin(\delta - \Delta)\cos 2\Phi)^{-3/2}. \end{aligned} \tag{25}$$

At $\Delta = \delta$, the dependences of V_∞ and P_c on Φ are of the same type as in Eq. (24) but with other values of anisotropy parameters:

$$H = (H_0 + g)/(1 + H_0 g), \quad G = (G_0 + g)/(1 + G_0 g), \quad g = -\sin\delta. \tag{26}$$

Since the migration equation retained its form, the spiral rotation frequency is $F(H_0, G_0)$. Its value taking into account Eq. (25) can be written

$$F(H_0, G_0) = (1 - g^2)^{3/4}F(H, G)/[(1 + H_0 g)^{1/2}(1 + G_0 g)^{3/2}]. \tag{27}$$

Using Eqs. (26) and (27), the frequency at any H and G will be expressed through the rotation frequency where the critical-nucleus radius is not anisotropic $F_0(H_0) = F(H_0, 0)$:

$$F(H, G) = (1 - G^2)^{1/4}(1 - HG)^{1/2}F_0((H - G)/(1 - HG)). \tag{28}$$

The paths in the HG plane, in which the transformation is performed, are plotted in Fig. 2.* It also is easy to find the shape of the spiral step. It depends periodically on time in contrast to an isotropic spiral rotating at a constant frequency.

---

*The possibility of finding a solution at $H = G$ was demonstrated in [12].

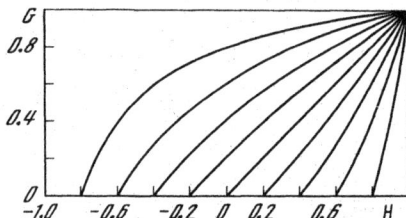

Fig. 2. Paths of the transformation of anisotropy parameters.

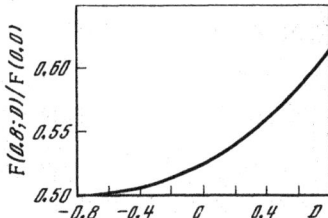

Fig. 3. Dependence of spiral rotation frequency on critical-
nucleus radius anisotropy parameter at $E = 0.8$.

If the anisotropy parameters are small, then

$$F(H, G) \approx (1 - G^2/4 - HG/2 - \alpha(H - G)^2/2)F_0(0),\qquad(29)$$

where $F_0(0) \approx 19v_0/\rho_0$ and the coefficient α is of the order of unity. We note that the rotation frequency can be obtained by this method for the functions $v_\infty(\varphi)$ and $\rho_c(\varphi)$, given in Eq. (24), where the direction of the rate anisotropy and critical-nucleus radius are different.

Numerical Solution of the Kinematic Equation. Let us examine migration of a spiral step at $v_\infty = (1 + E\sin n\varphi)v_0$ and $\rho_c = (1 + D\sin n\varphi)\rho_0$. We will first find F by examining Eq. (3) near the dislocation termination limited by the lowest order in s. At $s \to 0$, it follows from Eq. (4) that $u \to 0$ and $1 + \varphi'(0, t)\rho_c(\varphi) \to 0$. Therefore,

$$\dot{\varphi}(0, t) = u\varphi' + v' \approx v_\infty\partial_s\rho_c\varphi' \approx v(\varphi(0, t))/\rho_c(\varphi(0, t)).\qquad(30)$$

Integrating Eq. (30) for v_∞ and ρ_c given above, we obtain

$$F(E, D) = \frac{(1 - E^2)^{1/2}F(0, 0)}{1 - D(1 - (1 - E^2)^{1/2})/E}\ .\qquad(31)$$

The values F as functions of the anisotropy parameters were calculated by numerical integration of Eq. (3) for $n = 2$. The results agree qualitatively with those from Eq. (31). However, F depends less on the anisotropy parameter of the critical-nucleus radius. Thus, if the maximal values of $v(\varphi)$ and $\rho_c(\varphi)$ are not more than an order of magnitude greater that the minimal ones, then $F(E, D)$ differs from $F(E, 0)$ by less than 20%. We note that the same exponential dependence of F on H and G is characteristic of the model of Eq. (24). This is consistent with the exact expression Eq. (28). If the anisotropy is small, the present model and Eq. (24) are the same to the first power of anisotropy and $E = 2H$, $D = 2H/3$. Therefore, F depends less on D than on E.

The dependence of spiral rotation frequency on D is plotted in Fig. 3. The function $F(E, 0)$ is shown in Fig. 4. Their shapes suggest that the average angular critical-nucleus radius and the average geometric maximal and minimal rectilinear step migration rates should be used to find the spiral rotation frequency in this model. We obtain $p_\pm \approx [(1 \pm E)/(1 \mp E)]^{1/2}/(19\rho_0)$ for the extremal steepness of the growth hillock formed by the dislocation source. We note that the steepness does not depend on the average step migration rate but is determined only by its anisotropy and average critical-nucleus radius.

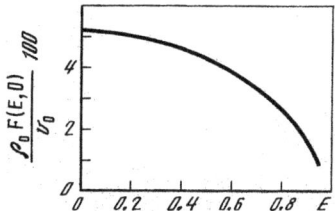

Fig. 4. Dependence of spiral rotation frequency on step rate anisotropy parameter at $D = 0$. At $E \to 1$, $2d\ln F_0(E)/d\ln (1 - E^2) = 0.85 \pm 2$.

The dependence of spiral rotation rate on the anisotropy parameters gives a nonquadratic (in contrast to the isotropic case) dependence of face growth rate on supersaturation σ if E and D change with a change of σ.

2. SPIRAL STEP KINETICS WITH NONEQUILIBRIUM IMPURITY ADSORPTION

Impurities adsorbing under nonequilibrium conditions on the crystal surface have a substantial effect on the growth process. It was noted in [13] that nonequilibrium adsorption of impurity can cause an unstable growth regime. A dislocationless crystal stopped growing in [14] in the presence of an impurity due to formation of two-dimensional nuclei on the surface. Another consequence of nonequilibrium adsorption is the enlargement of steps as they migrate along the crystal surface [15].

In the present section we will examine generation of steps by a dislocation source with nonequilibrium adsorption of impurities. The characteristic adsorption times τ_i are of the order of the spiral rotation period $T = F^{-1}$. This qualitatively changes the crystal growth rate as a function of the supersaturation σ. If the nonequilibrium is slight, the dislocation growth mechanism is $R \approx \sigma^2$ since $v_\infty = \beta\sigma$ and $\rho_c = \beta_c/\sigma$ [1].

Generation of Dislocation Steps with Nonequilibrium Impurity Adsorption. The step migration rate on the crystal surface depends on the impurity concentration. Moreover, it should be realized that the impurity concentration in this region is controlled by the time passing since the moment of formation of this region. We note that the spiral is stable to changes in shape resulting from the dependence of the migration rate on the curvature. This enables the spiral dynamics to be described qualitatively using a one-dimensional model. Let x_n be the distance from point of the dislocation termination to the nth step along a certain direction. When the step coordinate x_n reaches the value $L \approx 19\rho_c$, i.e., the spiral finishes the nth turn, the next step begins to migrate from the point $x = 0$. Its migration rate $V(\tau)$ at the given point x depends on the time $\tau(x)$ passing since the moment when the previous step was located at that point. Thus, the problem becomes the movement of a point along a closed line of length L. Its passage through $x = 0$ corresponds to generation of the next step. Let $t_n(x)$ be the step migration time on the nth spiral turn from $x = 0$ to x, $t_n(0) = 0$. Then

$$dt_{n+1}/dx = V^{-1}(t_{n+1}(x) - t_n(L) - t_n(x)). \tag{32}$$

We switch to a dimensionless variable by choosing the time unit τ_i, length L, and setting $\tau_i V(\tau)/(19\beta_c) = \sigma U(\tau)$. We define X and $t(X)$ as

$$X = n - 1 + x(t_n), \quad t(X) = \sum_{i=0}^{n-1} t_i(1) + t_n(x).$$

Equation (32) acquires the form

$$dt/dX = \frac{1}{\sigma^2} U^{-1}(t(X) - t(X-1)). \tag{33}$$

Here $t(n)$ is the time after which the spiral makes n turns. The face growth rate in dimensionless variables is $F = n/t(n)$. The initial condition for the functional-differential equation (33) is that the function $t(X)$ is set to the single interval $t(X) = t_0(X)$, $-1 \le X \le 0$.

Stability of the Steady-State Growth Regime. Steady-state spiral rotation, i.e., a constant growth rate F, corresponds to $t = TX$. Such a regime occurs if T is a solution of the equation $TU(T) = 1$. We will study the stability of the steady-state regime. Let us find a solution of Eq. (33) in the form $t = TX + \epsilon(X)$. To an approximation that is linear in ϵ,

$$\gamma^{-1}\dot{\epsilon} = \epsilon(X) - \epsilon(X - 1), \quad \gamma = -U'/U^2(T). \tag{34}$$

The total set of solutions to Eq. (34) has the form $\epsilon_\lambda(X) = C_\lambda \exp\{\lambda X\}$ [15], where λ satisfies the equation

$$\lambda = \gamma(1 - e^\lambda). \tag{35}$$

It can be demonstrated [16] that at $\gamma < 1$ Eq. (35) does not have solutions with $\mathrm{Re}\lambda > 0$, i.e., the steady-state solution is stable.

We note that the stability criterion of the steady-state regime Eq. (33) is different from the stability criterion of a fixed point for the discrete representation $T_{n+1} = U^{-1}(T_n)$ that was used in [13, 14], $|\gamma| < 1$. The one-dimensional discrete representation can be interpreted as an explicit difference plot of Eq. (33). The single intercept X is represented by one point. It is easy to confirm that the stability criterion for an arbitrary number of points at the single intercept agrees with $\gamma < 1$ for implicit difference plots.

Dependence of Growth Rate on Supersaturation. We will assume that the dependence of adsorbed impurity concentration on time is exponential and that the dependence of the step migration rate on impurity concentration is linear. Then $U(\tau) = 1 - I + I\exp\{-\tau\}$, $0 \le I < 1$. Here $1 - I$ is the ratio of step migration rate at equilibrium impurity concentration to the rate without impurity. At small concentrations $I < I^* = e^2/(1 + e^2)$ has a single stable steady-state solution for any values σ. However, there are two stable $F_-(\sigma)$ and $F_+(\sigma)$ ($F_- < F_+$) and one unstable $F_- < F_a < F_+$ solution at $I^* < I < 1$ in a certain supersaturation region. The rate F_+ corresponds to rapid spiral rotation where the impurity does not get adsorbed. This causes subsequent steps to migrate faster. For F_-, they migrate slower. Numerical integration of Eq. (33) indicates that the path under different initial conditions rapidly converges to one of the stable steady-state solutions. The steady-state growth rates are determined by the equation

$$F/\sigma^2 = 1 - I + I\exp\{-1/F\} \tag{36}$$

The various kinetic curves are plotted in Fig. 5.

The existence of two unstable regimes for generation of steps by a dislocation source means that hysteresis will be observed as the face growth rate changes with increasing and decreasing supersaturation. The probability of switching between F_+ and F_- must be small for hysteresis to be evident. This condition agrees with our supposition that there should be few fluctuations. Moreover, the rate of change of the controlling bifurcation parameter (supersaturation) should be sufficiently large or a single relation between the growth rate and σ will be observed. In this case, a sharp jump between F_+ and F_- will occur for a certain value $\sigma = \sigma_0$. The value σ_0 will be determined by a Maxwell-type rule for the van der Waals gas isotherm. The nonlinear dynamics of a domain wall were similarly analyzed in [17].

We also examined step generation by a dislocation source in the presence of impurities that affect step migration by changing the step advancement rate on the growing face. The impurity will have such an effect if the lifetime on the surface or the time for incorporating it into the crystal is small compared to the spiral rotation period. The proposed model is not applicable for describing the effect of impurities that stop step migration, i.e.,

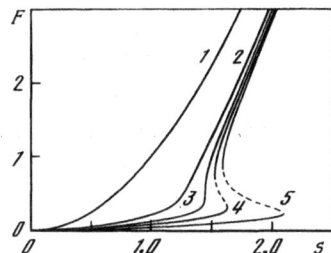

Fig. 5. Dependence of growth rate on supersaturation at various I: 0 (no impurities) (1), 0.8 (2), I^* (3), 0.92 (4), 0.96 (5). The dashed lines represent unstable solutions of Eq. (36).

where a "Cabrera cascade" is formed [7]. In this case, the position of the spiral center of the step fluctuates and is determined by the position of the impurities that have been adsorbed near the dislocation termination.

The model examined makes it easy to assess the effect of the decrease of concentration of crystallizing substance as a step passes. The step migration rate $V(\tau)$ will drop at values τ smaller than the characteristic time of supplying the substance to the surface. This does not change the characteristic features of the function $F(\sigma)$ since a fixed point with $dV/d\tau < 0$ is always stable. This produces the same successive bifurcations on changing σ.

Experimental observations of hysteresis as a function of face growth rate due to supersaturation are interesting. For water-soluble crystals, these can be made using the methods described in [11].

Conclusion. The information presented enables anisotropy parameters to be determined experimentally. The dependence of rectilinear step migration rate on angle can be calculated from the shape of the spiral step far from a dislocation using Eq. (23). The anisotropy of the critical-nucleus radius is controlled only by changes in the first turns of the spiral since the step radius of curvature then becomes much greater than ρ_c and the step migration rate does not depend on curvature. For example, we obtain the critical-nucleus radius by measuring the spiral radius of curvature at a dislocation termination at any time point. Thus, v_∞ and ρ_c can be determined from the shape of the spiral and the normal growth rate.

The effect of impurities on the spiral step rotation in our case produces a kinetic phase transition, analogous to a first-order thermodynamic phase transition. Similar transitions can occur for migration of the step itself. Apparently this occurs with a sharp change of step migration rate on the (100) face of KDP as a function of supersaturation [7]. Moreover, the kinetic transitions are evident in the morphology of the vicinal faces, for example, where macrosteps are generated or their structure is changed.

REFERENCES

1. W. K. Burton, N. Cabrera, and F. C. Frank, "Growth of crystals and their equilibrium surface structure," in: *Elementary Crystal Growth Processes* [Russian translation], Izd. Inostr. Lit., Moscow (1959), pp. 11-109.
2. A. A. Chernov, E. I. Givargizov, Kh. S. Bagdasarov, et al., *Modern Crystallography, Vol. 3: Crystal Formation* [in Russian], Nauka, Moscow (1980).
3. A. A. Chernov, L. N. Rashkovich, I. L. Smol'skii, et al., "Growth of KDP-group crystals from solution," in: *Growth of Crystals*, Vol. 1, Consultants Bureau, New York (1988).
4. K. Tsukamoto, "In situ observation of mono-molecular growth steps on crystals growing in aqueous solutions. 1," *J. Cryst. Growth*, **61**, No. 2, 199-209 (1983).
5. J. D. Weeks, "The roughening transition," in: *Ordering in Strongly Fluctuating Systems*, Plenum Press, New York (1980), pp. 293-323.
6. N. Cabrera and M. Levine, "Toward a dislocation theory of crystal vaporization," in: *Elementary Processes of Crystal Growth* [Russian translation], Izd. Inostr. Lit., Moscow (1959), pp. 152-165.
7. A. A. Chernov, L. N. Rashkovich, and A. A. Mkrtchyan, "Optical interference investigation of the surface processes of growth of crystals of KDP, DKDP, and ADP," *Kristallografiya*, **32**, No. 3, 737-754 (1987).

8.   G. Caginalp, "The role of microscopic anisotropy in the macroscopic behavior of a phase boundary," *Ann. Phys. (N.Y.)*, **172**, No. 1, 136-155 (1986).

9.   V. S. Zykov, "Kinematic steady-state regulation in an excited medium," *Biofizika*, **25**, No. 2, 319-322 (1980).

10.   K. Tsukamoto and I. Sunagava, "In situ observation of mono-molecular growth steps on crystals growing in aqueous solution. 2. Specially designed objective lens and Nomarski prism for in situ observation by reflected light," *J. Cryst. Growth*, **71**, No. 1, 183-190 (1985).

11.   V. P. Ershov, S. Yu. Potapenko, and N. V. Khlyunev, "Layered growth of a crystal face in the tangential flux of a solution," Inst. Appl. Phys., USSR Acad. Sci., Preprint No. 129, Gor'kii (1985).

12.   G. T. Avanesyan, "Effect of elliptical anisotropy in the tangential growth rate and of critical-nucleus shape on the growth spiral dynamics," in: Expanded Abstracts of the Seventh All-Union Conf. on Crystal Growth, Vol. 2 [in Russian], Moscow (1988), pp. 22-23.

13.   A. A. Chernov and B. Ya. Lyubov, "Questions on Crystal Growth Theory," in: *Growth of Crystals*, Vol. 5, Consultants Bureau, New York (1968).

14.   A. A. Chernov, V. F. Parvov, M. O. Kliya, et al., "Growth stoppages of bipyramid faces of dislocationless ADP crystals in the presence of an impurity," *Kristallografiya*, **26**, No. 5, 1125-1135 (1981).

15.   J. P. Van der Eerden and H. Müller-Krumbhaar, "Dynamic coarsening of crystal surfaces by formation of macrosteps," *Phys. Rev. Lett.*, **57**, No. 19, 2431-2433 (1986).

16.   G. M. Hale, *Theory of Functional-Differential Equations* [Russian translation], Mir, Moscow (1984).

17.   S. V. Gomonov, A. K. Zvezdin, and M. V. Chetkin, "Probabilistic description of nonlinear dynamics of domain walls," *Zh. Éksp. Teor. Fiz.*, **94**, No. 11, 133-139 (1988).

*Translator's Note: The citations for references 1 and 6 appear in journals. Apparently several journal articles were collected and published as a book in Russian. The original citations are as follows: (1) W. K. Burton, N. Cabrera, and F. C. Frank, "The growth of crystals and the equilibrium structure of their surfaces," *Trans. Roy. Soc. (London)*, **A243**, 299-358 (1951). (6) M. M. Levine, "Dislocation theory of evaporation of crystals," *Phil. Mag.*, 8th Ser., **1**, No. 5, 450-458 (1956).

HIGH-RESOLUTION TRANSMISSION ELECTRON MICROSCOPIC STUDY
OF EPITAXIAL LAYERS

N. A. Kiselev, V. Yu. Karasev, and A. L. Vasil'ev

High-resolution transmission electron microscopy (TEM) provides much information on epitaxial layers. The structure of layers grown on oriented substrates by various methods can be studied. The layers can be investigated at different stages of growth. The atomic structure of the interface can be characterized. The method of compensating the structural misfit of the layer with the substrate is revealed. The effect of several technical factors can be found.

Modern electron microscopy (EM) makes possible imaging of the lattice on an atomic level. The availability of microscopes with accelerating potentials up to 300-400 kV has been very important in developing studies of the lattice and, in particular, epitaxial layers and the interface. The point resolution d_1 of these instruments is greater than that of instruments with the usual accelerating potential. The resolution is known to be controlled by the minimal reciprocal vector u_1 ($d_1 = u_1^{-1}$) at which the imaginary part of the microscope transfer function $\exp[i\chi(u)]$ reverts to zero as long as the defocus Δf is set equal to the Scherzer value. The point resolution determines the size of details in the image that can be interpreted directly. Another characteristic of the microscope is the information-resolution limit caused by extinction of the transfer function. It indicates the minimal size of details for which information can be extracted from the image. The extinction in turn is caused by chromatic aberration of the instrument and divergence of the incident beam. The reciprocal lattice vector u_2 ($d_2 = u_2^{-1}$) at which the transfer function decreases e times compared to the maximal value is taken as the information limit d_2. For microscopes with an accelerating potential $V = 300\text{-}400$ kV, $d_2 \approx 1.4$ Å. The book by Spence [1] contains more detailed discussion of the Scherzer focus, resolution, and information limit.

Another advantage of the instruments with increased accelerating potential is the ability to work with a wider range of specimen thicknesses. This can sometimes be very important, for example, where the interfaces of similar materials are being investigated.

Widespread use of megavolt instruments to study epitaxial layers is not expected since the probability of radiational degradation and the movement of atoms by the beam are significant [2].

Despite the achievements of modern EM, columns of 6-30 atoms viewed on end rather than individual atoms are seen at lattice sites in photomicrographs in practically all cases. If these columns are too close together they cannot be resolved. For example, most images of the Si lattice are taken along [110]. In this projection the columns of atoms are separated pairwise by 1.36 Å and are not resolved in the overwhelming majority of cases. Photomicrographs of Si with separated columns were obtained by Hutchinson on a JEM4000EX microscope with

400 kV accelerating potential. The distance between columns was less than the Scherzer point-to-point resolution for this microscope (1.65 Å) and was close to the instrument information limit (1.42 Å). The optimal defocus [in order to transmit the (004) reflection] and crystal thickness were used [3, 4]. Therefore, an image of this type is somewhat artificial.

The columns of Si atoms along ⟨100⟩ are separated by 1.91 Å. This is greater than the instrument Scherzer resolution. Therefore, a true image of the column arrangement can be obtained. In particular, it was obtained during a study of Si on sapphire along Si⟨001⟩ [4].

Besides high resolution, a goniometer with two axes is required for TEM of epitaxial layers in order to align the specimen crystallographically. For studies of the interface, a goniometer is usually necessary to orient the interface plane along the beam axis.

Two types of specimens are typically used in studies of epitaxial layers and the interface. These are "in-plane" specimens where the substrate is removed partially or completely by chemical, usually jet, erosion or ion milling (Fig. 1a) and cross sections [5, 6] where the final treatment is also ion milling (Fig. 1b). As a rule, low-energy (1.5-7 keV) Ar^+ or Kr^+ beams are used. However, artifacts can arise during erosion of $A^{III}B^V$ and $A^{II}B^{VI}$ materials. These take the form of metallic islands on the surface, small defects such as packing defects and microtwins, and thick defect layers. Chew and Cullis [7] recommend that I^+ be used in these cases with cooled specimens in the ion milling chamber. An inclined part of the specimen is studied after polishing (Fig. 1b). A certain specimen thickness (from 30 to 200 Å) in the interfacial area is needed to obtain an image of the crystal lattice.

Several examples will be given of the study of epitaxial layers using the simplified classification of Hutchinson [8] based on the similarity or difference of materials (for example, by bond types) and lattice parameters.

"Similar" Materials with Small Lattice Mismatch. The structure GaAs/Al_xGa_{1-x}As(001), which has been studied in many laboratories, is an example of a molecular heterostructure of this type. The columns of Ga (or Al) and As atoms along the [110] direction are close to each other in this projection and are not usually resolved in photomicrographs obtained by organometallic vapor [9] or molecular-beam epitaxy (MBE) [10].

Visualization of the interface may become a problem in principle in this and similar structures. DeJong et al. [11] carried out calculations by a multilayer method for the [110] orientation for various thicknesses and defocuses at 200 kV accelerating potential.

The contrast between layers is very weak for a Scherzer focus about −80 nm and a specimen thickness 19 nm. This is evident in photomicrographs and calculations. Increasing the defocus by another −20 nm in both experiments and calculations produces contrast between layers.

Images of GaAs and AlAs with different contrasts were obtained by studying heterostructures prepared by MBE along ⟨001⟩ at a given thickness and defocus [12]. At thickness $t = 100$ Å, light spots in the AlAs image coincide with the structure projection whereas both light (As) and dark (Ga) spots correspond to columns of atoms for GaAs.

The approach described above was used by Hutchison et al. [13] to study CdTe/InSb(100) layers prepared by MBE. The small differences in scattering ability of the film and substrate (in the [110] orientation) were used. Calculations for a relatively large range of specimen thicknesses were made. The cross sections were milled by I^+ bombardment in order to avoid defects. Photomicrographs of thin (5 nm) crystals obtained at 200 kV and the Scherzer focus did not enable the interface to be distinguished. Images for 200 kV and a wide range of t and Δf were calculated using a multilayer method in hope of obtaining optimal windows in which the combination of these two parameters would enable the interface to be revealed. As it turned out, there is a substantial difference in the calculated images of the two materials at $t = 15$ and $\Delta f = -10$ nm. These conclusions were confirmed experimentally. The interface is clearly seen in the photomicrographs although the atomic structure of the layers under these conditions is given incorrectly.

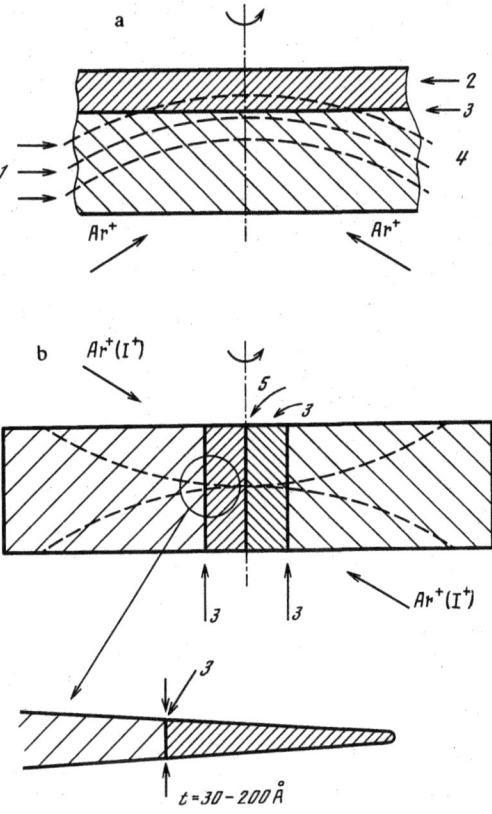

Fig. 1. Two principal specimen types for films on substrates: "in-plane" (a) and cross sections (b). a) Milling of specimens treated on one side; b) on two sides. The lower part of (b) shows a magnified view of the angled part of a specimen of thickness t = 30-200 Å used for TEM: 1) indentation boundary; 2) film; 3) interface; 4) substrate; 5) epoxy resin.

Homoepitaxy is an ideal example of the variant examined. The interface in this case in principle should be poorly visible. However, defects can be found if the process is destroyed at the interface. A cross section of a GaAs epitaxial layer grown on GaAs(100) by MBE is shown in Fig. 2. The packing defects and twins observed are typical of epitaxial interfaces with a large lattice mismatch (of which there is none in this case).

"Similar" Materials with Significant Lattice Mismatch. We will examine two examples of such structures. Heteroepitaxial InAs/GaAs structures prepared by MBE in the Institute of Semiconductor Physics of the Siberian Branch, Academy of Sciences of the USSR, were studied in the Institute of Crystallography of the Academy of Sciences of the USSR. These materials have a rather strong interatomic bond through the interface, the same type of bonds, and the same lattice type. However, the lattice mismatch is relatively large (of the order of 7.2%). Pintus et al. [14] proposed an explanation for the transition from two-dimensional growth to three-dimensional using the Stranskii—Krastanov mechanism during a study of the growth mechanisms of heteroepitaxial films. This explanation is based on an energy increase in the heteroepitaxial interface on introducing dislocation loops (DL).

These ideas were confirmed by high-resolution electron microscopy along ⟨100⟩ of cross sections of the InAs/GaAs heterosystem [15]. Multilayered heterosystems of alternating InAs and GaAs layers were grown on atomically pure surfaces of (001)GaAs by MBE. The InAs layer thicknesses varied between 0.6-40 nm. Those of GaAs were 6.0-40 nm. Cross sections of the heterostructures were treated mechanically and milled by 3 keV Ar⁺ to an angle of 15°. Photomicrographs were taken on a Philips EM-430 electron microscope at 300-kV accelerating potential. The lattice of the multilayered structure was imaged by passing nine diffracted electron beams through

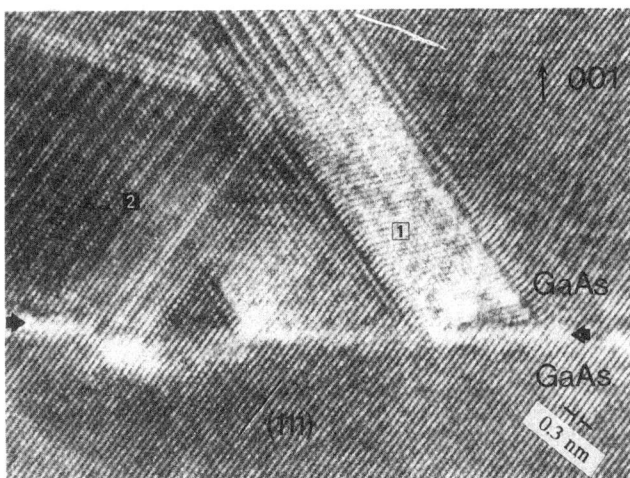

Fig. 2. Epitaxial GaAs film grown by MBE on GaAs substrate with process destruction. The interface is shown by short arrows. Twins are shown by numbers in squares.

the microscope aperture. These beams were the central, four 111 types, two 002 types, and two 220 types. The defocus was close to the Scherzer focus.

According to TEM, the elastically strained (pseudomorphous) state of InAs layers persists to a thickness of 1.0 nm. An image of the lattice of a cross section of the multilayered heterostructure with pseudomorphous InAs layers of thickness 0.6 nm separated by GaAs layers of thickness 6 nm is shown in Fig. 3. The pseudomorphism of the system studied was proved by analyzing multibeam images of the InAs layer crystal lattice. It can be seen that the InAs and GaAs lattices are joined without DL at the interfaces. Slight bending of the atomic ($\bar{1}11$) and (111) planes are observed in the heterointerfaces. This effect is caused by tetragonal distortions of the pseudomorphous InAs lattice. In this case the angle between the [$\bar{1}12$] and [001] directions in the elastically strained InAs lattice should be 3.3° less than the corresponding angle in a strained cubic lattice. This agrees in magnitude and direction with the observed local twist of atomic rows near the InAs layer. The local twist of atomic rows cannot be explained by differences in the electron scattering structure factor of the joined materials since the planes do not twist in thick InAs layers where the elastic deformation is compensated by DL.

In order to study the morphological instability of InAs layers as a function of their thickness, a multilayered structure in which the thickness of InAs layers was varied from 0.6 to 12 nm was grown in a single MBE process (Fig. 4). Increasing the critical thickness of the pseudomorphous InAs layer to 1 nm relaxes mismatch strains by generating DL. This transforms the initially smooth film into an island (Fig. 5). The insert in Fig. 5 shows a multibeam image of an isolated InAs island containing a dislocation loop in the interface. Further growth of the island film generates DL in the upper and lower boundaries with GaAs and generates partial DL with packing defects (cf. Fig. 5). After reaching 10-nm thickness, the InAs islands coalesce to form a continuous film (Fig. 6). The nature of the distortions of atomic planes near DL nuclei in Fig. 6 is consistent with these being edge-type dislocations with Burgers vectors (a/2) [110] and a Lomer dislocation at the interface. At certain heterojunction sites, partial dislocations associated with packing defects along {111} planes are observed. A schematic image of the three types of layers grown in a single MBE process is shown in Fig. 7.

In the CdTe/(100)GaAs system produced by organometallic vapor epitaxy, the lattice mismatch is about 15%. Cullis et al. [16] demonstrated that the epitaxial layer is ordered far from the interface whereas it is compensated at the interface itself by a quasiperiodic row of dislocations (they are located at a distance of about 28 Å). Although the interface is not planar it is clearly seen. The epitaxial layer is very ordered as it draws away from

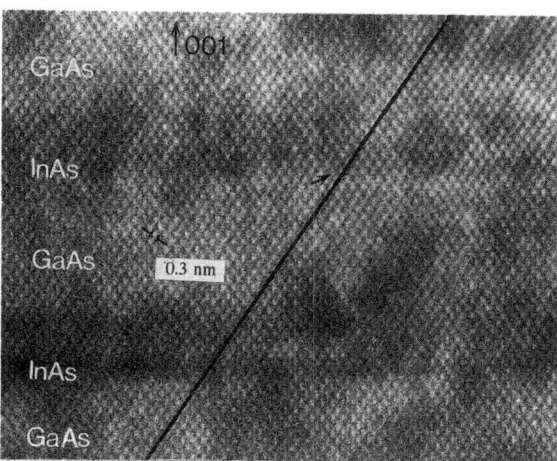

Fig. 3. Pseudomorphous InAs layers. The distortion of {111} planes in InAs layers is shown by arrows.

Fig. 4. Photomicrograph of three growth stages of InAs epitaxial layers of variable thickness grown in a single MBE process. 1) Continuous pseudomorphous film; 2) island films with increasing island thickness (the stage is given by the square bracket); 3) continuous film with dislocation loops at the interfaces (formed by coalescence of islands).

it. Thus, the lattice mismatch in this case does not produce packing defects and microtwins in the epitaxial layer although the interface is very distorted.

Layers of CdTe with the (111) plane parallel to the (100)GaAs substrate were grown in the Institute of Semiconductor Physics of the Siberian Branch, Academy of Sciences of the USSR, by MBE. Electron microscopy of cross sections performed in the Institute of Crystallography of the Academy of Sciences of the USSR demonstrated [17] that the interface along (110) is coherent due to the small lattice mismatch (about 0.7%) along the interface (along [11$\bar{2}$]). Thus, this example would be intermediate between the first and second types of interfaces.

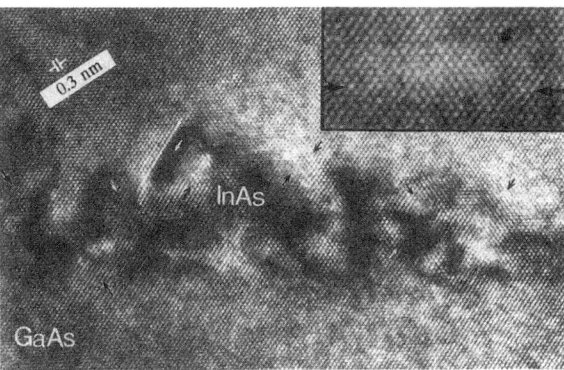

Fig. 5. Image of isolated island of InAs on GaAs(100). Dislocation loops (DL) are shown by small arrows. The interface of InAs with the GaAs epitaxial layer is shown by large arrows. The inset shows an InAs island at the initial growth stage after pseudomorphism is destroyed by introduction of a DL.

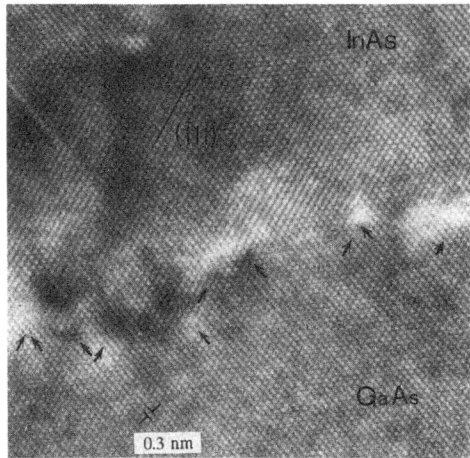

Fig. 6. Portion of a continuous InAs film with a DL (shown by arrows) at the interfaces.

A multibeam image of the cross section (0$\bar{1}$1) of the (111)CdTe film on the singular (100)GaAs surface is shown in Fig. 8. A high density of microtwins is observed at the interface. These are clearly visualized on the terrace of the {111} planes. The density of film portions in the mutually twinned position is approximately identical along the heterojunction. This is consistent with the equal probability of forming CdTe nuclei in twinned positions [17]. It is obvious that the higher islands absorb the lower ones to form a continuous layer (Fig. 9). One of the two neighboring orientations outlives the other at film thicknesses 3-30 nm. This explains the decrease of twin density far from the interface. At film thicknesses greater than 30 nm the amount of twinned interfaces decreases sharply. They are completely absent in separate film portions (Fig. 10). The increased number of twins in later stages of film growth is possibly due to destruction of the optimal growth process.

Different Materials with Small Lattice Mismatch. These materials include the metal silicides $CoSi_2$ and $NiSi_2$ or the fluorides CaF_2 and BaF_2 on Si, Ge, or GaAs substrates. For example, $CoSi_2$ is cubic. The mismatch for (111)$CoSi_2$/(111)Si is 1.4%. Solid-state epitaxy could be used to prepare pseudomorphous $CoSi_2$ layers of 9-Å thickness [18] on which Si layers were deposited using MBE. It is interesting to note that the latter were fault-free.

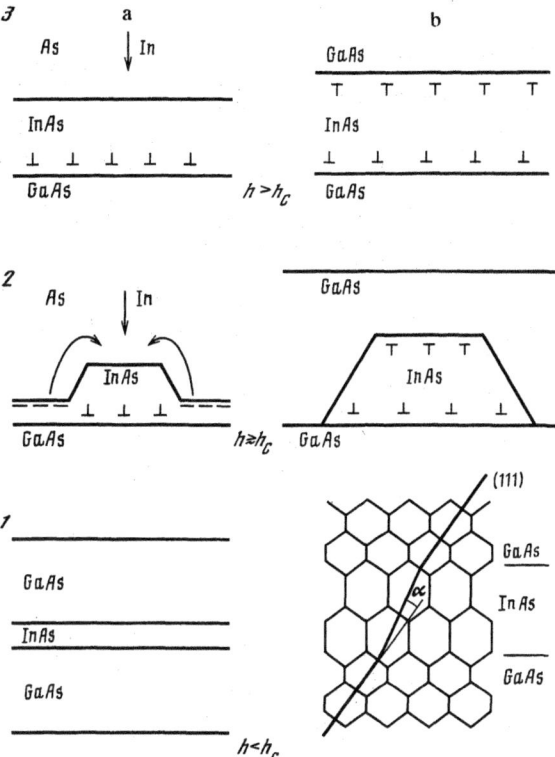

Fig. 7. Schematic image of the three stages of InAs growth of variable thickness: *h*) layer thickness; h_c) critical layer thickness for introducing DL. 1: a) Continuous pseudomorphous InAs layer between GaAs layers; b) diagram of tetragonally distorted InAs crystal lattice due to twisting of the {111} plane by angle α. 2: a) Rearrangement of continuous pseudomorphous film into an island during growth by introducing DL; b) isolated InAs island with DL at the interfaces. 3: a) Formation of a continuous InAs film by coalescence of islands; b) continuous InAs film with DL on the upper and lower interfaces.

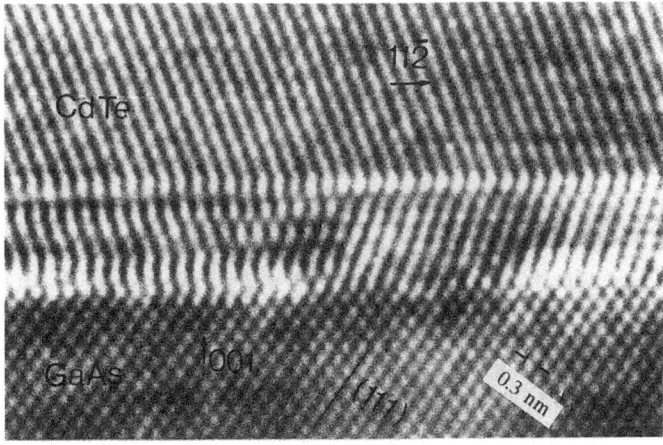

Fig. 8. Image of the interface of an epitaxial CdTe(111) film on GaAs(001) substrate.

The study of NiSi$_2$/Si(111) by Cherns et al. [19] is another example of the use of EM to study heterjunctions with a small mismatch. Interface models were constructed for two NiSi$_2$ film orientations, "normal" and "heterotwinned," due to the threefold symmetry along the $\langle 111 \rangle$ axis in the cubic lattice.

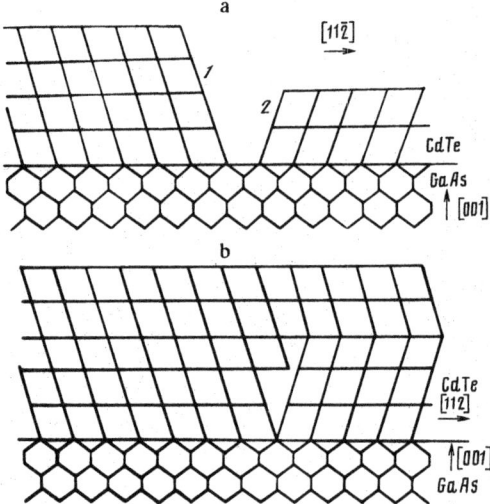

Fig. 9. Schematic image of stages of twin formation in a CdTe(111) layer on GaAs(001) substrate. a) Formation of nuclei in mutually twinned positions relative (1, 2); b) coalescence of nuclei.

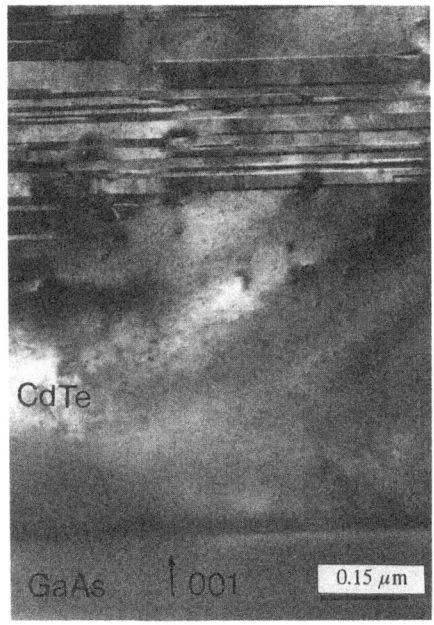

Fig. 10. Epitaxial (111)CdTe film on (001)GaAs substrate.

Films of the high-temperature superconductor (HTSC) $YBa_2Cu_3O_{7-\delta}$ on $SrTiO_3$ substrate is an interesting example of an epitaxial system of different structures with a small lattice mismatch. The HTSC has an ortho-rhombic lattice with constants $a = 0.382$ nm, $b = 0.389$ nm, and $c = 11.67$ nm or, depending on the oxygen content, a tetragonal lattice with $a = b = 0.386$ nm. However, $SrTiO_3$ has a perovskite cell with constant $a = 0.389$ nm, close to a and b of the HTSC. The unit cell of the HTSC consists of three cells with a structure similar to the perovskite cells. Each of these cells is similar in structure and constants to the substrate cell. Studies of the film interface HTSC/substrate are consistent with a strong dependence of structure and orientation on the film preparation technology.

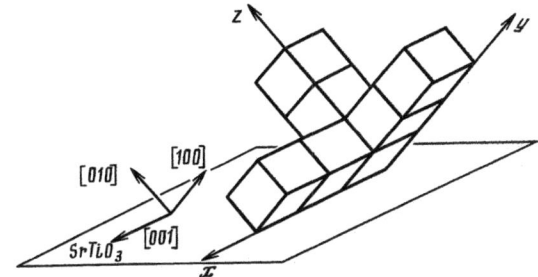

Fig. 11. Diagram of the growth of $YBa_2Cu_3O_{7-\delta}$ film on $SrTiO_3(110)$.

Films prepared by laser sputtering on single crystalline $SrTiO_3$ substrates are oriented near the interface with the c axis perpendicular to the interface [20] even if the interface is uneven. Such an orientation is obtained for films on polycrystalline Si, GaAs, and sapphire substrates with a ZrO_2 buffer layer. This suggests an orienting mechanism associated with anisotropic substrate deformation. Twins characteristic of the orthorhombic modification are seen. Judging from the reflected electron diffraction, the films are textured and mosaic on $SrTiO_3$. Electron microscopy of cross sections demonstrated that the grain boundary is coherent.

High resolution TEM enabled the structure, in particular the orientation of grains and the degree of intergrain boundary coherence and the presence of cracks, to be related to the electrophysical properties, in particular the critical current J_c.

The epitaxial films of $YBa_2Cu_3O_{7-\delta}$ prepared by magnetron sputtering [21] on $SrTiO_3(110)$ substrates are oriented with the c axis along one of the three substrate lattice axes $\langle100\rangle$, $\langle010\rangle$, $\langle001\rangle$. This is shown in Fig. 11. Analysis of electron diffraction patterns obtained from films near the interface demonstrated that the film structure is tetragonal. Twins and packing defects were not observed in the film.

Different Materials with Large Lattice Mismatch. Silicon on sapphire (SOS) has a large mismatch of the cubic and hexagonal lattices and is widely used. Silicon has covalent bonds. Its principal axes are orthogonal. Sapphire is a completely different material with ionic bonds and principal axes at an angle of 80°. Gas-phase deposition is the main method of preparing SOS. Specimens prepared by this method were studied by EM, as will be discussed later. Suitable specimens can also be prepared by MBE.

The study of SOS by high-resolution EM can take three directions (they are representative in general of EM studies of epitaxial layers):

1) The characteristics of layers between the Si and sapphire lattices, the characteristics of the interface, the method of compensating the mismatch [22-26].

2) The study of early stages of Si growth, defects at this stage, the effects of temperature and other factors [27].

3) The study of the influence of treatment in order to improve the quality of the layers. Solid-state epitaxy is at present the principal method of such treatment. In this case, it consists of ion implantation and subsequent rapid annealing using, for example, an electron beam [28].

Silicon is usually deposited on the sapphire $(10\bar{1}2)$ surface. The epitaxial orientation on this surface is Si(001). The lattice fit is better where $\langle100\rangle$ is oriented along $[10\bar{1}1]$ and $[1\bar{2}10]$ of sapphire. This was confirmed experimentally using electron diffraction. The lattice mismatch along $[100]//[10\bar{1}1]$ is 6%; $[110]//[2\bar{2}1]$, 9.3%; and $[010]//[1\bar{2}10]$, 12.3%.

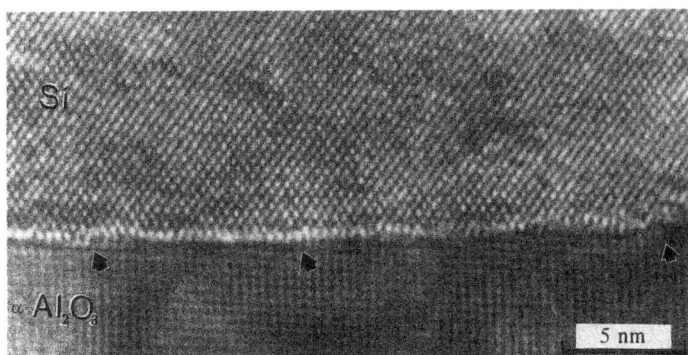

Fig. 12. Interface in SOS. Certain steps at the interface are shown by arrows.

Cross-sectional specimens of SOS are usually prepared along the (110)Si plane. The sapphire (01$\bar{1}$2) plane is parallel to it. Besides the lattice mismatch, the Si lattice is rotated relative to the sapphire lattice by 4°. As a result, the axes of Si ⟨110⟩ and sapphire ⟨02$\bar{2}$1⟩ do not coincide. It is difficult to obtain a good image of both lattices at the same time.

A part of SOS where the Si layer is practically fault-free and the interface is incoherent is shown in Fig. 12. The large mismatch observed in SOS should be compensated somehow. For the majority of epitaxial systems, as was demonstrated above, compensation is achieved by forming DL and a quasicoherent interface. Dislocation loops are not observed in the SOS system.

The Si layers in SOS are extensively twinned (Fig. 13). The twins are mainly first order T_1 with twinning planes parallel to {111} planes. If the mismatch of interplanar distances of the substrate and the Si matrix is +0.37 Å, then it is −0.21 Å for T_1 and −0.34 Å for T_2. Due to a small disorientation the placement of the {111} planes is nonsymmetric. The (1$\bar{1}$1) planes tilted to the right make an angle of 52.5° with the substrate. The (111) planes are tilted to the left by 57°. As a result, the mismatch of these two systems is 13.5 and 7.4%, respectively. Ponce and Aranovich [25] propose that this is due to the predominance of twinning planes parallel to ($\bar{1}$11) that are controlled by strain compensation. Twins oriented in this direction are much longer than in the opposite direction.

The frequently observed moire picture ($T_1 + M$) characterized by a banded structure with 9.5 Å periodicity is shown in Fig. 13. The twins in SOS are proposed to compensate forces arising due to the mismatch. However, there can be rather large portions that are free of twins. Therefore, it is unlikely that the twins are the only method of compensating the mismatch.

Before the high-resolution studies of SOS it was proposed that an intermediate layer or amorphous sublayer of several hundred angstroms thickness might be found at the interface. The first high-resolution studies demonstrated that there was no amorphous layer and that the intermediate layer, if it exists, cannot be thicker than 2-3 atomic layers [22-25].

Electron microscopy reveals no distortion of the Si lattice at the interface (the planes are not curved). Moreover, 2-3 atomic layers at the interface differ in contrast from both Si and sapphire [22-26]. They can be interpreted as an intermediate structure. However, it is possible that this contrast is due to interfacial irregularities or steps on it in the direction of the electron beam.

The nature of the chemical bonds at the interface indicates [26] that bonds arise by restructuring the last layer of Al in sapphire and forming Al—O—Si chains with O degrees of freedom greater than in sapphire. These chains are used to explain the lack of interfacial coherence. Nuclei of (001)Si should arise on them. The chains are proposed to form an intermediate layer consisting of one or several types of alumoinsilicates.

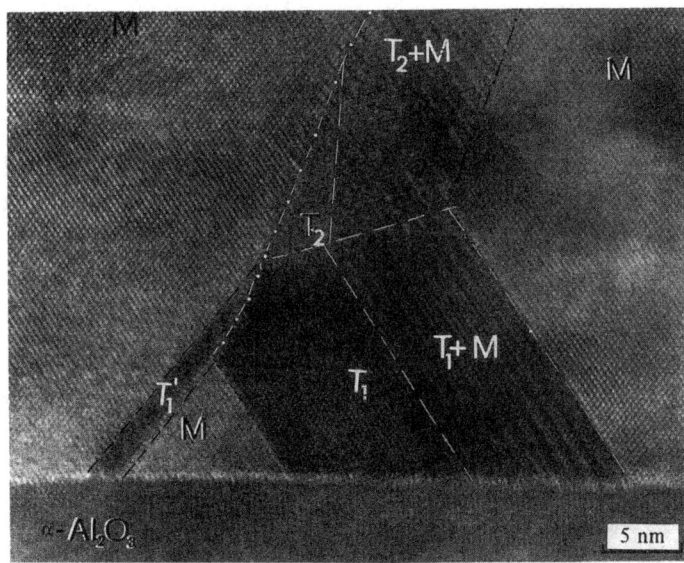

Fig. 13. Image of the SOS interface: M is the Si matrix lattice; T_1 and T_1' are first order twins; T_2 are second order twins; $T_1 + M$ and $T_2 + M$ are superposition patterns of twins on the matrix. Coherent and incoherent boundaries of twins are shown by dashed and dash-dot lines, respectively.

Solid-state epitaxy of SOS has been studied by electron microscopy. For example, implantation of Si ions (10^{15} Si$^+$/cm^2, 190 kV) in Si has been studied. This makes the whole material amorphous with the exception of the thin subsurface layer [27]. Fault-free Si is formed by subsequent electron beam annealing. The amorphous region of several nanometers thickness persists near the interface.

Paus et al. [27] have investigated the early growth of SOS using cross sections. The SOS grows by the Volmer—Weber mechanism. Islands where Si is more or less rolled into a sphere with the center above the interface are typical of the early growth stage. Twins that are not in contact with the substrate are interesting. This indicates that their appearance is not due to differences in the thermal expansion coefficients of Si and sapphire. Multiply twinned fault-free twins are also found. Twinning along planes parallel to {111} observed even in the earliest stages indicates that the twinning does not occur as the islands coalesce. The number of twins decreases with increasing temperature. However, the substrate begins to erode at 1050°C. In the later stages (940°C) the islands become facetted and coalesce with each other.

Epitaxial films of TiSi$_2$ are important for microelectronics technology for a number of reasons. They have high thermal stability, low specific resistivity, etc. Two phases are found, the metastable C-49 and the low-ohmic C-54. Both are orthorhombic. The usual method of preparing the epitaxial films is annealing a Ti layer. The interface morphology in this case depends strongly on the substrate orientation [29]. The layer defectiveness depends on the purity of the substrate and the vacuum conditions. It is noteworthy that for an oriented Si(111) substrate and substrate purified by heating at 900-1000°C in 10^{-2} Pa vacuum with subsequent Ti deposition and annealing under high vacuum, the Si(111)/C-54 TiSi$_2$ interface is even and sharp. For Si substrates with 001 orientation, the boundary is uneven with broken triangular faceting along the {111} planes in the form of saw-teeth. This indicates that the (111) plane is preferred for epitaxy. It was found that the principal TiSi$_2$ orientation is (202) although analysis of the cross-sectional images and diffraction patterns from specimens in the plane and cross sections reveals a number of other epitaxial relations due to a difference in lattice types and parameters.

Catana et al. [30] discuss the "common lattice" resulting from coincidence of sites on superimposing the primitive lattices for various epitaxial relations. Packing defects and "microtwins" in the TiSi$_2$ film are investigated.

Conclusion. The following conclusions can be reached about the possibilities of high-resolution transmission electron microscopy for studying epitaxial layers.

1. Columns of atoms or pairs of columns of atoms can be resolved by studying the lattices of the layer and substrate. Columns of atoms are usually resolved if $V \geq$ 200 kV in studies of a diamondlike lattice along $\langle 100 \rangle$. In studies along $\langle 110 \rangle$, columns can be resolved if $V \geq 400$ kV. However, they are more distinct for $V \geq$ 600 kV.

2. High-resolution TEM enables the mutual orientation of the epitaxial layer and substrate to be found. Specimens "in plane" (diffraction simultaneously from the layer and substrate) and cross sections are used.

3. High-resolution TEM enables the following information about the interface to be obtained (mainly from investigation of cross sections):

(a) the transition sharpness, i.e., the number of atomic layers forming the boundary. Computer modelling of the image is necessary to estimate the transition sharpness. These data are valuable in particular in that the presence of an intermediate layer of several atomic planes thickness can in principle be found;

(b) the planarity, i.e., the type of interface, wavy or smooth. The dimensions of the irregularities, the roughness and steps.

(c) the atomic structure of the interface and the degree of coherence. These studies must be conducted with computer modelling of the interface in order to increase the reliability. In the majority of cases artifacts can be avoided.

(d) the method of lattice constant mismatch compensation can be found. It was mentioned above that high-resolution EM can at present reveal the following compensation methods: lattice distortions (pseudomorphous layer); the presence of DL, microtwins, packing defects, and an intermediate layer;

(e) the appearance and characterization of the "common lattice."

4. Grain boundaries in the epitaxial layer are special. In a number of cases, for example for HTSC, the grain boundaries determine the physical properties of the epitaxial films.

5. Early growth stages can be studied uniquely using EM. This enables the layer formation mechanism to be investigated. The study of early growth stages is also very effective where *in situ* EM methods are used, in particular, use of reflection EM. However, this question requires special examination.

REFERENCES

1. J. C. H. Spence, *Experimental High-Resolution Electron Microscopy*, Oxford Univ. Press, New York (1981).
2. A. Bourret, "New Development in the HREM technique," *Inst. Phys. Conf. Ser.*, No. 93, Vol. 1, 165-170 (1988).
3. J. L. Hutchison, "Below the 'Scherzer resolution' limit: fact and artifact," *Ultramicroscopy*, 9, 191-196 (1982).
4. J. L. Hutchison, T. Honda, and E. D. Boyes, "Atomic imaging of semiconductors," *JEOL News, [Ser.] Electron Opt. Instrum.*, 23, No. 3, 9-13 (1986).
5. M. S. Abrahams and C. J. Buiocchi, "Cross-sectional specimens for transmission electron microscopy," *J. Appl. Phys.*, 45, 3315-3326 (1974).
6. J. C. Bravman and R. Sinclair, "The preparation of cross-section specimens for transmission electron microscopy," *J. Elect. Microsc. Tech.*, 1, 53-61 (1984).
7. N. G. Chew and A. G. Cullis, "The preparation of transmission electron microscopy specimens from compound semiconductors by ion milling," *Ultramicroscopy*, 23, No. 2, 175-198 (1987).
8. J. Hutchison, "Advances in HREM studies of semiconductor interfaces," *Inst. Phys. Conf. Ser.*, No. 87, 1-8 (1987).

9. M. R. Leys, M. P. A. Viegers, and G. W. Hooft, "Metal-organic vapor-phase epitaxy with a novel reaction and characterization of multilayer structures," *Philips Tech. Rev.*, **43**, 133-142 (1987).

10. B. A. Joyce and C. T. Foxon, "Molecular beam epitaxy of multilayer structures with GaAs and $Al_xGa_{1-x}As$," *Philips Tech. Rev.*, **43**, 143-153 (1987).

11. A. F. DeJong, H. Bender, and W. Coene, "Actual comparison of experimental and simulated lattice images of the GaAs/AlAs interface," *Ultramicroscopy*, **21**, 373-378 (1987).

12. H. Ichinose, Y. Ishida, and H. Sokaki, "Lattice imaging analysis of GaAs/AlAs heterointerface by [100] illumination," *JEOL News, [Ser.] Electron Opt. Instrum.*, **26**, No. 1, 8-11 (1988).

13. J. L. Hutchison, W. C. Waddington, A. G. Cullis, and N. G. Chew, "High resolution imaging of the CdTe/(100)InSb interface: a lattice-matched heteroepitaxial structure," *J. Microsc. (Oxford)*, **142**, 153-162 (1986).

14. S. M. Pintus, S. I. Stenin, A. I. Toropov, and E. M. Trukhanov, *Morphological Stability and Growth Mechanism of Heteroepitaxial Films* [in Russian], Inst. Fiz. Prob., Sib. Otd., Akad. Nauk SSSR, preprint No. 5-86, Novosibirsk (1986).

15. V. Yu. Karasev, N. A. Kiselev, E. V. Orlova, et al., "HREM of InAs and GaAs-based multilayered heterosystems," *Inst. Phys. Conf. Ser.*, No. 100, 33-38 (1989).

16. A. G. Cullis, N. G. Chew, J. L. Hutchison, et al., "HREM studies of II—VI heteroepitaxial layers," *Inst. Phys. Conf. Ser.*, No. 76, 29-34 (1985).

17. S. A. Dvoretski, A. K. Gutakovski, V. Yu. Karasev, et al., "Twinning in CdTe(111) films of (100)GaAs substrates," *Inst. Phys. Conf. Ser.*, No. 3, Vol. 2, 407-408 (1988).

18. H. Von Känel, J. Henz, M. Ospelt, and P. Wachter, "Growth of high-quality CoSi/Si superstructures on Si(111)," *Superlattices Microstruct.*, **4**, No. 1, 27-32 (1988).

19. D. Cherns, J. C. H. Spence, G. R. Antis, and J. L. Hutchison, "Atomic structure of the $NiSi_2$/(111)Si interface," *Phil. Mag. A*, **46**, 849-862 (1982).

20. N. A. Kiselev, A. L. Vasiliev, O. V. Uvarov, et al., "High resolution electron microscopy of superconducting films $YBa_2Cu_3O_{7-\delta}$," *Inst. Phys. Conf. Ser.*, No. 93, Vol. 2, 223-229 (1988).

21. M. Tomita, Y. Hayashi, H. Takaoka, et al., "Cross-sectional TEM observation of $YBa_2Cu_3O_{7-\delta}$," *Jpn. J. Appl. Phys.*, **27**, No. 4, L636-L638 (1988).

22. M. S. Abrahams, J. L. Hutchison, and G. R. Booker, "Direct observation of silicon—sapphire heteroepitaxial interface by high resolution transmission electron microscopy," *Phys. Status Solidi A*, **63**, K3-K6 (1981).

23. J. L. Hutchison, G. R. Booker, and M. S. Abrahams, "Transmission high resolution electron microscopy studies of silicon—sapphire epitaxial layers structure," *Inst. Phys. Conf. Ser.*, No. 60, 139-146 (1980).

24. A. L. Vasil'ev, A. L. Golovin, K. M. Manafov, et al., "High resolution electron microscopy and x-ray diffraction studies of epitaxial layers of silicon on sapphire," *Poverkhnost*, No. 1, 123-132 (1987).

25. F. A. Ponce and J. Aranovich, "Imaging of the silicon-on-sapphire interface by high resolution transmission electron microscopy," *Appl. Phys. Lett.*, **38**, No. 6, 439-441 (1981).

26. F. A. Ponce, "Fault-free silicon at the silicon/sapphire interface," *Appl. Phys. Lett.*, **41**, No. 4, 371-373 (1982).

27. K. C. Paus, J. C. Barry, G. R. Boomes, et al., "Investigation of the early growth of epitaxial silicon-on-sapphire using high resolution transmission electron microscopy," *Inst. Phys. Conf. Ser.*, No. 76, 35-40 (1985).

28. D. J. Smith, L. A. Freeman, R. A. McMahon, et al., "High resolution electron microscopy of Si-implanted and electron-beam annealed silicon-on-sapphire," *Inst. Phys. Conf. Ser.*, No. 67, 83-88 (1983); 35-40 (1985).

29. A. Catana, M. P. Heintze, P. E. Shmid, and P. Stadelmann, "Structural study of the TiSi/Si(111) interface by high resolution electron microscopy," *Inst. Phys. Conf. Ser.*, No. 93, Vol. 2, 401-402 (1988).

30. A. Catana, M. P. Heintze, P. E. Shmid, and P. Stadelmann, "Investigation of epitaxial titanium silicide thin films by high resolution electron microscopy," *Inst. Phys. Conf. Ser.*, No. 87, 529-534 (1987).

STRUCTURAL RECONSTRUCTION OF ATOMICALLY-CLEAN SILICON SURFACE DURING SUBLIMATION AND EPITAXY

A. V. Latyshev, A. L. Aseev, A. B. Krasil'nikov, and S. I. Stenin

The elucidation of elementary processes occurring on a surface during sublimation and epitaxy is a fundamental problem for the physics of semiconductors. The determination of the nature of superstructure reconstructions, the structure of surface defects, the kinetics of adatoms, and the mechanisms of interaction with troughs are some of these processes [1-3]. The solution of these problems is especially important for the development of molecular-beam epitaxy (MBE), which requires an atomically pure surface with the required structure and micromorphology. This is necessary to control the flux density for adatoms and to determine the formation mechanisms of submonolayer coatings and structural reconstructions during preparation of epitaxial films [4-7].

These processes must be studied using methods that can achieve high spatial resolution during *in situ* experiments. Scanning tunneling microscopy (STM), ultrahigh-vacuum reflective electron microscopy (UHV-REM), and low-energy electron diffraction (LEED) are the most promising of the modern methods according to a review of existing methods for studying the structure of pure surfaces [3]. Reflective electron microscopy (REM) and LEED have lower resolution than STM. However, they are much more suitable for *in situ* experiments [8, 9].

In the present work, results are presented from a study of structural reconstruction on atomically pure Si(111) and -(001) surfaces using REM during sublimation, superstructural reconstruction, and homoepitaxy.

1. Ultrahigh-Vacuum Reflective Electron Microscopy. Questions about the determination of the spatial resolution and contrast in REM are discussed in detail in [10-13]. The resolution perpendicular to the electron beam was demonstrated not to differ substantially from that attained in transmission electron microscopy. The main reason for poor resolution in REM (up to 1 nm) is due to increased chromatic aberration. An examination of the mechanisms for forming contrast in REM [11, 12] showed that it is a superposition of diffraction and phase contrast due to a shift of the phases on defocus (Fresnel contrast).

It should be noted that the magnification (and therefore the resolution) parallel to the electron beam $M_|$ is less than that perpendicular to it $M_\perp$ by 20-30 times. This is caused by the small angle at which the beam glances off the studied surface. The difference in magnifications compresses the images in the direction of the electron beam. The first results from computer treatment of photomicrographs to remove the scale difference are presented in [14].

A differential cryogenic pump attached to the microscopy column, supplemental nitrogen screens, and continuous pumping with a getter are used to attain the UHV in REM [15]. The apparatus enables the sample to be heated by passing an electric current through the crystal. It also ensures that the vacuum is not compromised

Fig. 1. Image of the Si(111) surface.

when the crystal is rotated in the Bragg and azimuthal planes and locked. Controlled deposition of various substances on the surface, measurement of the intensities of diffracted beams using a semiconducting electron detector built into the microscope, and analysis of the gas composition in the vacuum system using a mass spectrometer are possible.

Samples of Si of dimensions $0.3 \times 1 \times 8$ mm were subjected to thermal oxidation after polishing. The thermal oxide was removed before the samples were placed into the electron microscopy column. Subsequent heating to 900-1000°C under ultrahigh vacuum removed the natural oxide and formed epitaxial SiC particles of dimensions 5-10 nm on the surface. The particles are clearly visible in REM images and give additional reflections in the diffraction pattern. At temperatures above 1000°C, the SiC particles are removed from the surface. The criteria for complete cleaning of the Si surface in REM are, first, the absence of additional diffraction reflections and, second, the presence of reversible phase transitions $7 \times 7 \leftrightarrow 1 \times 1$ for Si(111) and $1 \times 2 \leftrightarrow 1 \times 1$ for Si(001). These transitions are observed in the diffraction pattern and in surface images as the temperature is changed. The third criterion is that monoatomic steps without stoppage centers can be directly observed on the surface. The adequacy of these criteria is confirmed by reproducing effects observed experimentally by REM in UHV on MBE apparatuses equipped for chemical analysis of the surface.

A Si(111) surface containing a system of steps and a dislocation termination with a Burgers vector component perpendicular to the surface is shown in Fig. 1. The dislocation termination also lies on a contour of one of the steps. Since the differences in intensity of this and the remaining steps are the same, all steps of this surface have a monoatomic height 0.314 nm. Changing the diffraction conditions rotates the dislocation contrast and a part of the step surfaces. This is consistent with the conclusions of [10-13] about the predominance of diffraction contrast in the image of monoatomic steps.

2. Reconstruction of Systems of Monoatomic Steps during Sublimation onto Si(111). Contradictory data on the micromorphology of the Si(111) surface during sublimation can be found in the literature. According to [16], the data of which were obtained by scanning electron microscopy (SEM), the principal surface relief after sublimation consists of steps with a height of tens of interatomic distances. Only monoatomic steps were observed in [17] using REM. Therefore, we studied experimentally the kinetics of formation of the stepped surface (111) at various sublimation temperatures and the factors controlling formation of various types of surface micromorphology.

It was found that the Si surface during sublimation in certain temperature ranges contains a system of monoatomic steps separated by the same distance from each other $d_0 \approx a_{111}/\tan \alpha$, where a_{111} is the interplanar distance for (111) and α is the angle of deviation from the singular (111) plane. The monoatomic steps advance during sublimation at a rate directly proportional to the distance between steps (Fig. 2). The rate of step advance is an Arrhenius function of temperature with an activation energy 4.2 ± 0.2 eV, close to the enthalpy of sublima-

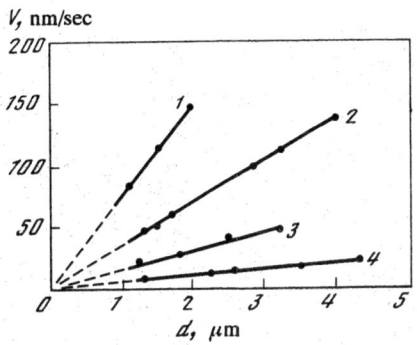

Fig. 2. Rate of advance of monoatomic steps as a function of distance between neighboring steps at various crystal sublimation temperatures: 1) 1200; 2) 1170; 3) 1130; 4) 1090°C.

Fig. 3. REM image of a stepped Si surface with two-dimensional sublimation islands (indicated by arrows).

tion of Si [18]. In agreement with the theory of [19], the rate of advance of a system of parallel steps on a crystal surface is defined by

$$v = 2\sigma\lambda_s\nu\exp(-W/kT)\mathrm{th}(d/2\lambda_s),$$

where λ_s is the diffusion length of an adatom; ν is the atomic vibration frequency; W is the heat of sublimation; $\sigma = (P_0 - P)/P$ is the vapor phase undersaturation; P and P_0 are the current and equilibrium vapor pressures; and k is Boltzmann's constant. The rate of step advance is proportional to d for a distance between steps $d \ll \lambda_s$. Experimental data (cf. Fig. 2) indicate that this condition is fulfilled in the studied sublimation temperature range. Thus, the steps advance by diffusional exchange of adatoms between monoatomic steps. If $d \gg \lambda_s$ (for example, by rapidly reducing the sublimation temperature), then the rate of step advance does not depend on the distance between steps. Under these conditions, two-dimensional sublimation islands bounded by a monoatomic step were found on the terraces (Fig. 3). Thus, the absence of two-dimensional islands in our experiments is consistent with the presence of diffusional exchange between monoatomic steps under the given sublimation conditions.

The system of equidistant monoatomic steps (Fig. 4a) rearranges into a system of step echelons consisting of portions with a low step density and pile-ups of monoatomic steps [20] on heating by direct current in certain temperature ranges. The monoatomic steps in pile-ups are separated by distances much less than d_0 (Fig. 4b). Changing the imaging conditions of the echelons demonstrated that the contrast from them is diffractive, like from the monoatomic steps. It was demonstrated that the temperature ranges at which echelons form depend on the direction of the electric current heating the crystal. If the direction is the same as that of the monoatomic step advance during sublimation then the equidistant steps rearrange into echelons at 1050-1250 and greater than

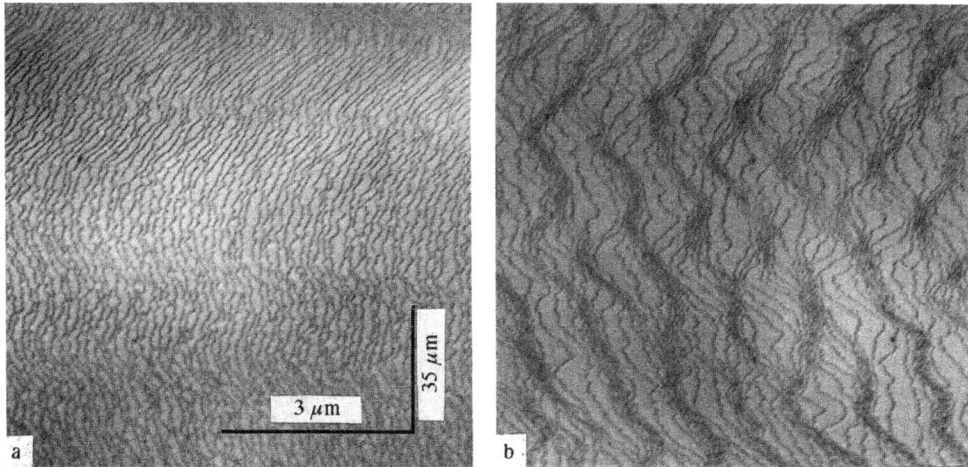

Fig. 4. Structure of the stepped (111) surface at sublimation temperatures 1270 (a) and 1180°C (b). Heating by direct current. Directions of current and step advance are the same.

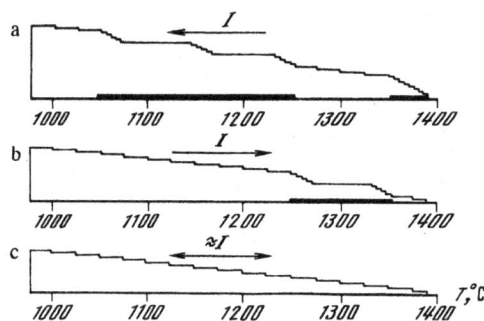

Fig. 5. Temperature ranges of monoatomic step rearrangements on a Si crystal surface on heating by passage of electric current. Directions of electric current and step advance during sublimation are the same (a) and opposed (b). Heating by alternating current (c).

1350°C (Fig. 5a). If the directions are opposed, then echelons of steps form between 1250-1350°C (Fig. 5b). Heating the crystal with 50 Hz alternating current forms a system of equidistant monoatomic steps with $d \approx d_0$ at any sublimation temperature (Fig. 5c).

The evenly spaced steps rearrange reversibly into echelons. If the sample temperature is changed beyond the limits of the temperature range at which echelons form, they transform into a regular step system. The rate of echelon destruction depends on temperature. The lower the temperature the greater the time required to form the regular step system.

The echelon formation processes were studied on a surface containing centers of step stoppage. The stoppage centers are contaminant particles that were formed by degradation of the vacuum to above 10^{-4} Pa. These experiments demonstrated that the change in distance between steps due to stopping at the particles splits the monoatomic steps into pairs (Fig. 6). For example, stoppage of step 1 on particles A and B increases the distance between step 1 and step 2, which is higher. This causes the distance between steps 2 and 3 to decrease. Thus, the system of equidistant monoatomic steps is unstable toward fluctuations of the distances between them in temperature ranges at which echelons form (cf. Fig. 5).

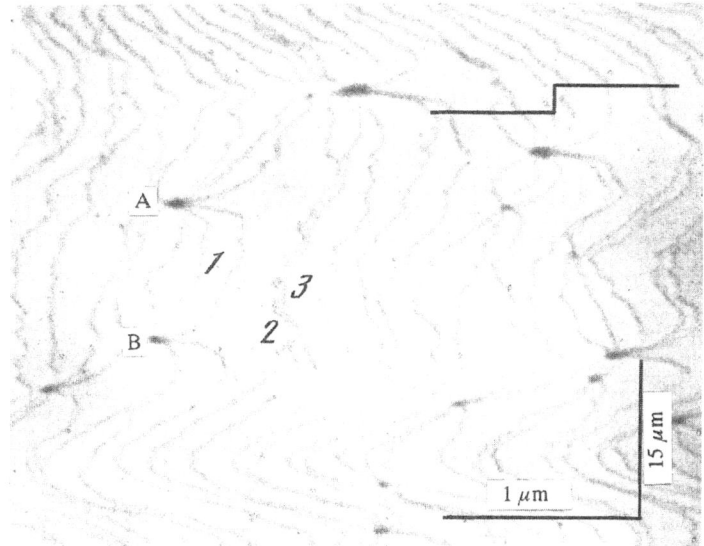

Fig. 6. REM image of the Si(111) surface with contaminant particles (*A*, *B*) after heating at temperature ranges at which echelons of steps form.

The experimental data obtained by us indicate that the system of diffusionally related monoatomic steps is unstable relative to fluctuations in the distance between them. As a result, the monoatomic steps coalesce into a macrostep with a height greater than the interplanar distance. Within the framework of existing theories, this effect is caused by generation of kinematic waves in the system of diffusionally related steps [1, 21, 22]. In agreement with the examination carried out in [23] and the calculations of [24], the reversible fluxes of adatoms approaching a step from the two adjoining terraces must be different in order for echelons to form. The data obtained by us for the temperature ranges at which echelons form and the dependence on heating current direction indicate that the theoretical model of echelon formation on the Si(111) surface should take into account the following factors.

The first factor is the presence of electrically charged adatoms on the Si surface. It can also be assumed that charged adatoms of a metallic impurity that block incorporation of adatoms into the step are present on the surface. However, a study of echelon formation processes in UHV apparatuses combined with Auger analysis of the chemical composition of the surface did not find metallic impurities within the detection limits of the Auger method. Thus, it should be accepted that there is an effective adatom charge that in general is determined by the relation [2]

$$Z_{eff} = e(Z_0 - nC_n l_n + pC_p l_p),$$

where Z_0 is the degree of atomic ionicity; $C_{n, p}$ and $l_{n, p}$ are the scattering cross section and the path length of electrons (n) and holes (p), respectively; and e is the electron charge. It can be seen that the effective adatom charge at certain parameters can change sign and correspondingly change the flux of adatoms to the step from the lower and higher terraces due to drift of adatoms in the direct electric field applied to the sample [20]. The stability of the equidistant steps on heating in alternating current is possibly related to the fact that the tendency to form echelons that appears in the first half-period changes to a tendency to decomposition during the second half-period (a transition from Fig. 5a to Fig. 5b and back again).

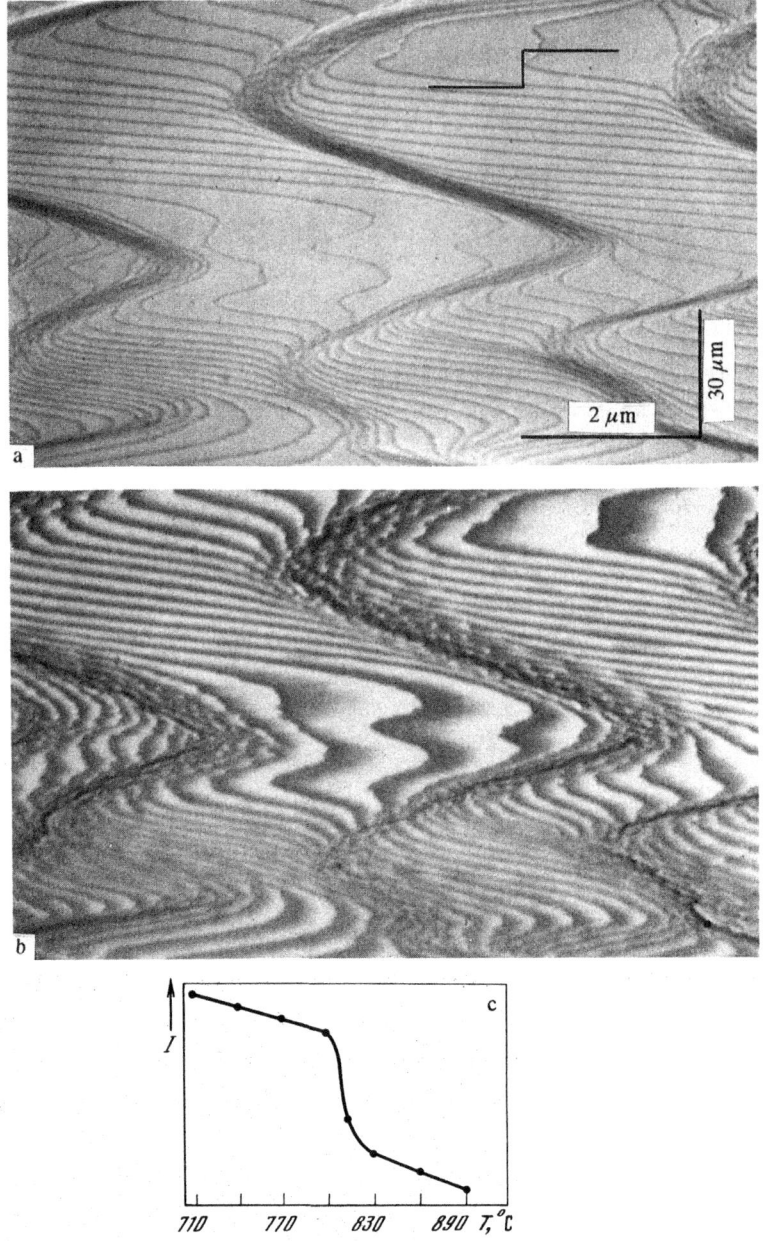

Fig. 7. Structural transition 1 × 1 (a) → 7 × 7 (b) and the change of intensity of the electron beam reflected from the Si(111) surface during the phase transition 7 × 7 ↔ 1 × 1 (c).

The second factor is the existence of elastic forces between monoatomic steps that prevent coalescence of separate steps on decreasing the distance between them during echelon formation. The existence of elastic deformations in the echelons is confirmed indirectly by the presence of diffractive contrast in the REM images of the echelons.

The third factor is related to the existence of various energy barriers for incorporating adatoms into the monoatomic step from the lower and higher terraces. The barrier to incorporation from the lower terraces is less. This follows from the fact that the step advance rate is observed to be more sensitive to a change of distance to the neighboring step on the lower terrace (see above).

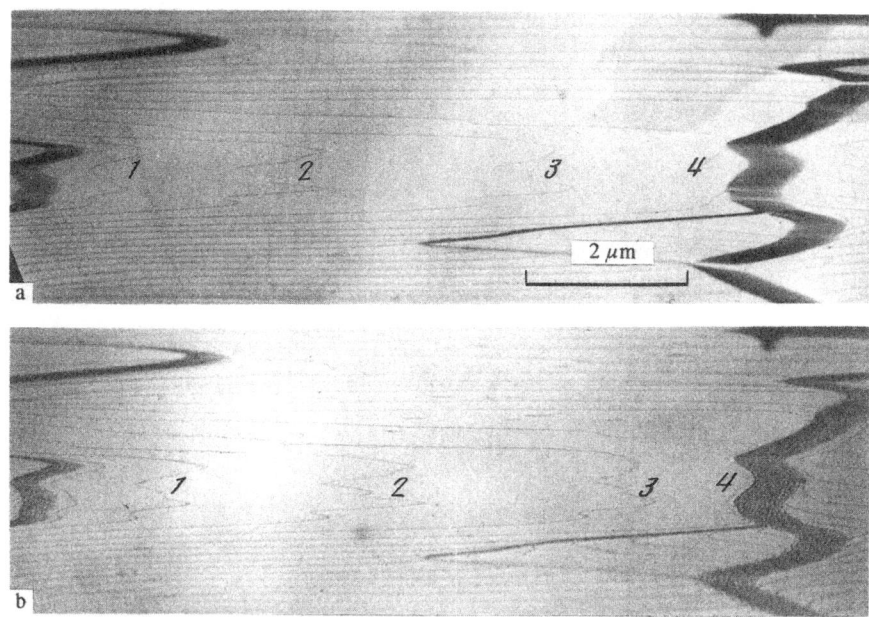

Fig. 8. Image of the same surface for various surface structures: 1×1 (a) $\rightarrow$ 7×7 (b).

The results of the present work indicate that Si sublimation is accompanied by formation of both a system of evenly spaced monoatomic steps and of steps with a height up to $10\text{-}10^3$ interatomic distances. The formation of echelons depends on temperature and the direction of crystal heating current. Thus, the contradictory data from various works that monoatomic steps are the principal element of surface micromorphology [17] and that the larger surface relief elements of Si(111) predominate on sublimation [16] are due to the different crystal heating conditions in these works.

3. Structural Reconstructions of the Si(111) Surface during the Phase Transition $1 \times 1 \leftrightarrow 7 \times 7$. The atomic positions in the unit cell of the superstructure 7×7 were determined using STM [25]. Pashley et al. [26] proposed that atoms migrate during reconstruction of 1×1 into 7×7. This should affect the behavior of steps. Thus, is seemed interesting to study the features of kinetically reversible reconstruction of 1×1 into the super-structure 7×7 by REM.

Figure 7 presents REM images of sequential stages of the structural transition $1 \times 1 \rightarrow 7 \times 7$ on a surface consisting of almost singular portions with small step and echelon densities. The starting surface at $T \geq 830°C$ (Fig. 7a) is characteristic of 1×1. The dark regions near monoatomic steps on the side of the upper terraces are superstructural domains 7×7 (Fig. 7b). Many kinks appear on the steps during the transition $1 \times 1 \rightarrow 7 \times 7$ such that the steps acquire a zig-zag form. Kinks in steps were generated earlier in [8].

The transition $7 \times 7 \leftrightarrow 1 \times 1$ is accompanied by a reversible shift of monoatomic steps (Fig. 8) [27]. The step shift during a superstructural transition is seen most clearly in portions of the surface with a low step density and for steps bounded by two-dimensional islands. The steps shift toward the lower terraces during the transition $1 \times 1 \rightarrow 7 \times 7$. They shift toward the higher terraces during the reverse transition $7 \times 7 \rightarrow 1 \times 1$. The diameter of islands changes as the steps bounded by two-dimensional islands shift. The shift of the monoatomic step is 0.2-0.3 of the width of the adjoining terraces. The number of atoms absorbed or released by steps during the phase transition is estimated from the shift of the steps as $4 \cdot 10^{14}$ at·cm^{-2}. If an amount of the order of 0.3 mono-layers of Si were deposited on a Si surface with superstructure 7×7, the phase transition $7 \times 7 \rightarrow 1 \times 1$ did not shift the monoatomic steps. However, the growth islands did disappear. This means that about $4 \cdot 10^{14}$ at·cm^{-2} were involved in the surface reconstruction. It should be noted that the shift of monoatomic steps on formation of

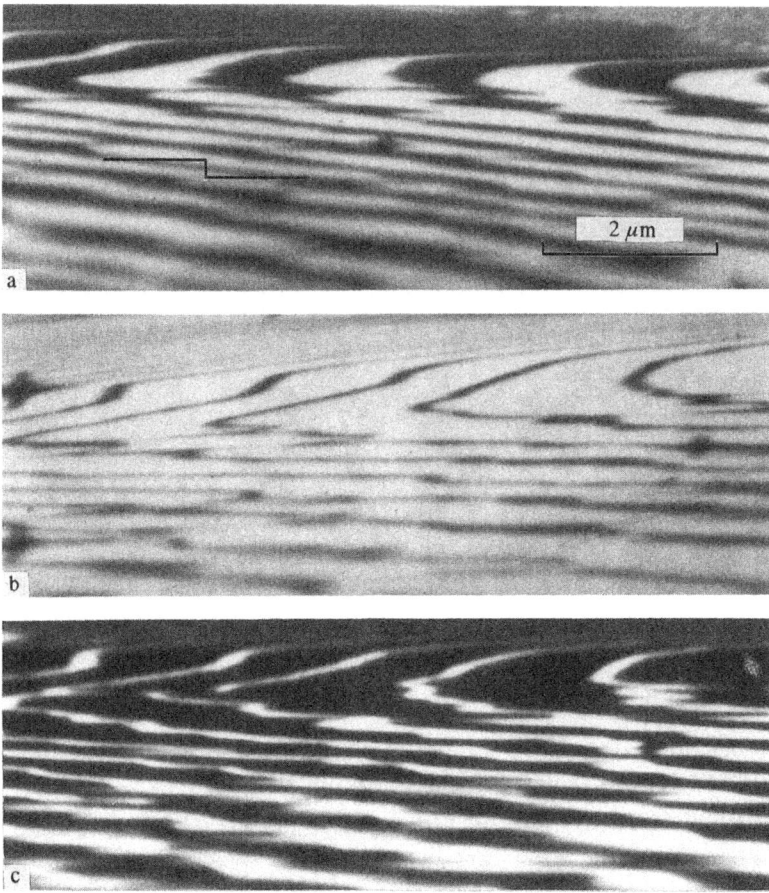

Fig. 9. REM images of the Si(001) surface obtained for superstructural reflection 1×2 with the electron beam in the $\langle 110 \rangle$ direction. Crystal heated by alternating (a) and direct (b, c) currents with the current direction coincident (b) or opposed to (c) the direction of step advance during sublimation.

superstructure is determined largely by the width of the lower terraces adjoining the steps. This is confirmed by the existence of various energy barriers to incorporation of adatoms into the step from the lower and higher terraces (cf. Sec. 2).

According to [25, 28], the unit cell of the 7×7 structure contains 12 atoms in the uppermost layer and one vacancy in the second. If this model is used, then 1×1 is formed by removing 12 atoms from the first 7×7 surface layer and building them into a monoatomic step or placing them into the first upper layer of the superstructure at the expense of the atoms from the step. According to experiments, the shift of steps during the transition $7 \times 7 \to 1 \times 1$ corresponds to generation of adatoms by steps such that the first superstructure layer is filled during formation of 1×1. The step shift should be about 0.8 of the terrace width instead of that observed in the experiment (0.2-0.3) in order to fill this layer completely. The discrepancy in the shift values can be explained by assuming that the surface of 1×1 is partially filled with a degree of filling 0.5. This conclusion is confirmed by measurements of the intensity of reflected electrons (Fig. 7c). A decrease in the reflected intensity is usually related to development of microrelief [29]. The data of Fig. 7c indicate that the Si(111) surface at $T > 830°C$ is atomically rough. The possible existence of atomically rough crystal surfaces was examined earlier in [1]. It can be proposed that the filled upper layer of 1×1 consists of small clusters of adatoms that are not resolved by REM.

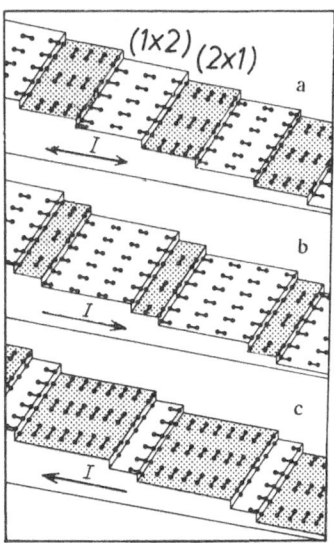

Fig. 10. Diagram of the structure of the Si(001) surface during sublimation. Heating conditions for a-c are the same as in Fig. 9a-c.

Additional experimental studies of the migration of adatoms and their interaction with fluxes at $T > 830°C$ are needed to confirm this conclusion.

4. Monoatomic Steps on the Si(001) Surface. According to [2, 30-35], steps of both diatomic and monoatomic height can exist on Si(001). Features of the superstructural reconstruction of this surface due to dimerization causes domination of either the superstructure for the surface with diatomic steps or superstructures 2×1 and 1×2 on the surface with monoatomic steps. The kinetics of the behavior of monoatomic steps during sublimation of crystals of given orientation were studied to determine the factors characterizing the relative areas of superstructures 2×1 and 1×2.

It was found that a system of equidistant monoatomic steps that advance during sublimation without changing the distance between them forms on the pure surface if the crystal is heated by alternating current. The terraces between steps have alternately bright and dark contrast (Fig. 9a). This is caused by alternation of superstructures 1×2 and 2×1 for a step height $a/4$ (Fig. 10a).

The monoatomic steps approach pairwise if the crystal is sublimed by heating with direct current at $T > 900°C$ [36]. The minimal step distance in the vapor reaches 0.1 μm at an average distance 0.5-2 μm in the equidistant system. If the direction of the direct current and that of the step advance during sublimation coincide, then almost the whole surface has the superstructure 1×2 and the dimers are perpendicular to the monoatomic steps. The portions between steps in the pair have superstructure 2×1 and the dimers are parallel to the steps (Figs. 9b and 10b). If the current and the step advance directions are opposed, then the surface has mainly the superstructure 2×1 (Figs. 9c and 10c).

The transition from one type of surface to the other on changing the direct current polarity is reversible. It occurs by decreasing the advance rate of the lower step in the pair and by increasing the advance rate of the upper step during sublimation. For any current direction at temperatures below 900°C the system of equidistant monoatomic steps with alternating superstructures 1×2 and 2×1 on terraces (the type in Fig. 9a) is stable.

The ratio of surfaces occupied by superstructures 1×2 and 2×1 under these heating conditions is determined firstly by anisotropic surface diffusion due to orientation of dimers into various domains and secondly by the presence of an effective adatom charge. The assumption of an effective adatom charge agrees with results

1 μm

Fig. 11. REM image of the Si(001) surface containing two-dimensional sublimation islands.

obtained earlier for the influence of the direction of the current used to heat the crystal on echelon formation processes of monoatomic steps on the pure Si(111) surface [20]. In fact, asymmetric islands are observed on Si(001) containing two-dimensional $a/4$ sublimation islands. The islands are elliptical (Fig. 11). The long axis of the ellipse is perpendicular to the orientation of dimers within the island. The asymmetric shape of the two-dimensional islands (the ratio of the short and long axes is 0.7-0.8) indicates that preferred directions of adatom diffusion exist on Si(001) containing superstructural domains.

Under these conditions the mechanism of paired approach of monoatomic steps consists of the following. The presence of effective adatom charge increases the reverse flux of adatoms from steps of these terraces for which the dimerization direction is parallel to the step line. Thus, a reverse flux arises on the upper steps of the 2×1 terraces for the experimental situation in Fig. 10b. This decreases the rate of shift of these steps relative to the lower steps of these terraces. For the case shown in Fig. 10c, the rate of shift of the lower steps of the 2×1 terraces decreases. The steps do not join into pairs on heating with alternating current. This is due to the rapid (50 Hz) change of the reverse flux of adatoms to the upper and lower steps due to drift in the alternating field. Thus, the change of relative advance rates of monoatomic steps causes paired approach of steps. This changes the relative areas of superstructures 2×1 and 1×2.

5. Initial Stages of Homoepitaxial Film Formation. Clarifying the role of monoatomic steps of the substrate as a trough for adatoms is one of the principal problems during a study of silicon homoepitaxy. The starting Si(111) surface in our experiments had a system of monoatomic steps that shifted toward the lower terraces during growth at $T > 830°C$, in agreement with a step-layer growth mechanism.

An image of the surface with monoatomic steps at various distances from each other is shown in Fig. 12. The parts of the monoatomic step A and B (Fig. 12a) have upper and lower terraces of approximately the same total width adjoining the step. However, part A has a lower terrace of about twice the width of that of part B, which is about twice the width of the upper terrace. Deposition of Si on this surface shifts the steps (Fig. 12b). The shift of part A is about twice that of part B. Thus, the shift of the monoatomic step during growth is determined mainly by the width of the lower terrace adjoining the step. It follows that the probability of incorporating adatoms into the step is greater from the lower terrace than the upper. The difference in the energy barriers for incorporating adatoms into the step is 0.2-0.6 eV according to estimates of adatom fluxes from the upper and lower terraces.

It was found that there is a critical flux of atoms onto the surface. Two-dimensional growth islands form on terraces between monoatomic steps if this flux is exceeded. The critical flux of atoms necessary to form the two-dimensional growth islands is controlled by the substrate temperature and the width of terraces between steps. Thus, for $d \approx 2 \mu m$ and $T = 870°C$, the critical flux is $(0.5-1) \cdot 10^{15}$ at $\cdot cm^{-2} \cdot sec^{-1}$.

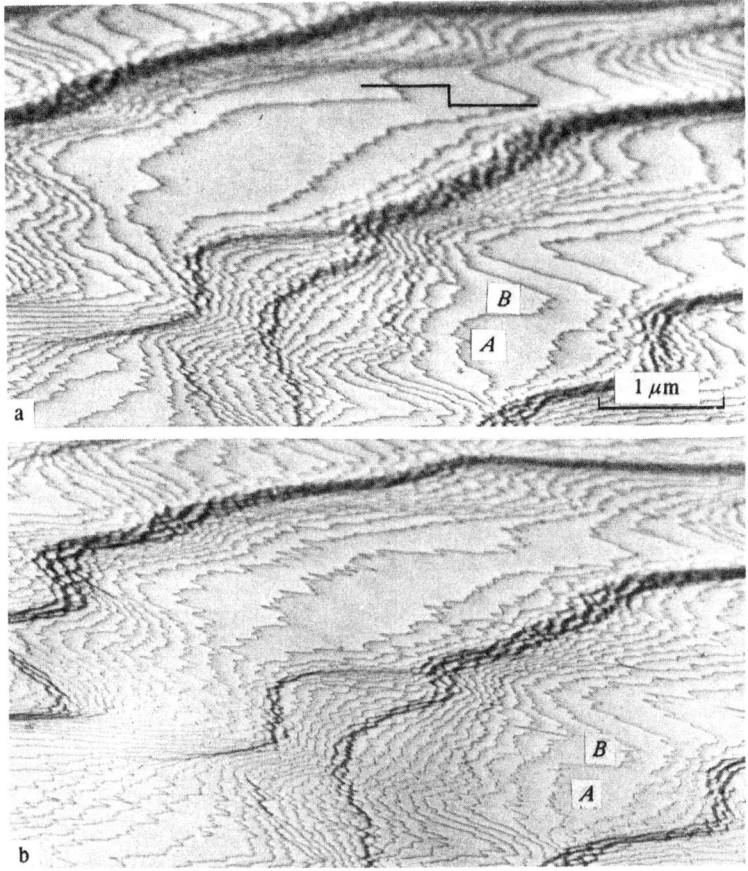

Fig. 12. Image of Si surface with monoatomic steps before growth (a) and after deposition of 0.25 Si monolayers at $T = 830°$C.

Sequential stages in formation and enlargement of the two-dimensional growth islands during homoepitaxy are shown in Fig. 13. Silicon was deposited at $T = 850$°C onto the starting surface (Fig. 13a). This formed two-dimensional growth islands (Fig. 13b). Further deposition of Si is accompanied by island growth (Fig. 13c) with subsequent coalescence with neighboring islands and the monoatomic step adjoining it (Fig. 13d and e). The surface with an almost filled monolayer, except for small parts appearing as two-dimensional negative islands of various shapes that were formed by coalescence of two-dimensional growth islands, is shown in Fig. 13e. This represents a shift of the monoatomic step by the width of a terrace and formation of one Si monolayer. Further growth is accompanied by periodic changes of the density and size of the surface islands (cf. Fig. 13b and f). A change of the density and size of the islands causes intensity oscillations of the reflected electron beam (Fig. 14). One oscillation corresponds to the time for formation of one Si monolayer. This agrees with the model of intensity oscillations proposed in [26] and based on the idea that the intensity of the reflected electron beam decreases when the surface is partially filled and increases when the monolayer is completed. The maximal number of oscillations observed in our experiments exceeds $5 \cdot 10^2$ at $T = 400$°C.

Deposition of tens of monolayers of Si at $T < 830$°C leads to the appearance of dark—bright contrast on the terraces between monoatomic steps. This contrast is caused by two-dimensional growth islands (Fig. 15) the size of which decreases and the density of which correspondingly increases as T is lowered. Zones free of islands are found near the monoatomic steps. The dimensions of these zones increase with increasing T. However, the width of the zone does not depend on the distance between steps. Thus, the monoatomic step is an effective trough for adatoms on the surface within a region with a width related to the adatom diffusion length. The

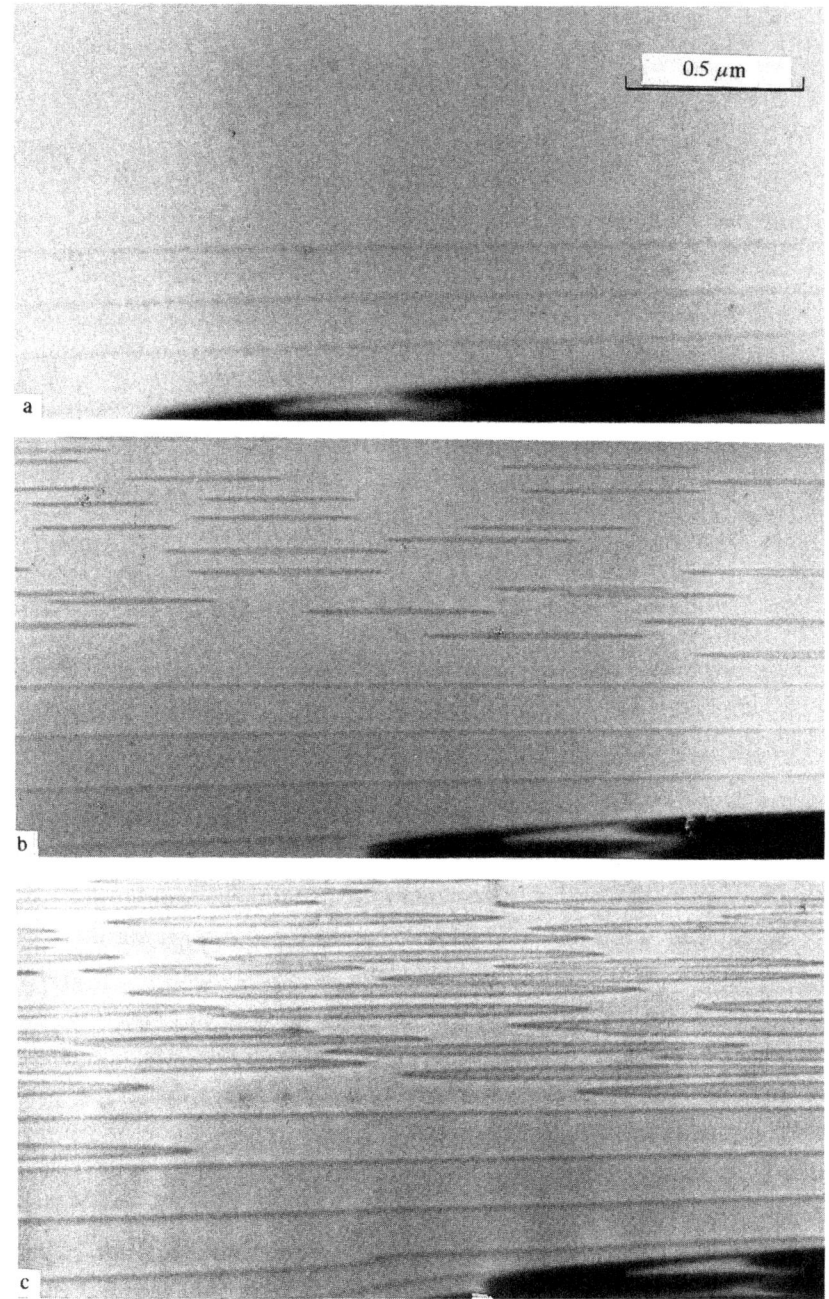

Fig. 13. Change of size of two-dimensional growth islands on the surface after deposition of Si: 0.0 (a), 0.1 (b), 0.3 (c), 0.6 (d), 0.9 (e), 1.2 (f) monolayer at $T = 850°C$.

temperature dependence of the width of the zones free of islands enables the activation energy of surface diffusion to be estimated for the adatoms. The value obtained, 1.3 ± 0.2 eV, is close that given in [37].

It was found that the size of the zones depleted in islands that adjoin the step from the upper terrace is greater than that of zones approaching the step from the lower terrace (cf. Fig. 15). An analysis of the image contrast of the crystal surface obtained from the superstructure reflection 7×7 demonstrated that the zone free from islands that directly adjoins the monoatomic step on the upper terrace has the 7×7 structure. However, the zone on the side of the lower terrace has the 1×1 structure (Fig. 16). This means that the 7×7 structure is more stable during growth on the parts of the upper terraces adjoining the steps. This conclusion agrees with the data obtained in Sec. 3. It can be assumed that the high stability of the 7×7 structure in these surface regions is

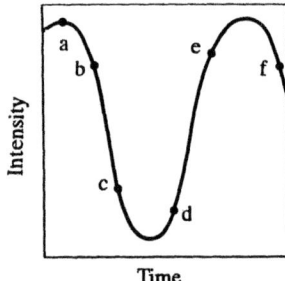

Fig. 14. Change of reflected electron beam intensity from the surface during growth. Points a-f correspond to surface images in Fig. 13a-f.

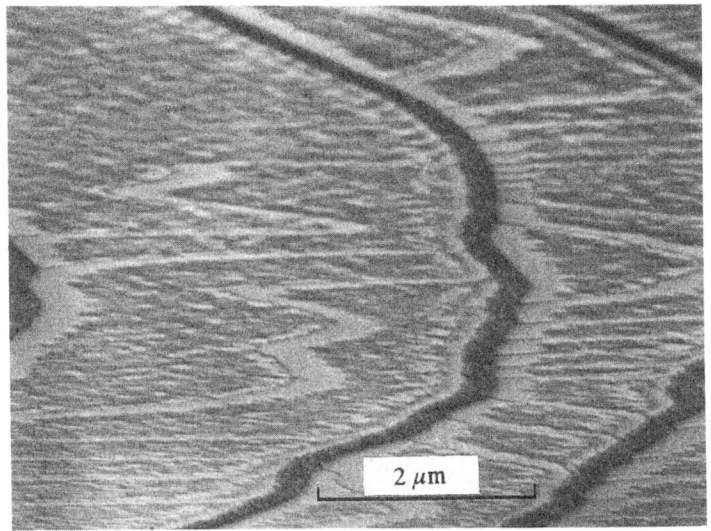

Fig. 15. Surface after deposition of 0.5 monolayer of Si at $T = 730°$C.

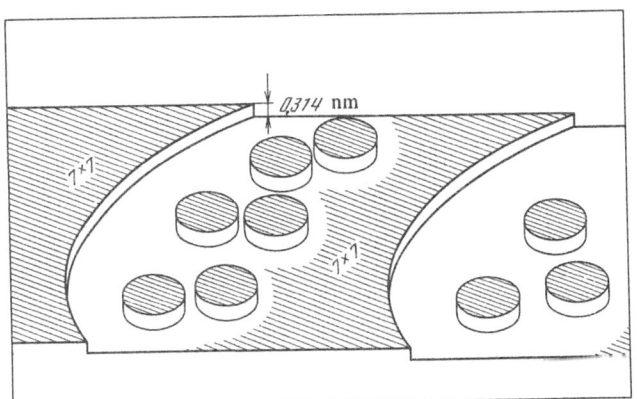

Fig. 16. Schematic diagram of islands on a Si surface during homoepitaxial growth at $T < 830°$C.

accompanied by a shift of adatoms from these regions to that with the less stable 7×7 structure or to the two-dimensional islands. The increased mobility of adatoms with increasing T increases the size of the zones on the steps with 7×7 structure. This is observed experimentally. The experimental data obtained indicate that the required superstructure reconstruction during epitaxy on Si(111) changes the formation kinetics of two-dimensional growth islands in surface regions adjoining the monoatomic steps of the substrate.

REFERENCES

1. A. A. Chernov, *Modern Crystallography*, Vol. 3 [in Russian], Nauka, Moscow (1980).
2. B. A. Nesterenko and O. V. Snitko, *Physical Properties of Atomically Clean Semiconductor Surfaces* [in Russian], Naukova Dumka, Kiev (1983).
3. I. F. Lyuksyutov, A. G. Naumovets, and V. L. Pokrovskii, *Two-dimensional Crystals* [in Russian], Naukova Dumka, Kiev (1988).
4. Y. Ota, "Silicon molecular beam epitaxy," *Thin Solid Films*, **106**, 1-136 (1983).
5. I. G. Neizvestnyi, A. V. Rzhanov, S. I. Stenin, and V. N. Shumskii, "Molecular-beam epitaxy as a method for fabricating modulated semiconducting structures," in: *Problems in Crystallography* [in Russian], Nauka, Moscow (1987), pp. 190-215.
6. A. V. Rzhanov and S. I. Stenin, "Molecular epitaxy: Status, problems, and prospects for development," in: *Growth of Semiconducting Crystals and Films* [in Russian], Nauka, Novosibirsk (1984), pp. 5-34.
7. A. V. Rzhanov, S. I. Stenin, and B. Z. Ol'shanetskii, "Methods of monitoring the surface state and problems of molecular beam epitaxy," *Mikroélektronika*, **9**, No. 4, 292-301 (1980).
8. K. Yagi, "Reflection electron microscopy," *J. Appl. Crystallogr.*, **20**, 147-160 (1987).
9. A. L. Aseev, A. V. Latyshev, and S. I. Stenin, "Structural reconstructions on atomically clean semiconductor surfaces studied using ultrahigh vacuum reflection electron microscopy," in: *Problems of Electronics Materials Science* [in Russian], Nauka, Novosibirsk (1986), pp. 109-127.
10. P. E. Højlund Nielsen and J. M. Cowley, "Surface imaging using diffracted electrons," *Surf. Sci.*, **54**, 340-354 (1976).
11. N. Osakabe, Y. Tanishiro, K. Yagi, and G. Honjo, "Image contrast of dislocations and atomic steps on (111) silicon surface in reflection electron microscopy," *Surf. Sci.*, **102**, 424-442 (1981).
12. Y. Uchida and G. Lehmpfuhl, "Observation of double contours of monoatomic steps on single crystal surfaces in reflection electron microscopy," *Ultramicroscopy*, **23**, 53-60 (1987).
13. H. Shuman, "Bragg diffraction imaging of defects at crystal surfaces," *Ultramicroscopy*, **2**, 361-369 (1977).
14. D. Katzer, M. Taege, and A. V. Latyshev, "REM-Untersuchung von Stufenbewegungen auf Si(111) Oberflächen," Veröffentlichungen zur 12. Tagung "Elektronenmikroskopie," Dresden (1988), pp. 101-105.
15. A. A. Kroshkov, É. A. Baranova, O. A. Yakushenko, et al., "Apparatus for differential ultrahigh vacuum evacuation of an electron microscope," *Prib. Tekh. Eksp.*, No. 1, 199-202 (1985).
16. Y. Ishikawa, N. Ikeda, M. Kenmochi, and T. Ichinokawa, "UHV-SEM observations of cleaning process and step formation on silicon (111) surfaces by annealing," *Surf. Sci.*, **159**, 256-264 (1985).
17. N. Osakabe, Y. Tanishiro, K. Yagi, and G. Honjo, "Reflection electron microscopy of clean and gold deposited (111) silicon surfaces," *Surf. Sci.*, **97**, 393-408 (1980).
18. R. E. Honig, "Sublimation studies of silicon in the mass spectrometer," *J. Chem. Phys.*, **22**, 1610-1611 (1954).
19. W. K. Burton, N. Cabrera, and F. C. Frank, "The growth of crystals and the equilibrium structure of their surfaces," *Trans. R. Soc. (London) A*, **243**, 299-358 (1951).
20. A. V. Latyshev, A. L. Aseev, A. B. Krasil'nikov, et al., "Behavior of monoatomic steps on silicon (111) surface during the sublimation under conditions of electric current heating," *Dokl. Akad. Nauk SSSR*, **300**, 84-88 (1988).
21. M. J. Lighthill and G. B. Whitman, "On kinematic waves," *Proc. R. Soc. (London) A*, **229**, 281-345 (1955).
22. F. C. Frank, "On the kinematics theory of crystal growth from solution," in: *Growth and Perfection of Crystals*, Wiley, New York (1958), pp. 411-419.
23. R. N. Schwoebel and E. J. Shipsey, "Step motion on crystal surfaces," *J. Appl. Phys.*, **37**, 3682-3686 (1966).
24. C. Leenwen, R. Rosmalen, and P. Bennema, "Simulation of step motion on crystal surfaces," *Surf. Sci.*, **44**, 213-236 (1974).
25. G. Binnig, H. Rohrer, C. Gerber, and E. Weibel, "7×7 reconstruction on Si(111) resolved in real space," *Phys. Rev. Lett.*, **50**, 120-123 (1983).
26. M. D. Pashley, K. W. Haberern, and W. Friday, "The effect of cooling rate on the surface reconstruction of annealed silicon (111) studied by scanning tunneling microscopy and low-energy electron diffraction," *J. Vac. Sci. Technol.*, A, **6**, No. 2, 488-492 (1988).
27. A. V. Latyshev, A. L. Aseev, and S. I. Stenin, "Anomalous behavior of monoatomic steps during the structural transition 1×1 → 7×7 on atomically clean silicon (111) surface," *Pis'ma Zh. Eksp. Teor. Fiz.*, **49**, No. 9, 448-450 (1988).
28. K. Takayanagi, Y. Tanishiro, M. Takahashi, S. Takahashi, "Structural analysis of Si(111) 7×7 by UHV transmission electron diffraction and microscopy," *J. Vac. Sci. Technol.*, A, **3**, 1502-1506 (1985).
29. J. H. Neava, P. J. Dobson, and B. A. Joyce, "Reflection high-energy electron diffraction oscillations from vicinal surfaces - a new approach to surface diffusion measurements," *Appl. Phys. Lett.*, **47**, 100-102 (1985).
30. R. J. Hamers, R. M. Tromp, and J. E. Demuth, "Scanning tunneling microscopy of Si(001)," *Phys. Rev. B: Condens. Matter*, **34**, 5343-5357 (1986).
31. B. Z. Olshanetsky and A. A. Shklyaev, "LEED studies of vicinal surface of silicon," *Surf. Sci.*, **82**, 445-452 (1979).

32.   P. E. Wierenga, J. A. Kubby, and J. E. Griffith, "Tunneling images of biatomic steps on Si(001)," *Phys. Rev. Lett.*, **59**, 2169-2172 (1987).

33.   D. E. Aspnes and J. Ihm, "Biatomic steps on (001) silicon surfaces," *Phys. Rev. Lett.*, **57**, 3054-3057 (1986).

34.   T. Nakayama, Y. Tanishiro, and K. Takayanagi, "Monolayer and bilayer high steps on Si(001) 2×1 vicinal surface," *Jpn. J. Appl. Phys.*, **26**, L1186-L1188 (1987).

35.   N. Inoue, Y. Tanishiro, and K Yagi, "UHV-REM study of changes in the step structures on clean (100) silicon surfaces by annealing," *Jpn. J. Appl. Phys.*, **26**, L293-L295 (1987).

36.   A. V. Latyshev, A. B. Krasil'nikov, A. L. Aseev, and S. I. Stenin, "Influence of electric current on the area ratio of the domains 2×1 and 1×2 on clean silicon (001) surface during sublimation," *Pis'ma Zh. Eksp. Teor. Fiz.*, **48**, No. 9, 484-487 (1988).

37.   R. F. C. Farrow, "The kinetics of silicon deposition on silicon by pyrolysis of silane," *J. Electrochem. Soc.*, **121**, 899-907 (1974).

Part II

MOLECULAR-BEAM EPITAXY

MOLECULAR-BEAM EPITAXY OF SILICON

S. I. Stenin, B. Z. Kanter, and A. I. Nikiforov

Molecular-beam epitaxy (MBE) of silicon is attracting increasing scientific and practical interest due to the low film growth temperatures and the accurate control of their thickness, composition, and doping level. These capabilities enable thin multilayered structures with strictly controlled composition and very sharp doping profiles to be obtained with high reproducibility [1-3]. Moreover, the number of heterostructures available by this method can be substantially increased by combining MBE of silicon with epitaxy of dielectrics, metal silicides, and Ge—Si solid solutions. The potentialities of MBE were used as a basis for construction of new instruments and for improvement of the characteristics of those already existing. Progress in this area suggests that MBE will enable in the future a new fundamental basis for microelectronics to be found. In particular, three-dimensional integrated circuits will become feasible.

In the present work, the experimental status of MBE of silicon is analyzed. The following questions are examined: preparation of substrates, doping, and the use of intensity oscillations of diffracted electrons. Possible instrumental uses of multilayered structures obtained by MBE are listed in the conclusion.

Preparation of Silicon Substrate Surface. Standard methods of cleaning the silicon surface by chemical etching before placing the substrate in the MBE chamber prevent rapid adsorption of hydrocarbons and moisture on the surface. Surface hydrocarbons react with Si to form SiC on heating in vacuum. The carbide particles act as stoppers for growth steps and help the generation of defects. In earlier works, heating in ultrahigh vacuum to a temperature near the melting point was used for cleaning. At this temperature the natural oxide and SiC are removed. However, this process is accompanied by redistribution of the impurity near the substrate surface (for example, cf. [4]).

Depletion of the subsurface layer is equivalent to expansion of the transitional region between the film and substrate. Accumulation of an impurity causes a heavily doped layer to appear at the interface. Generation of a layer enriched in boron was demonstrated in [5] for growth of an undoped homoepitaxial film of Si on the substrate with a B concentration of $4 \cdot 10^{14}$ cm^{-3} (cf. Fig. 1, 2). The substrate was cleaned at 1250°C for 3 min before growth. This led to accumulation of B on the surface due to its lower vapor pressure compared with Si. Profile 1 in Fig. 1 corresponds to an undoped Si film on the substrate with a B concentration 10^{19} cm^{-3}. Films of both types were obtained at a substrate temperature $T_s = 800$°C. The charge carrier concentration distribution was determined by a C—V method. Dashed lines denote profiles corrected for the Debye length. The concentration transition on the heavily doped substrate (curve 1) is sharp. The background doping level is $2 \cdot 10^{14}$ cm^{-3} and does not depend on the substrate doping level. Apparently this is due to weak B segregation on the growth surface.

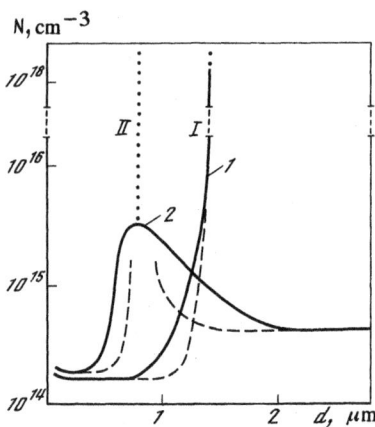

Fig. 1. Charge carrier distribution in the film on p^+-substrates (1) and p-substrates (2). The numbers I and II denote the interface for the corresponding structures.

These effects and the inability to use high-temperature cleaning on substrates with multilayered structure necessitated development of methods of cleaning the Si surface at temperatures near the epitaxy temperature (700-800°C).

Two basic directions in the development of such methods can be identified. The first is based on removal of adsorbed and oxidized layers by ion bombardment with subsequent annealing of radiation defects. The drawbacks of this method are the difficulty of cleaning substrates of large diameter, the high purity inert gas required, and the possible transport of contaminants onto the substrate as the ions impinge on the parts of the vacuum chamber.

The substrate cleaning method at temperatures near 800°C proposed in [6] has been used most widely in MBE. The substrate was subjected to several oxidation cycles with subsequent removal of the oxide before placing into the chamber. This removed metal and hydrocarbon impurities and then created a thin protective oxide film. The cleaning was performed under ultrahigh vacuum at 800°C. The SiO_2 is reduced to SiO which desorbs from the surface. After cleaning, the Si surface is free from contaminants according to Auger spectroscopy. According to reflective high-energy electron diffraction (RHEED) the surface has superstructure characteristic of the clean surface. However, secondary-ion mass spectrometry indicates that carbon is present at a level of 10^{18} cm^{-3} at the film—substrate interface [7].

Cleaning in a flux of Si is used to reduce the desorption temperature of the protective oxide and to decrease the defect density [8]. The cleaning temperature decreases by 50-80 K with Si deposition. The defect density is reduced by about an order of magnitude. The optimal Si layer thickness, equal approximately to half of the oxide thickness, occurs for the (111) orientation. The (100) surface is less critical to the amount of deposited Si.

Desorption of the protective SiO_2 film was studied in detail by Auger spectroscopy in [9]. The activation energy for thermal desorption was 3.54 ± 0.2 eV. The sublimation temperature of an oxide 2.5 nm thick could be lowered from 850-900°C to 700°C at a Si flux $3 \cdot 10^{13}$ cm$^{-2} \cdot$sec^{-1}. In this case the activation energy of desorption decreased to 0.84 ± 0.2 eV.

Methods of Doping and Forming Structures with a Given Impurity Distribution Profile. A wide range of impurity concentrations and sharp concentration transitions must be ensured in doping to fabricate multilayered structures based on Si. The impurity concentration in the bulk of the growing film is determined by the atom density on the crystal surface, the temperature, and the height of the activation barrier for incorporating impurity from the surface into the bulk. The impurity atom density on the surface is a function of the impurity flux, the growth rate, the substrate temperature, and the height of the activation barrier of desorption. Therefore, studies of processes accompanying doping are an important inherent part of MBE.

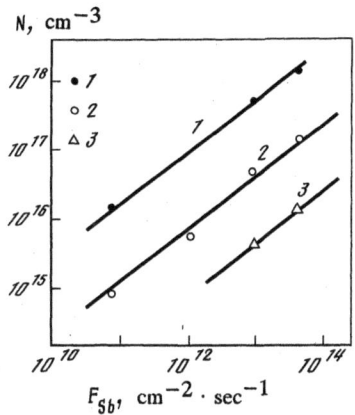

Fig. 2. Dependence of doping level on substrate temperature and Sb flux. $T_s = 700$ (1), 800 (2), 900°C (3).

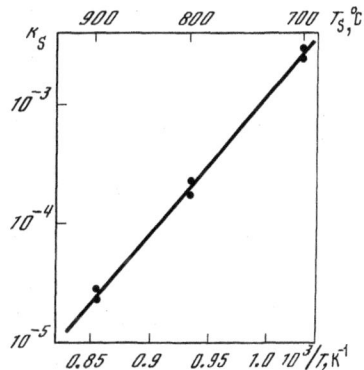

Fig. 3. Dependence of sticking coefficient of Sb to the Si(111) surface on substrate temperature.

Antimony is used at present as the principal donor impurity. Gallium and recently boron are used as acceptors. Doping with Sb is studied in most detail in [10]. It is important that Sb from the source arrives at the growth surface as Sb_4 tetramers. Partial or complete dissociation into monomers during surface adsorption is a function of temperature. A high cluster concentration in the adlayer can cause heavily doped (more than 10^{18} cm^{-3}) films to have a high defect density.

A plot of the Sb doping level during MBE of Si vs. Sb flux and T_s (Fig. 2) shows that the doping level depends on both the dopant flux and the substrate temperature. The temperature effect is due to a strong dependence of the sticking coefficient of Sb to the Si(111) surface on T_s. This dependence is plotted in Fig. 3. The activation energy of the sticking coefficient of Sb determined from the graph is 2.3 eV. These results agree well with both the sticking coefficients obtained in [10] and their calculated dependence of doping level on dopant fluxes and substrate temperature.

In order to understand the doping process better, the behavior of dopant on the surface must be known. This explains the interest in studies of superlattice structures and desorption processes on enriching the Si surface with dopant. Formation conditions of two-dimensional ordered phases on the Si(111) surface stabilized by Sb at 600-800°C are studied in [11]. It is demonstrated by RHEED that Sb forms ordered two-dimensional dopant phases Si(111)—$\sqrt{3} \times \sqrt{3}$—Sb and Si(111)—$5\sqrt{3} \times 5\sqrt{3}$—Sb (denoted further as $\sqrt{3}$—Sb and $5\sqrt{3}$—Sb). Regions of existence of the ordered phases stabilized by a constant Sb flux onto the surface, i.e., where the desorbed flux is equal

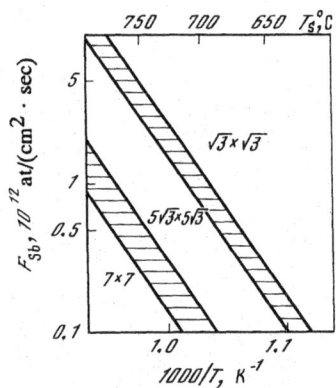

Fig. 4. Regions of existence on the Si(111) surface of two-
dimensional ordered phases stabilized by Sb flux.

to the incident, are plotted in Fig. 4. The streaked regions correspond to regions of coexistence of the two phases. After the flux is stopped at these temperatures, Sb desorption causes the sequential superlattice transitions $\sqrt{3}$—Sb $\rightarrow$ $5\sqrt{3}$—Sb $\rightarrow$ 7×7.

Two-dimensional ordered phases on the Si(111) surface are also formed during isothermal annealing of Sb film with coverages from 0.2 to 2 monolayers deposited at $T_s = 300°C$. The dependence of the residence time of the phases on temperature (Fig. 5) was used to determine the activation energy of superstructural transitions $\sqrt{3}$—Sb $\rightarrow$ $5\sqrt{3}$—Sb and $5\sqrt{3}$—Sb $\rightarrow$ 7×7. The corresponding values are 1.8 ± 0.1 eV and 2.1 ± 0.1 eV. These are apparently the activation energies of Sb desorption. The difference in activation energies may represent the difference in bond energies of Sb to the Si(111) surface for various ordered two-dimensional phases.

Doping with Sb has much in common with introduction of other dopants such as Ga and In [12, 13]. Of course the parameters characterizing this system will be quantitatively different. Like for Sb, desorption of In depends on its structural state on the Si surface. This produces an activation energy difference of about 0.3-0.4 eV [13].

Boron is singled out among the widely used dopants. It should have the greatest activation energy of desorption from the Si surface. It is difficult to use in pure form due to a low vapor pressure and, consequently, a high vaporization temperature. Nevertheless, vaporization of elemental B is used in MBE of Si since it has a sticking coefficient that does not depend on substrate temperature and is near unity [14]. This enables sharp concentrational transitions of less than 10 nm to be prepared for an order of magnitude change of dopant concentration. The compounds B_2O_3 [15, 16] and HBO_2 [17] have been proposed for depositing B on the growth surface. The incorporation of B into the growing crystal is just as effective as for vaporization of elemental B but the source temperature can be reduced considerably. The mechanism of depositing B in the adsorption layer apparently consists of the following. A molecule containing dopant reacts with Si atoms on the substrate. The B atom released by the chemical reaction is incorporated into the film. The oxygen forms SiO, which desorbs from the surface.

The effectiveness of the incorporation is easily seen from the dependence of the hole concentration on the ratio of B and Si fluxes for films prepared at a substrate temperature 800°C (Fig. 6) [16]. For concentrations from 10^{16} to 10^{20} cm^{-3}, the doping level is determined only by the B and Si fluxes with the condition that the B sticking coefficient is unity (solid line). The deviation from linearity at B concentrations greater than 10^{20} cm^{-3} is apparently due to formation of complexes containing B in the film when the limiting B solubility in Si at this temperature is exceeded. The doping level at a constant ratio of B and Si fluxes does not depend on T_s between 600-850°C. This indicates that the B sticking coefficient is constant in this range.

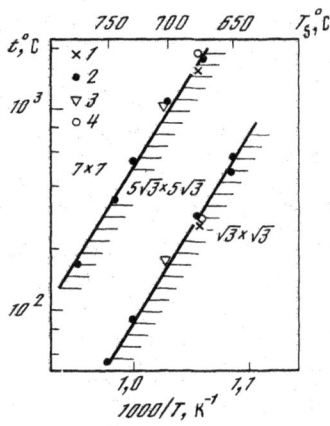

Fig. 5. Dependence of the residence time of two-dimensional phases on the Si(111) surface on temperature with Sb coverage from 0.2 to 2 monolayers. (1) $\theta_0 = 0.2$; (2) $\theta_0 = 0.4$; (3) $\theta_0 = 0.6$; (4) $\theta_0 = 2$.

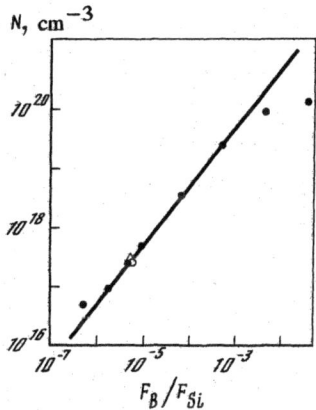

Fig. 6. Dependence of B doping level on ratio of B and Si fluxes. $T_s = 600$ (○), 800 (●), 850°C (△).

Radiation-stimulated incorporation of dopants in MBE has been studied recently. The surface is treated with external electrons [18] or Si ions [14, 19]. Electron irradiation increases the Sb sticking coefficient by about two orders of magnitude between 700-900°C. The maximal doping level for a perfect film increases from $5 \cdot 10^{17}$ to $3 \cdot 10^{19}$ cm^{-3}. Moreover, electron irradiation sharply reduces the amount of Sb on the growth surface. Thus, conditions are favorable for decreasing the width of concentrational transitions. The picture is similar for bombardment with Si ions. Another method of increasing the doping effectiveness is to use partially ionized dopant fluxes. This decreases the amount of dopant required on the surface and enables sharp concentrational transitions to be obtained.

One of the interesting applications of doping from molecular beams is the fabrication of doped δ-layers, i.e., layers in which the dopant is distributed practically in one atomic layer of the crystal. A δ-layer of Sb on Si(100) was prepared by combining solid-phase epitaxy (SPE) of a thin amorphous Si film (2-3 nm) under which was the δ-layer, with subsequent epitaxy of Si from a molecular beam [20] and by SPE alone [21]. The width of the δ-layer according to electron microscopy of a cross section is less than 2-3 nm. Formation of discrete quantum sublevels of two-dimensional electron gas confirms the sharpness of the transition. The values of the quantum sublevels calculated theoretically agree well with those observed experimentally for a δ-layer width less than 3 nm.

A δ-layer of Sb on Si(111) was fabricated using a combination of SPE and MBE. These structures were prepared on a Si MBE apparatus [22] fitted with a RHEED system and a sample changer for five substrates of 60-mm diameter. The distribution profile of electrically active dopant in a δ-layer containing $2 \cdot 10^{13}$ cm^{-2} Sb that was prepared at $T_s = 700°C$ is presented in Fig. 7. The δ-layer was formed on an epitaxial film of 1.3-μm thickness doped with Sb to a concentration of 10^{16} cm^{-3}. An undoped layer of 150-nm thickness was grown over the δ-layer. According to four-probe layered analysis the width of the δ-layer is less than the minimal thickness of the etched layer, i.e., 10 nm. The carrier concentration changes by an order of magnitude within 5 nm in the profile. This indicates that segregation and diffusion of Sb are suppressed during formation of the structure.

Another important doping application is fabrication of modulated n—i—p—i superlattices [23-25]. The first of these works studies the formation process and the influence of diffusion. The other two investigate electrophysical properties. A hole mobility at room temperature of about 1000 cm$^2 \cdot$V$^{-1} \cdot$sec^{-1} was detected for the p—i—p—i structure with a concentration in the p-region of 10^{18} cm^{-3} and a period of 30 nm. This is an order of magnitude greater than the mobility in the bulk for the same hole concentration [27]. The mobility increases according to $T^{-2.2}$ as the temperature is lowered. This is characteristic of the material itself. The value at 30 K is 40,000 cm$^2 \cdot$V$^{-1} \cdot$sec^{-1}. This is twice the bulk value. The doped superlattice described in [28] has a carrier lifetime of 1.2 sec, i.e., close to that in Si itself.

Use of RHEED to Study Growth Processes and to Monitor Thickness. The composition of the substrate surface and the growing film during MBE are known to be effectively monitored by RHEED. However, the detection of oscillations of reflected electron beams during growth of an epitaxial film by a two-dimensional-layered mechanism has recently provided the impetus for increased use of RHEED. Observation of this effect enables layered growth of a film to be monitored. Thus, it can be used to measure film thickness and growth rate and to study surface processes during epitaxy.

A minimally roughened surface must be used to obtain stable intensity oscillations. This is achieved by growing a buffer layer with subsequent annealing. As demonstrated by reflective electron microscopy (REM) [26, 27], a clean Si surface has terraces separated by monoatomic steps. A similar picture is repeated periodically during growth and corresponds to the RHEED intensity maximum. An intensity minimum occurs when the surface of the terraces is covered by two-dimensional nuclei over about one half of the area (cf. [26], Figs. 13 and 14).

Oscillations of RHEED intensity are observed during Si epitaxy over a wide range of growth temperatures, including ambient, on all three singular planes (100), (111), and (110) [28]. The plane (100) has the widest range of temperatures. The growth model of the (100) face proposed by Sakamoto et al. [29] is based on alternation of the domains 1×2 and 2×1 after each monolayer of height $a_0/4$, where a_0 is the interplanar distance. Islands extended along a certain direction are characteristic of each type of domain. This model was confirmed simultaneously by observed intensity changes of three reflections: the trivial one and superstructural ones with domains 1×2 and 2×1. The intensity of the latter changes out-of-phase and their period is two times less than the trivial one.

The oscillation period during growth of Si(111) and Si(110) films is equal to the time for growth of one monolayer of thickness $a_0/3$ and $a_0\sqrt{2}/4$, respectively.

Conclusion. The principal advantages of MBE of Si, namely the low epitaxy temperature, strict control of the dopant composition, and low background level of dopant, determine the possible applications. Two directions must be identified: fabrication of modulated superlattices for UHF and optomicroelectronics and of structures with increased speed, efficiency, and power due to thinner active regions and sharp concentrational transitions.

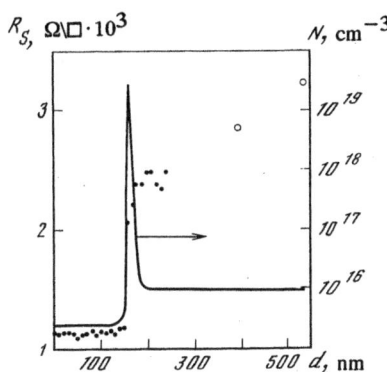

Fig. 7. Layered analysis of a δ-layer of Sb with concentration $2 \cdot 10^{13}$ cm^{-2} at a depth of 150 nm. Points are data obtained after anodic oxidation; circles, after chemical etching; solid line, dopant concentration.

Use of modulated Si superlattices has not yet been reported. A structure with alternating δ-layers can also provide a basis for fabrication of superlattices. The first application of the δ-layers was their use as the channel in an FET with a Schottky barrier and a MOSFET [20, 21]. The characteristics of these devices attest to their promise in spite of the nonoptimal geometry.

The high speed integrated circuit of [30] is a prominent example of how the device parameters can be improved using MBE. The working frequency could be increased from 900 MHz to 2.8 GHz by changing an epitaxial layer of 2.5-μm thickness prepared by gas-phase epitaxy to a layer of 0.9-μm thickness grown using MBE. The thickness of the active layer and the concentrational periods were decreased.

Another successful application of Si MBE is fabrication of avalanche—drift diodes (IMPATT) operating in the millimeter range. A complicated n- and p-type doping profile with sharp transitions is necessary for this class of UHF devices. The IMPATT obtained was used as a quasi-Read double-drift diode at 100 GHz and has 910 mW power with an efficiency of 11% [31]. This is close to the theoretical values for this structure. Integrated use of a quasi-Read double-drift IMPATT at 75 GHz has 200 mW power with an efficiency of 4.5% [32]. The output power for such use is an order of magnitude greater than the value for IMPATT fabricated using other technologies. Finally, planar IMPATT oscillators were first prepared using MBE [33]. This diode, placed on a high-ohmic substrate (10 kΩ), has a double-drift structure and an output power of 1 mW at 76 GHz.

As seen from the examples given, MBE of Si enables fabrication of devices and structures with unique characteristics.

REFERENCES

1. Y. Shiraki, "Silicon molecular beam epitaxy," *Prog. Cryst. Growth Charact.*, **12**, 45-66 (1986).
2. J. C. Bean, "Silicon molecular beam epitaxy," *J. Cryst. Growth*, **81**, No. 1/3, 411-420 (1987).
3. A. V. Rzhanov and S. I. Stenin, "Molecular epitaxy: Status, problems, and prospects for development," in: *Growth of Semiconducting Crystals*, Part 1 [in Russian], Nauka, Novosibirsk (1984), pp. 5-34.
4. A. V. Kozhukhov, B. Z. Kanter, and S. I. Stenin, "Behavior of gallium in the subsurface layer of silicon on heating in ultrahigh vacuum," in: Abstracts of Papers of the Seventh Conf. on Growth Processes and Synthesis of Semiconducting Crystals and Films, Vol. 3 [in Russian], Novosibirsk (1986), p. 90.
5. A. V. Rzhanov, S. I. Stenin, O. P. Pchelyakov, and B. Z. Kanter, "Molecular beam epitaxial growth of germanium and silicon films: Surface structure, film defects, and properties," *Thin Solid Films*, **139**, No. 2, 169-175 (1986).
6. A. Ishizaka, K. Nakagawa, and Y. Shiraki, "Low-temperature surface cleaning of silicon and its application to silicon MBE," in: Coll. Papers of the Second Int. Symp. on Molecular Beam Epitaxy and Related Clean Surface Techniques, Jpn. Soc. Appl. Phys., Tokyo (1982), pp. 183-186.

7. Y. H. Xie, Y. Y. Wu, and K. L. Wang, "Deep level defect study of molecular beam epitaxially grown silicon films," *Appl. Phys. Lett.*, **48**, No. 4, 287-289 (1986).

8. T. Tatsumi, N. Aizaki, and H. Tsuya, "Advanced techniques to decrease defect density in molecular beam epitaxial silicon films," *Jpn. J. Appl. Phys.*, **24**, No. 4, L227-L229 (1985).

9. D. C. Streit and F. G. Allen, "Thermal and Si-beam assisted desorption of SiO_2 from silicon in ultrahigh vacuum," *J. Appl. Phys.*, **61**, No. 8, 2894-2897 (1987).

10. R. A. Metzger and F. G. Allen, "Evaporative antimony doping of silicon during molecular beam epitaxial growth," *J. Appl. Phys.*, **55**, No. 4, 931-940 (1984).

11. B. Z. Kanter, A. I. Nikiforov, and S. I. Stenin, "Formation of two-dimensional ordered phases on the Si(111) surface during sputtering of Sb and isothermal annealing," *Pis'ma Zh. Tekh. Fiz.*, **14**, No. 21, 1963-1968 (1988).

12. F. G. Allen, S. S. Iyer, and R. A. Metzger, "Dopant incorporation studies in silicon molecular beam epitaxy (Si MBE)," *Appl. Surf. Sci.*, No. 11/12, 517-527 (1982).

13. J. Knall, J. E. Sundgren, G. V. Hansson, and J. E. Greene, "Indium overlayers on clean Si(100)—2×1: surface structure, nucleation, and growth," *Surf. Sci.*, **166**, 512-538 (1986).

14. R. A. A. Kubiak, W. Y. Leong, and E. H. C. Parker, "Enhanced sticking coefficients and improved profile control using boron and antimony as coevaporated dopants in Si-MBE," *J. Vac. Sci. Technol., B*, **3**, No. 2, 592-595 (1985).

15. R. M. Ostrom and F. G. Allen, "Boron doping in Si molecular beam epitaxy by coevaporation of B_2O_3 on doped silicon," *Appl. Phys. Lett.*, **48**, No. 3, 221-223 (1986).

16. A. I. Nikiforov, B. Z. Kanter, and S. I. Stenin, "Doping of boron from B_2O_3 during molecular beam epitaxy of silicon," in: Abstracts of the All-Union Conf. on Physical and Physical Chemical Principles of Microelectronics [in Russian], Moscow (1987), pp. 286-287.

17. T. Tatsumi, H. Hirayama, and N. Aizaki, "Boron heavy doping for Si molecular beam epitaxy using a HBO_2 source," *Appl. Phys. Lett.*, **50**, No. 18, 1234-1236 (1987).

18. F. A. D'Avitaya, S. Delage, and E. Rosencher, "Silicon MBE: recent developments," *Surf. Sci.*, **168**, 483-497 (1986).

19. H. Jorke, H.-J. Herzog, and H. Kibbel, "Secondary implantation of Sb into Si molecular beam epitaxy layers," *Appl. Phys. Lett.*, **47**, No. 5, 511-513 (1985).

20. H. P. Zeindl, T. Wegehaupt, I. Eisele, et al., "Growth and characterization of a delta-function doping in Si," *Appl. Phys. Lett.*, **50**, No. 17, 1164-1166 (1987).

21. A. A. van Gorkum, K. Nakagawa, and Y. Shiraki, "Controlled atomic layer doping and ALD MOSFET fabrication in Si," *Jpn. J. Appl. Phys.*, **26**, No. 12, L1933-L1936 (1987).

22. B. Z. Kanter, N. T. Moshegov, A. I. Nikiforov, et al., "Ultrahigh vacuum apparatus for molecular beam epitaxy of Si," *Prib. Tekh. Eksp.*, No. 2, 171-173 (1988).

23. D. C. Streit, E. D. Ahlers, and F. G. Allen, "Sharp profiles and low-temperature diffusion of Ga and Sb in silicon modulation-doped superlattices," *J. Vac. Sci. Technol., B*, **5**, 752-756 (1987).

24. K. Nakagawa and Y. Shiraki, "Anomalous mobility enhancement in Si doping superlattices," *Surf. Sci.*, **714**, 646-650 (1986).

25. K. H. Teo, J. N. McMullin, F. Weichman, et al., "Measurement and mechanism of free carrier recombination in a silicon doping superlattice," in: Workbook of the Fifth Int. Conf. on MBE, Sapporo, Japan (1988), pp. 248-251.

26. A. V. Latyshev, A. L. Aseev, A. B. Krasil'nikov, and S. I. Stenin, "Structural reconstruction on atomically clean silicon surface during sublimation and epitaxy," Growth of Crystals, Vol. 18, preceding article.

27. M. Ishikawa and T. Doi, "Observation of Si(111) surface topography changes during Si molecular beam epitaxial growth using microprobe reflection high-energy electron diffraction," *Appl. Phys. Lett.*, **50**, No. 17, 1141-1143 (1987).

28. T. Sakamoto, N. V. Kawai, and M. T. Nakagawa, "Intensity oscillation of reflection high-energy electron diffraction during silicon molecular beam epitaxial growth," *Appl. Phys. Lett.*, **47**, No. 6, 617-619 (1985).

29. T. Sakamoto, T. Kawamura, S. Nago, et al., "RHEED-intensity oscillation of alternating surface reconstructions during Si MBE growth on single-domain Si(001)—2×1 surface," *J. Cryst. Growth*, **81**, No. 1/3, 59-64 (1987).

30. E. Kasper and K. Wörner, "High speed integrated circuit using silicon molecular beam epitaxy (Si MBE)," *J. Electrochem. Soc.*, **132**, No. 10, 2841-2846 (1985).

31. J. F. Luy, E. Kasper, and W. Behr, "Semiconductor structures for 100 GHz silicon IMPATT diodes," in: Proc. 17th Europ. Microwave Conf., Italy (1987), pp. 820-825.

32. J. Buechler, E. Kasper, J. F. Luy, et al., "70 GHz integrated silicon oscillator," *Electron. Lett.*, **24**, No. 15, 977-978 (1988).

33. J. F. Luy, K. M. Strohm, and J. Buechler, "Silicon monolithic millimeter wave IMPATT oscillators," in: Proc. 18th Europ. Microwave Conf., Sweden (1988), pp. 382-387.

MOLECULAR EPITAXY OF A_3B_5 COMPOUNDS

Yu. O. Kanter and A. I. Toropov

Molecular-beam epitaxy (MBE) is a relatively new method for preparing epitaxial films. It recently has been studied and developed extensively. Multilayered films with unique characteristics can now be obtained. The most important achievement of MBE is the possibility of fabricating so-called modulated semiconducting structures (MSS). A particular case of these are superlattices (SL) consisting of thick layers of one or several monolayers. Superlattices are actually a new class of materials with a regular band structure. This holds great promise for developments in microelectronics.

The principles of MBE and the properties of epitaxial films prepared by this method (mainly GaAs and $Al_xGa_{1-x}As$) have been reviewed [1-4]. In the present work special attention is paid to physicochemical problems of MBE structures based on A_3B_5 compounds. Results from studies of the morphological instability of the planar crystallization front are presented. Questions of SL characterization and quantum well structures are discussed.

1. Apparatus for Molecular-Beam Epitaxy of A_3B_5 Compounds. The low growth rates inherent to MBE and the necessity to prepare films with limiting parameters impose stringent demands on the film growth apparatus. The growth chamber is surrounded with liquid-nitrogen-cooled cryoshrouds to reduce the amount of background impurities. Ultrahigh vacuum is normally required for various analytical methods such as quadrupole mass spectrometry, Auger electron spectroscopy (AES), relfective high-energy electron diffraction (RHEED), and automatic ellipsometry. These are used to monitor and study growth mechanisms and film properties. The number of analytical attachments in each chamber is optimized to the tasks undertaken and the materials studied.

It is noteworthy that a characteristic of the Group V elements (As, P, Sb) is their ability to exist in the vapor as two- or four-atom clusters. According to the literature, the type of cluster in the molecular beam has a great influence on the properties of the growing films [5]. For example, use of As_2 dimers increases the sticking coefficient of As, improves the electrophysical and optical properties of the films, and in particular decreases the concentration of deep levels and increases the lifetime of charge carriers. We will consider three methods of producing dimers: vaporization from dissociating A_3B_5 compounds [1], use of high-temperature sources for cracking As_4 [6], and use of gas sources with arsine or phosphine decomposition [7, 8].

Recently [7-9] the trend has been to combine various epitaxial growth technologies with MBE technologies. In particular, multicomponent gas sources (chemical molecular epitaxy) are widely used. This enables the merits of vapor phase and organometallic depositions to be used while retaining the ability to control precisely the growth process.

2. Preepitaxial Preparation of Substrates. The importance of preparing substrates and cleaning surfaces for production of high-quality films by MBE has been emphasized in the literature [1-3]. This problem becomes particularly pivotal for growth of structures with a thin active region, for example, SL. The usual procedure is to clean A_3B_5 substrates [10] by polishing, chemical etching, and formation of a protective oxide film that protects the surface from contaminants. A substrate prepared this way was until recently stuck by In to a Mo partner. Modern MBE apparatuses use direct radiational heating instead of the traditional liquid-metal contact [11]. This enables several steps associated with sample sticking to be eliminated and the reproducibility and uniformity of temperature over the area to be increased. The correct substrate surface temperature is set using known superstructural transitions as identified by RHEED [12].

The protective oxide layer is removed before epitaxy by heat treatment of the substrates in the MBE apparatus. The best results are obtained by using two-step treatment [13]. The first step involves heating in a preliminary preparation chamber to 200-500°C. It is then annealed in the growth chamber in a flux of Group V element. Traces of carbon, the concentration of which is 0.5-5% of a monolayer, are detected using AES and electronic spectroscopy for chemical analysis (ESCA) [10]. Until now the most reliable method of preparing atomically clean substrate surfaces was to grow homoepitaxial buffer layers immediately before growth of the device structures.

3. Production of Highly Uniform Film Properties Using MBE. Difficulties arise during growth of A_3B_5 compounds and especially solid solutions based on them due to the spatial nonuniformity and temporal instability of the molecular beams [15, 16]. In [15] the influence of the geometric configuration of the source substrate on film uniformity with depth was examined. An effusion cell with internal diameter A and crucible cone angle θ_0 was placed at an angle θ to the axis perpendicular to the substrate. The molecular-beam axis passes through the substrate center at a distance L from the top of the crucible. The level of vaporized substance is held below the aperture not only for the Group III elements but also for the Group V elements. It is recommended that the substrate rotate. In [15] GaAs and n-$Al_xGa_{1-x}As$ were grown on GaAs substrates of 75-mm diameter at $L = 250$ mm, $\theta_0 = 5.5°$, and $A = 35$ mm. The diameter of the region of uniform thickness and Si doping level was 70 mm.

The flux instability created by changing the shutter setting due to a change of the source heat treatment must be considered when fabricating SL and sharp heterojunctions. It can be seen in Fig. 1a that the initial film growth rates V_i depend on the time t_c that the shutter is in the closed position [17]. During growth, the rate V, the dependence on time of which is plotted (Fig. 1b) from the data of [18], decreases by almost 1.5 times. Thus, the ratio of fluxes V/III, which is the most important parameter of molecular-beam film growth, changes significantly. A steady state (relaxation time of the source temperature) is achieved after 3 min whereas typical growth rates are one monolayer per second. Therefore, two sources held at different temperatures [19] and shutters placed 30 mm from the source [20] are used if the composition has to be changed drastically.

4. Preparation of Atomically Smooth Surfaces and Interfaces. The problems with preparation of uniform films examined above involve control of the density of molecular beams and are not actually related to growth mechanisms. Microdefects caused by physicochemical processes can arise during MBE on the growth surface. The most prevalent types of morphological defects in A_3B_5 films are growth hillocks and oval defects. Growth hillocks were observed during epitaxy of GaAs, $Al_xGa_{1-x}As$ [21, 22, 27, 28], $In_xGa_{1-x}As$ [23], InAs [14, 24, 25], and GaP [26]. Contaminants on the substrate surface [2] or segregations of Group III elements [22] cause the hillocks. Miller et al. [27, 28] have proposed another explanation. They suggest that impurities in the residual atmosphere such as carbon degrade the smooth surface of, in particular, $Al_xGa_{1-x}As$ and disrupt the layered growth mechanism. They also proposed that C has different solubility in GaAs and $Al_xGa_{1-x}As$. As a result, the micromorphologies are different under identical growth conditions.

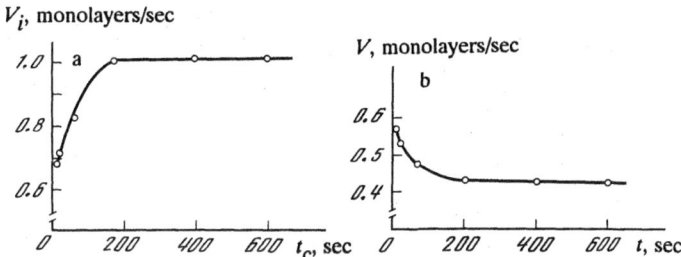

Fig. 1. Dependence of initial growth rate V_i on time for which the source is closed t_c (a) and of film growth rate V on growth time (b).

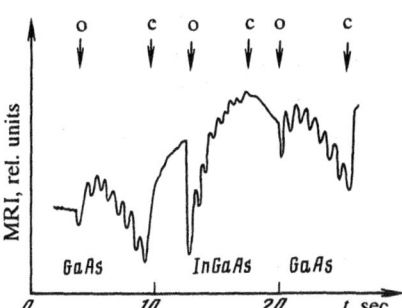

Fig. 2. Dependence of specular RHEED intensity (SRI) on growth time of the GaAs/In$_{0.12}$Ga$_{0.88}$As superlattice. Shutter positions of In and Ga are shown by arrows: o = open, c = closed.

The oval defects are caused by a local increase of Group III element on the growing surface [14, 22]. They appear in films prepared by MBE due to spraying of Ga (In, Al) [22] from the effusion cell and deposition of metal globules onto the substrate. Stall et al. [22] used thermal decomposition of Ga(CH$_3$)$_3$ in a ceramic tube instead of metallic Ga in order to avoid this effect. The oval defect density was reduced from 10^5 to $2 \cdot 10^2$ cm^{-2}. However, they could not be completely eliminated. This suggests that metal globules form on the growth surface where the adatoms of the Group III element are in excess due to repulsion by the growing step.

Oval defects were observed during growth of InAs films [14] only if the surface was enriched in In. If the films were grown with a high enrichment of As, a characteristic cellular structure formed and oval defects were not observed.

Smooth films of the majority of A$_3$B$_5$ compounds grow under very selective conditions [29]. Numerous oscillations of RHEED intensity can be observed (Fig. 2). These are caused by a periodic change of roughness in the film surface growing by a two-dimensional (2M) nucleation and growth mechanism.

A general theory of specular RHEED intensity (SRI) oscillations has not yet been developed. However, existing concepts about the nature of the oscillations enable them to be used widely to fabricate epitaxial structures. The behavior of SRI has been examined in a number of works using diffraction theory [30] or computer modelling of the growth process [31]. The SRI oscillation theory developed in [32] enables the mechanism of monolayer filling to be determined and the density of two-dimensional islands and their rate of lateral expansion to be calculated from the shape of individual oscillations.

The oscillation period is determined in the overwhelming majority of cases by the flux of Group III elements. The ratio of fluxes of Group III and V elements were measured directly neglecting the growth chamber geometry and the sensitivity of the flux sensors used by observing the SRI oscillations caused by As.

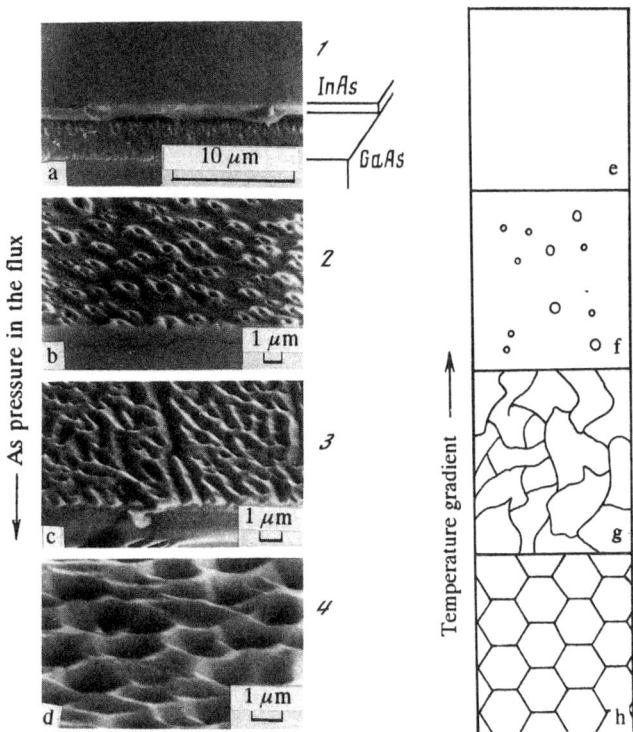

Fig. 3. Dependence of InAs surface film micromorphology on ratio of fluxes P_{As_4}/P_{In} (a-d) and characteristic structures arising from a loss of stability of the planar growth front (e-h): 1) Smooth surface; 2) pustules; 3) irregular cellular structure; 4) regular cellular structure.

It should be noted that the SRI increases smoothly, asymptotically approaching the constant value I_0 characteristic of the given material, when growth is stopped between alternating heterolayers. Such intensity behavior is due to a decreased roughness in the epitaxial film caused by surface mass transport. It is understandable that similar stoppages while the substrate temperature is raised enable an atomically smooth film surface and extremely sharp interfaces between SL layers to be prepared.

5. Supermonolayer Adsorptive Coating Model. The review of experimental data presented above indicates that the range of conditions suitable for growth of smooth A_3B_5 films is very limited. A change of one of the growth parameters, for example, the substrate temperature T_s or the ratio of Group III and V element fluxes, causes the morphology of the growing surface to be unstable. It was found using *in situ* control [14] that the diffuse background in RHEED patterns increases if the conditions deviate from optimal. Kanter et al. [14] proposed that a supermonolayer adsorptive coating (SMAC) is formed on the growing surface. The SMAC is a film of a solution of InAs in In for In-stabilized growth. The SMAC is enriched in As for As-stabilized growth. Quantitative estimates of the SMAC thickness made in [14] based on a measurement of the intensity of reflections and the diffuse background in RHEED patterns gave a value of 1-5 monolayers. The SMAC thickness is minimal under optimal growth conditions and increases as the excess component accumulates.

A similar conclusion was used in [35, 36] to explain film growth features from molecular beams. Herman [36] proposes the existence of a quasi-gas layer near the surface, the thickness of which is 5-8 monolayers.

It is noteworthy that the SMAC model is very similar to the model of [37] that describes the vapor—liquid—crystal growth mechanism. The existence of thick adsorptive layers during vapor-phase epitaxy has been pointed out, for example, by Chernov [38]. The SMAC model also includes concentrational supercooling near the growth surface. The obvious correlation between the morphology of film surfaces grown from molecular beams and crystals grown from the melt then becomes understandable. As an illustration, photomicrographs of InAs film surfaces [14] and a schematic drawing [39] of the surface shape of the crystal growing under different conditions are shown in Fig. 3.

Thus, oval defects or growth hillocks are not caused exclusively by a specific type of defects in films grown by MBE but by morphological defects, known in the theory and practice of crystal growth [39], caused by accumulation of an impurity or excess component near the growth surface. In this case it can be concluded [39] that reducing the rate should cause smooth films of InAs to grow from molecular beams even under nonoptimal conditions. In fact, it was demonstrated in [14] that the film surface develops relief during MBE of InAs at temperature T_s = 450°C, growth rate V = 0.8 μm/h, and ratio of pressures P_{As_4}/P_{In} = 10. Decreasing V to 0.1 μm/h at constant T_s and P_{As_4}/P_{In} enabled smooth films to grow.

The SMAC model can also be used to explain impurity and main-component segregation effects. These questions will be examined further.

6. Control of Electrical Properties of A_3B_5 Films Grown from Molecular Beams. Films with a minimal ($\leq 10^{14}$ cm^{-3}) level of background impurity must be prepared to fabricate epitaxial structures for devices. The level of background impurity can be reduced by using ultrahigh vacuum, high-purity materials, and high-vacuum construction parts. All modern MBE systems are equipped with cryoshrouds to decrease outgassing from the walls of the growth chamber. However, these requirements are insufficient to prepare films with high electrophysical properties. It has been demonstrated often [1, 2, 24] that the charge carrier concentration and mobility in A_3B_5 films depend strongly on the growth conditions and primarily on T_s and the ratio of fluxes of Group III and V elements.

As an illustration, we use data of [40] that were obtained in our laboratory and are characteristic of the majority of A_3B_5 systems. Dependences of concentration n and mobility μ of charge carriers in heteroepitaxial films of 1-μm thickness on the ratio P_{As_4}/P_{In} at T_s = 450°C are plotted in Fig. 4. The dependences presented have deep extrema in the region P_{As_4}/P_{In} = 30-40. A deviation from this interval increases n from $5 \cdot 10^{15}$ to $7 \cdot 10^{17}$ cm^{-3}. Analogous functions were obtained on homoepitaxial InAs films of thickness 4 μm but the value n_{min} in this case was $4 \cdot 10^{14}$ cm^{-3}.

The region in which films with the best electrophysical parameters grow corresponds to the transition between superstructures 2 × 4 As ↔ 4 × 2 In. The excess component does not accumulate in the SMAC under these conditions. Smooth films are obtained. If the conditions deviate from optimal the background impurity or intrinsic point defects are captured more efficiently. The data obtained by us agree well with results of [29] in which practical use is discussed for superstructural transition diagrams on the growth surface for growing InAs and GaAs. The range of conditions for growing smooth films of GaAs is much wider than for InAs.

7. Growth of III—III—V Solid Solutions. Such solid solutions as $Al_xGa_{1-x}As$, $Al_xIn_{1-x}As$, and $In_xGa_{1-x}As$ are commonly used to fabricate microsemiconducting structures (MSS). The forbidden band width and the band gap at heterojunctions can be changed in a controlled manner by varying the composition of the solid solutions used. During growth of III—III—V compounds and during deposition of the binary compounds problems associated with morphological instability of the planar growth shape and the control of concentrations of electrically active defects arise [38]. Moreover, a specific problem with the growth of solid solutions is the compositional nonuniformity in the growth direction. This may be caused by instability with time of the molecular beams, segregation of components, and revaporization of one of the Group III elements. Repulsion of In during growth of $In_xGa_{1-x}As$ has been observed [41, 42] not only at high T_s [43] but also at low T_s = 400°C. Segregation is evident where the streaks in the RHEED pattern transform into point reflections after the $In_xGa_{1-x}As$ film stops growing. This indicates morphological rearrangements in the film surface. Following the hypothesis about In repulsion, Kanter et al. [41] concluded that a SMAC as a liquidlike phase with a high In and low Ga content forms on the growth surface. As a result, the solid solution $In_xGa_{1-x}As$ with $x \approx 1$ crystallizes on the surface of the epitaxial film after growth stops. This hypothesis is confirmed by transmission electron microscopy (TEM) and x-ray photoelectron

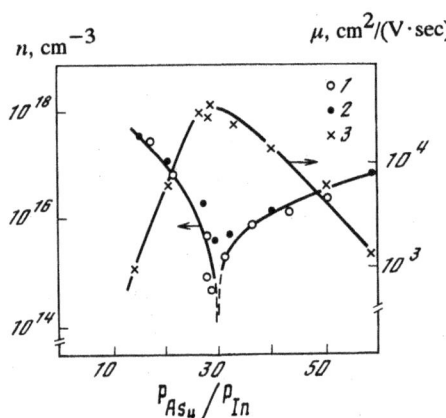

Fig. 4. Effect of ratio of fluxes of Group III and V elements on concentration (2) and mobility (3) of electrons in undoped films of InAs on GaAs. Concentration of electrons in homoepitaxial films of InAs (1).

spectroscopy [42]. It is noteworthy that this effect (segregation) is observed only during growth with metal enrichment. Increasing the pressure in a As_4 flux with constant T_s and fluxes of Ga and In suppresses segregation.

Favorable conditions for eliciting segregation are not only stabilization of the growth surface by a metal (i.e., low ratio of fluxes of Group V and III elements and high T_s) but also mechanical strains in the growing film. For example, sputtering of films of $Al_{0.48}In_{0.52}As$ on InP and $Al_{0.70}In_{0.30}As$ on GaAs was studied in [44]. In the first case the lattice constant of the heterosystem is matched. In the second, films of thickness up to 10 nm are pseudomorphous and strained. As it turned out, incorporation of In into the crystal is hindered. All growth conditions with the exception of the level of mechanical strains were identical. This suggested [44] stabilization of the film composition through the influence of elastic strains. This effect became especially noticeable at high T_s where revaporization of In adatoms is significant. The critical temperature for In according to [44] is 500°C and for Ga, 630°C.

Thus, As-stabilized conditions must be maintained during growth of MSS based on III—III—V solid solutions. The influence of mechanical strains must be considered during fabrication of strained SL. The epitaxy temperature should not exceed the limit beyond which the Group III atoms begin to revaporize.

8. Heterostructure Characterization. As noted, one of the most important achievements of MBE is the ability to fabricate a new type of semiconducting materials, SL. The ability to control their band structure is one of their excellent properties. Like ordinary crystalline solids, SL should have translational symmetry (as a rule, one of the directions is isolated) and do not contain structural defects since the properties of the electronic subsystem are determined by the crystal lattice. In this respect, the most important question in fabricating SL (and in general MSS of another type also) is the study of their atomic structure and the reasons that defects arise.

The defect nature of MSS is studied by TEM, x-ray diffraction, and Rutherford backscattering (RBS). We will examine several possible uses of these methods to study the structural perfection of strained SL (SSL) $In_xGa_{1-x}As/GaAs$ grown in our laboratory.

The RBS spectra (random and axial) of 1.6 MeV He^+ ions obtained from SSL containing 13 periods of the composition 20 monolayers GaAs and 20 monolayers $In_{0.2}Ga_{0.8}As$ grown on a $In_{0.1}Ga_{0.9}As$ buffer layer (BL) of 100-nm thickness are shown in Fig. 5. The normalized yield χ, the ratio of flux intensities of dechannelized and randomly scattered ions, is also shown in Fig. 5. The thicknesses of the BL, the total thickness of the SL (marked in Fig. 5), and the mole fraction x of In in the SL are determined from these data. The value $\chi_{(100)}^{SSL} = 4\%$ measured near the surface of the epitaxial structure is less than $\chi_{(100)}$ for GaAs substrate. This is consistent with high structural perfection of the SSL. The increased yield of dechannelized ions along the [110] axis is due to a periodic

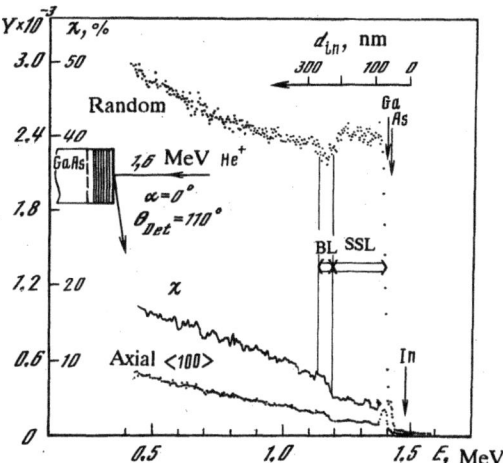

Fig. 5. Rutherford backscattering spectra obtained from SSL $In_{0.2}Ga_{0.8}As/$ GaAs: Y is the signal intensity; χ is the normalized yield of ions; d_{in} is the depth of the analyzed region; θ_{Det} is the detector collection angle.

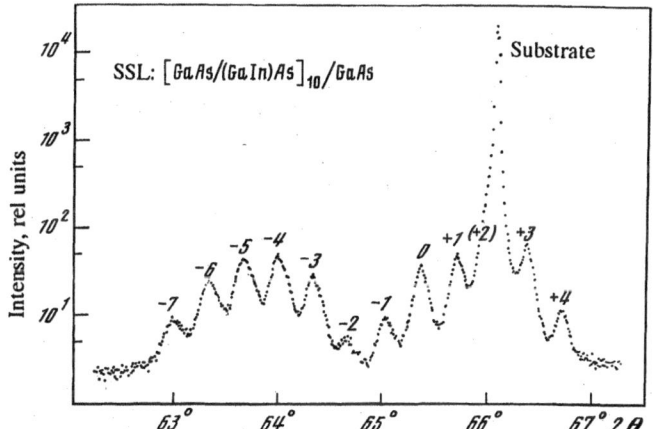

Fig. 6. X-ray diffraction spectra obtained from SSL $In_{0.23}Ga_{0.77}As/GaAs$. Numbers denote the principal peak (0) and satellites due to superstructure.

change of channel direction at each interface as a result of tetragonal distortions in the SSL. Comparison of the experimental spectra with model calculations found [45] that the mismatch of lattice constants is completely compensated by elastic deformations.

The SSl were analyzed using x-ray diffraction by recording oscillation curves near principal reflections. The composition of the solid solution, the strain level, the degree of perfection of the SL, etc., were determined by comparing the experimental and theoretical spectra. The x-ray diffraction spectrum obtained in [45] from SSL consisting of GaAs layers and the solid solution $In_{0.23}Ga_{0.77}As$ is shown in Fig. 6. The angular distribution between satellites near the principal reflection (400) that are caused by superstructure enables the average period of the SL $\Lambda = d_{GaAs} + D_{InGaAs} = 30$ nm to be determined. The presence of more than 12 satellites $n = \pm 1, 2, ...$ around the principal reflection "0" indicates that the SL is highly perfect. However, the halfwidth of the peaks is greater than that calculated. This broadening may be due to ternary compound composition fluctuations as a result of, for example, In segregation.

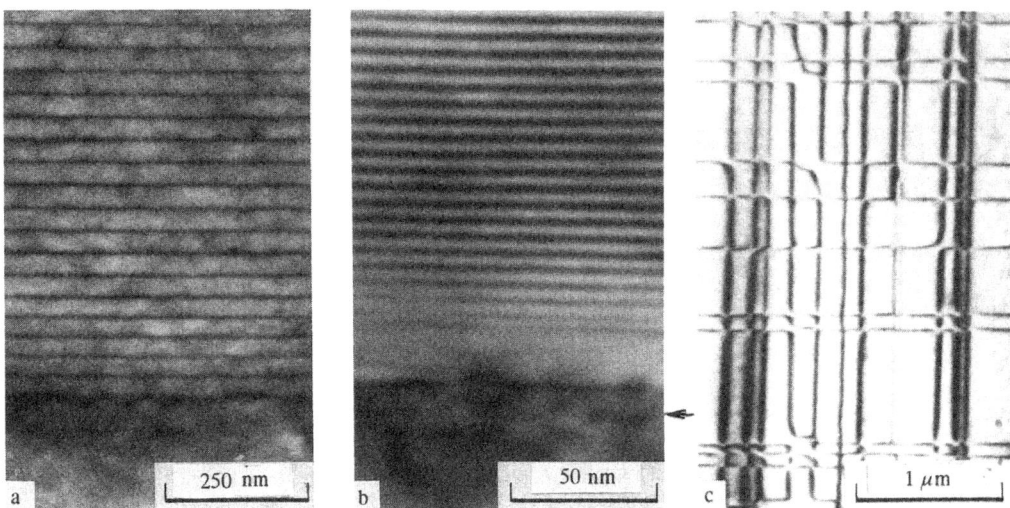

Fig. 7. TEM images of cross-sectional samples of SSL InAs/GaAs (a), $In_{0.2}Ga_{0.8}As$/GaAs (b), and networks of DL at the interface (c).

The structural defects can be visualized by TEM analysis of the SSL. Electron micrographs of cross-sectional samples of SSL InAs/GaAs and $In_{0.2}Ga_{0.8}As$/GaAs are shown in Figs. 7a and b. It can be seen from the images that the SL is strictly periodic and dislocations do not emerge. Moreover, the planar TEM image at the interface between the SL and the GaAs substrate (Fig. 7c) shows a network of orthogonal dislocation loops (DL) along the directions [110] and [1$\bar{1}$0]. The DL are Lomer edge dislocations for heteroepitaxy of InAs on GaAs, as was demonstrated in [41]. Their distinguishing feature is a rather strict ordering and periodicity. This enables the DL network to be viewed as a planar SL with a period of about 6 nm. The existence of such a SL of DL in the system InAs/GaAs was confirmed by observing electron diffraction from it [41].

Thus, these study methods enable the technical regimes of growing SSL to be controlled reliably. However, other methods such as Raman scattering and photoluminescence must be used for complete characterization of MSS.

REFERENCES

1.   A. Y. Cho, "Growth of III—V semiconductors by molecular beam epitaxy and their properties," *Thin Solid Films*, **100**, 291-317 (1983).

2.   K. Ploog, "Molecular beam epitaxy of artificially layered semiconductor structures — basic concept and recent achievements," in: *Physics, Fabrication and Application of the Multilayered Structures* (NATO Adv. Study Inst. Lect. Notes), Ile de Bendor (1987), Part 1, pp. 77-109.

3.   A. V. Rzhanov and S. I. Stenin, "Molecular epitaxy," in: *Growth of Semiconducting Crystals and Films* [in Russian], Nauka, Novosibirsk (1984), Part 1, pp. 5-34.

4.   S. I. Stenin, "Molecular beam epitaxy of semiconductor, dielectric and metal films," *Vacuum*, **36**, No. 7/8, 419-426 (1986).

5.   B. A. Joyce, "Effect of arsenic species (As_2 or As_4) on the crystallographic and electronic structure of MBE-grown GaAs(001) reconstructed surfaces," *Surf. Sci.*, **133**, 267-278 (1983).

6.   B. S. Krusor and R. Z. Bachrach, "Two-stage arsenic cracking source with integral getter pump for MBE growth," *J. Vac. Sci. Technol.*, B, **1**, No. 2, 138-141 (1983).

7.   W. T. Tsang, "Chemical beam epitaxy of InP and GaAs," *Appl. Phys. Lett.*, **45**, No. 11, 1234-1236 (1984).

8.   Y. Kawaguchi, H. Asahi, and H. Nagai, "MBE growth of high-quality InP using triethylindium as an indium source," *Jpn. J. Appl. Phys.*, **23**, No. 9, L737-L739 (1984).

9.   E. Tokumitsu, Y. Kudou, M. Konagai, and K. Takahashi, "Metalorganic molecular-beam epitaxial growth and characterization of GaAs using trimethyl- and triethyl-gallium sources," *Jpn. J. Appl. Phys.*, **24**, No. 9, 1189-1192 (1985).

10.   R. P. Vasquez, B. F. Lewis, and F. J. Grunthaner, "Cleaning chemistry of GaAs(100) and InSb(100) substrates for molecular beam epitaxy," *J. Vac. Sci. Technol.*, B, **1**, No. 3, 791-794 (1983).

11. L. P. Erickson, G. L. Carpenter, D. D. Seibel, et al., "MBE film growth by direct free substrate heating," *J. Vac. Sci. Technol., B,* **3**, No. 2, 536-537 (1985).

12. V. V. Preobrazhenskii, D. I. Lubyshev, and V. P. Migal', "Temperature dependence of GaAs and InAs(100) surface structure reconstructions," in: Expanded Abstracts of the Seventh All-Union Conf. on Crystal Growth: Symposium on Molecular-Beam Epitaxy, Vol. 4 [in Russian], Moscow (1988), pp. 145-146.

13. M. Heiblum, E. E. Mender, and L. Osterling, "Growth by molecular beam epitaxy and characterization of high purity GaAs and AlGaAs," *J. Appl. Phys.,* **54**, No. 12, 6982-6988 (1983).

14. Yu. O. Kanter, A. I. Toropov, A. V. Rzhanov, et al., "Micromorphology of epitaxial films of InAs during growth from molecular beams on GaAs substrates," *Poverkhnost,* No. 9, 83-87 (1986).

15. J. Saito and A. Shibatomi, "Highly uniform GaAs and AlGaAs epitaxial layers grown by molecular beam epitaxy," *Fujitsu Sci. Tech. J.,* **21**, No. 2, 190-197 (1985).

16. J. Massies, "Mismatch and electron mobility in MBE $Ga_xIn_{1-x}As$ epitaxial layers on InP substrates," *Appl. Phys. A,* **32**, 27-30 (1983).

17. Yu. O. Kanter, M. A. Revenko, and A. A. Fedorov, "Measurement of InAs film growth rates from molecular beams using RHEED intensity oscillations," *Pis'ma Zh. Tekh. Fiz.,* **13**, No. 18, 1127-1130 (1987).

18. T. Sakamoto, H. Funabashi, K. Ohta, et al., "Phase-locked epitaxy using RHEED intensity oscillation," *Jpn. J. Appl. Phys.,* **23**, No. 9, L657-L659 (1984).

19. P. A. Maki, S. C. Palmateer, A. R. Calawa, and B. R. Lee, "Elimination of flux transients in molecular beam epitaxy," *J. Vac. Sci. Technol., B,* **4**, No. 2, 564-567 (1986).

20. T. Mizutahi and K. Hirose, "High mobility GaInAs thin layers grown by molecular beam epitaxy," *Jpn. J. Appl. Phys.,* **24**, No. 2, L119-L121 (1985).

21. F. Alexandre, L. Goldstein, G. Leroux, et al., "Investigation of surface roughness of molecular beam epitaxy $Ga_{1-x}Al_xAs$ layers and its consequences on $GaAs/Ga_{1-x}Al_xAs$ heterostructures," *J. Vac. Sci. Technol., B,* **3**, No. 4, 950-955 (1985).

22. R. A. Stall, J. Zilko, V. Swaminathan, and N. Schumaker, "Morphology of GaAs and $Al_xGa_{1-x}As$ grown by molecular beam epitaxy," *J. Vac. Sci. Technol., B,* **3**, No. 2, 524-527 (1985).

23. H. Saito, J. O. Borland, H. Asahi, et al., "Hillock defects in InGaAs/InP multilayers grown by MBE," *J. Cryst. Growth,* **64**, No. 3, 521-528 (1983).

24. R. A. A. Kubiak, E. H. C. Parker, S. Newstead, and J. J. Harris, "The morphology and electrical properties of heteroepitaxial InAs prepared by MBE," *Appl. Phys. A,* **35**, 61-66 (1984).

25. M. Kano, M. Nogami, Yu. Matsushima, and M. Kimata, "Molecular beam epitaxial growth of InAs," *Jpn. J. Appl. Phys.,* **16**, No. 12, 2131-2137 (1977).

26. S. L. Wright, H. Kroemer, and M. Inada, "Molecular beam epitaxial growth of GaP on Si," *Appl. Phys.,* **55**, No. 8, 2916-2927 (1984).

27. R. C. Miller, W. T. Tsang, and O. Munteanu, "Extrinsic layer at Al_xGa_{1-x}—As—GaAs interfaces," *Appl. Phys. Lett.,* **41**, No. 4, 374-376 (1982).

28. P. M. Petroff, R. C. Miller, A. C. Gossard, and W. Wiegmann, "Impurity trapping interface structure and luminsecence of GaAs quantum wells grown by molecular beam epitaxy," *Appl. Phys. Lett.,* **44**, No. 2, 217-219 (1984).

29. S. M. Newstead, R. A. A. Kubiak, and E. H. C. Parker, "On the practical applications of MBE surface phase diagrams," *J. Cryst. Growth,* **81**, No. 1/4, 49-54 (1987).

30. J. M. Van Hove, C. S. Lent, P. R. Pukite, and P. I. Cohen, "Damped oscillations in reflection high-energy electron diffraction during GaAs MBE," *J. Vac. Sci. Technol., B,* **1**, No. 3, 741-746 (1983).

31. S. Clarke and D. D. Vvdeensky, "Growth kinetics and step density in reflection high-energy electron diffraction during molecular beam epitaxy," *J. Appl. Phys.,* **63**, No. 7, 2272-2283 (1988).

32. D. Kashchiev and Yu. O. Kanter, "Oscillations of specular beam intensity in reflection diffraction from the surface of growing epitaxial film: A theoretical study," *Phys. Status Solidi A,* **110**, No. 1, 61-76 (1988).

33. J. H. Neave and B. A. Joyce, "Dynamics of film growth of GaAs by MBE from RHEED observations," *Appl. Phys. A,* **31**, 1-8 (1983).

34. B. F. Lewis, R. Fernandez, A. Madhukar, and F. J. Grunthaner, "Arsenic-induced intensity oscillations in reflection high-energy electron diffraction measurements," *J. Vac. Sci. Technol., B,* **4**, No. 2, 560-563 (1986).

35. M. A. Herman, "Quasi-gas transition layers occuring in MBE growth of microdevices and superlattices," *Superlattices Microstruct.,* **2**, No. 4, 345-348 (1986).

36. M. A. Herman, "The problem of a near surface quasi-gas transition layer in MBE," *Cryst. Res. Technol.,* **2**, No. 11, 1413-1420 (1986).

37. L. S. Palatnik and Yu. F. Komnik, "On the kinetics of metal condensation in vacuum," *Dokl. Akad. Nauk SSSR,* **124**, 808-814 (1959).

38. A. A. Chernov, "Crystallization processes," in: *Modern Crystallography,* Vol. 3 [in Russian], Nauka, Moscow (1980), pp. 176-189.

39. R. Laudise and R. Parker, "Kinetics of crystal growth," in: *Growth of Crystals* [Russian translation], Mir, Moscow (1974), pp. 126-132.

40. I. S. Bzinkovskaya, Yu. O. Kanter, V. A. Kolosanov, et al., "Electrical and optical characteristics of InAs films grown from molecular beams," in: Abstracts of Symp. Mol. Beam Epitaxy, Frankfurt on Oder (1987), p. 20.

41. Yu. O. Kanter, A. K. Gutakovsky, A. A. Fedorov, et al., "Study of the growth mechanism of modulated structures in the InAs—GaAs system," *Thin Solid Films,* **163**, No. 1/3, 497-502 (1988).

42. M. A. Revenko, S. V. Rubanov, and A. A. Fedorov, "Segregation of In during growth of InAs films by MBE," in: Expanded Abstracts of the Seventh All-Union Conf. on Crystal Growth: Symp. on Mol. Beam Epitaxy, Vol. 4 [in Russian], Moscow, 1988, pp. 67-68.

43. D. V. Morgan, H. Onho, E. C. Wood, et al., "Ion beam analysis of molecular beam epitaxy InAlAs/InGaAs layer structures," *J. Electrochem. Soc.,* **128**, No. 11, 2419-2424 (1981).

44.   F. Turco and J. Massies, "Strain-induced In incorporation coefficient variation in the growth of $Al_{1-x}In_xAs$ alloys by molecular beam epitaxy," *Appl. Phys. Lett.*, **51**, No. 24, 1989-1991 (1987).
45.   R. Flagmeyer, K. Lankeit, T. Baumbach, et al., "Characterization of $In_xGa_{1-x}As/GaAs$ strained layer superlattices by ion backscattering," *Phys. Status Solidi A*, **107**, No. 1, K19-K24 (1988).

EPITAXY OF SOLID SOLUTIONS AND MULTILAYERED STRUCTURES
IN THE SYSTEM Cd—Hg—Te

Yu. G. Sidorov and S. I. Chikichev

Solid solutions $Cd_xHg_{1-x}Te$ (CMT) are one of the principal IR microphotoelectronics materials. Quantum-sized multilayered CdTe—HgTe structures are viewed as the most promising for fabricating photodetectors using "band engineering." With respect to the structural and electrophysical parameters of the CMT films, all common epitaxial methods give the same excellent results. The carrier mobility in n-type $Cd_{0.2}Hg_{0.8}Te$ reaches $\mu_{77} \geq 2 \cdot 10^5$ cm^2·V/sec at a concentration $n \leq 10^{15}$ cm^{-3}. Therefore, liquid-phase epitaxy, metalorganic chemical vapor deposition (MOCVD), sublimational—diffusional gas-phase epitaxy, and molecular-beam epitaxy (MBE) are equally capable of satisfying the demands of traditional photoelectronics in high-quality CMT layers. The single method of preparing quantum-sized structures and superlattices (SL) until very recently was MBE. The advantage of MBE for growing SL based on CMT is the low growth temperature compared to other epitaxy methods. The interdiffusion coefficients of HgTe and CdTe are relatively large. Interdiffusion of the layer components can destroy the SL. If the interdiffusion coefficients of HgTe and CdTe given in the literature are used [1], the calculated thickness of the diffusion layers at 200°C is 0.1-1.0 nm. Superlattices based on CMT can be grown by MBE at 200°C. This growth temperature cannot be exceeded substantially without destroying the sharpness of the composition change at the boundaries of the SL layers. In MOCVD, use of photoactivation [2], precracking [3], flame excitation of the reaction mixture [4], and nontraditional organometallic compounds [5] enabled the CdTe growth temperature to be lowered to 250-200°C (and even to 85°C for HgTe). Thus, ultrathin layers could be synthesized [6]. The development of a hydride MBE method using organometallic compounds, like that already used for A_3B_5, seems logical.

Superlattices are artificial materials with properties close to ideal for application in IR technology. The band width and effective carrier mass for transfer perpendicular to the layers can be varied widely independently of each other. This property, which is lacking in alloys, makes possible fabrication of more perfect instruments than those using alloys [7]. The thickness and composition of the layers in SL do not need to be held as accurately as in alloys to set the required forbidden band width. The calculated tunnelling currents in SL are much less than in alloys with the same forbidden band width. The diffusional currents in p-material in SL are calculated to be less than in alloys [7].

Superlattices of CdTe—HgTe were proposed in 1979 [8] and were grown in 1982 on CdTe substrates [9]. At present several research groups have reported growth of SL of CdTe—HgTe on various substrates such as CdZnTe and GaAs [10,11].

More than 100 layers are known. The thickness is of the order of 10 nm. Secondary ion mass spectrometry (SIMS) and Auger profiling reveal periodic structure in SL that consists of HgTe and CdTe layers with layers of variable composition 4-5-nm thick [12]. In fact, this thickness corresponds to the depth resolution of the profiling methods used. High-resolution transmission electron microscopy of cross-sectional samples of SL reveals sharp boundaries between HgTe and CdTe layers [10].

Superlattices $Hg_{1-x}Zn_xTe$—CdTe with $x = 0.06$-0.15 and $Hg_{1-x}Mn_xTe$—CdTe with $x = 0.02$-0.12 have been grown [13]. High hole mobility is observed in SL based on HgZnTe. Thus, $Hg_{0.85}Zn_{0.15}Te$—CdTe SL has a hole mobility 20,000 $cm^2 \cdot V/sec$ at 25 K [13].

It can be demonstrated by analyzing the epitaxy conditions of A_3B_5 that certain MBE characteristics of these compounds obey thermodynamic laws [14]. Results of MBE of A_2B_6 such as CdTe, HgTe, and HgCdTe also showed that there are no kinetic limitations and a quasithermodynamic approach is suitable [15]. Certain results from a thermodynamic MBE model of A_2B_6 compounds are given in [16, 17]. The MBE model is based on a balance of fluxes to the growing surface. An approximation is used in which the flux of the vaporizing substance J_i has the equilibrium value.

The incident flux is determined by the source:

$$S_i = p_i(2\pi m_i kT)^{-\frac{1}{2}}.$$

At equilibrium $J_i = S_i = p_{iequil}(2\pi m_i kT)^{-1/2}$, i.e., in agreement with the approximation used, it is possible to determine independently J_i since the equilibrium vapor pressures of the components are calculated uniquely. Here p_i is the vapor pressure and m_i is the molecular (atomic) mass of the component. For the binary compound AB vaporizing with formation of the monatomic component A and the n-atomic component B_n, the vaporization rate of the components is

$$J_A J_B^{1/n} = k(T)(2\pi m_A kT)^{\frac{1}{2}}(2\pi m_B kT)^{\frac{1}{2}} \equiv G(T),$$

where $k(T)$ is the equilibrium constant of the dissociation reaction. The deposition rate is naturally

$$v = S_A - J_A = nS_B - nJ_B,$$

We obtain a system of equations relating the deposition rate to the vapor pressures from the sources. Actual substrate temperatures above which the growth rate decreases sharply are found from a comparison with experiment. The validity of the thermodynamic examination means that elements A and B react on the surface rather quickly so that thermal equilibrium can be established. The system of equations becomes complicated for deposition of a three-component substance, for example, $Hg_{1-x}Cd_xTe$:

$$(1-x)v = S_{Hg} - J_{Hg},$$
$$xv = S_{Cd} - J_{Cd},$$
$$v = 2S_{Te} - 2J_{Te},$$
$$J_{Hg} J_{Te}^{\frac{1}{2}} = (1-x)G_{HgTe}(T),$$
$$J_{Cd} J_{Te}^{\frac{1}{2}} = xG_{CdTe}(T).$$

Calculations show that the growth rate up to 200°C is controlled by the Te flux; the composition (x), by the ratio of Cd and Te fluxes (Fig. 1). Vaporization of Te must be considered at the higher temperatures. Predictions of the calculation are confirmed quantitatively by experiments. Thus, conditions ensuring the given growth rate and the required layer composition can be calculated. The limits beyond which the required material cannot in principle be obtained can also be determined. The sticking coefficient can be calculated from the model

$$C_{Hg} = (S_{Hg} - J_{Hg})/S_{Hg} = (1-x)v/S_{Hg}.$$

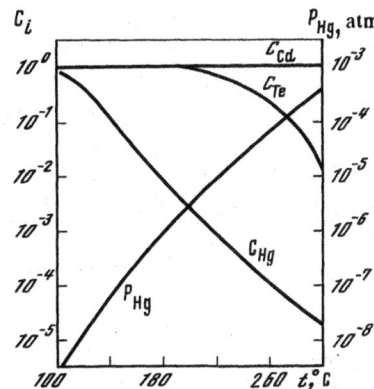

Fig. 1. Calculations of incorporation coefficients of components and minimal Hg vapor pressures over the surface of the growing CMT film using the data of [16].

According to Fig. 1, the sticking coefficient can change. For example, it lies between 10^{-2} and 10^{-3} at 180°C and decreases with increasing temperature of the Hg source. This agrees with experiment. An important consequence of the low Hg sticking coefficients is the increased consumption of 100 g/h. The thermodynamic model can predict the principal features of the process. However, it is useless for a detailed examination of surface processes due to the assumed equilibrium on the surface. In fact, equilibrium is not attained or else there would not be the resulting growth or sublimation. The deviation from equilibrium on the surface depends on the deposition rate, the temperature, and other factors, for example, the surface orientation. Studies of the Hg sticking coefficient to various CdTe planes during MBE of CMT films [18] clearly demonstrate the effect of kinetic factors on the MBE process. The required Hg flux for growth on CdTe(111)B is an order of magnitude less than for (111)A at 185°C. The Hg sticking coefficient on CdTe(100) has an intermediate value. The orientational dependence of the Hg sticking coefficient is explained by the different bonding of surface atoms. The CdTe(111)B surface is composed of Te atoms. An Hg—Te bond is formed on adsorption of a Hg atom. Weaker Hg—Cd bonds form on CdTe(111)A, composed of Cd atoms. A large flux of Hg atoms is needed to give the concentration of adsorbed Hg atoms on (111)A necessary to grow CMT. The Hg vapor pressure over HgTe at 450 K, corresponding to growth of structurally perfect HgTe, is about 10^{-1} Pa [19]. This does not correspond to MBE conditions. The high vapor pressure of Hg necessitates specific changes in the construction of the MBE apparatus. An Hg vaporizer with good reproducibility, high feed rate, and a system for Hg collection and removal is used [20].

A diagram of the CMT MBE apparatus is shown in Fig. 2. Superlattices CdTe—HgTe are grown using three effusion cells. Elemental Hg and Te are used as sources for growth of HgTe layers. The deposition rate of CdTe and HgTe layers is held constant at about 2 monolayers/sec. The growth temperatures of CdTe and HgTe are identical during growth of SL. The highest structural perfection was obtained at 473 K. However, the optimal growth temperatures for CdTe and HgTe are different. Studies of MBE of CdTe films uniquely determine the optimal epitaxy temperature of CdTe to be 520-540 K. This growth temperature gives the highest structural perfection and the best electrophysical quality of the CdTe films [21]. The HgTe growth temperature cannot be raised to 520-540 K due to the prohibitively high Hg vapor pressure over HgTe at these temperatures (~1 Pa). Reducing the growth temperature to 473 K degrades the structural perfection of the CdTe layers. Thus, one of the problems with increasing the structural perfection of the CdTe—HgTe SL leads to a search for ways to grow perfect CdTe layers at reduced temperatures.

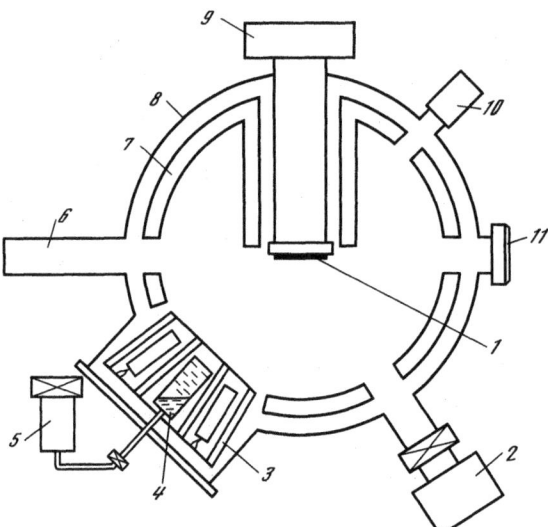

Fig. 2. Diagram of the MBE apparatus for growing CMT films: 1) Substrate; 2) demercurization system; 3) vaporization system with cryoshrouds; 4) Hg vaporizer with diaphragm system; 5) Hg feed system; 6) diffractometers; 7) vacuum chamber cryoshrouds; 8) vacuum chamber; 9) multistepped manipulator; 10) mass spectrometer; 11) screen.

The theory of MBE is developed better for A_3B_5 than for A_2B_6. Methods for reducing the epitaxy temperature, varying the intensity of the molecular beams, and illuminating the surface during growth are suggested in [22] on the basis of a study of the mechanism of autoepitaxial growth on GaAs(100) substrates. Definite stages occur in formation of alternating layers of Ga and As during growth on the GaAs(100) plane. The presence of Ga or As in several of these stages is highly undesirable. Therefore, it is reasonable to assume that pulsed Ga and As flux intensities synchronized with the monolayer growth cycle can increase the structural perfection or reduce the epitaxy temperature. Sequential deposition of Ga and As monolayers was carried out experimentally. This produced high-quality GaAs at temperatures below 200°C. The effect of light can cause certain bonds to rupture during growth resulting in a decreased need for thermal activation and the possibility of reducing the growth temperature. The compound CdTe was grown by MBE by sequential deposition of Cd and Te layers in [23, 24]. It was possible to prepare high-quality CdTe layers. However, data on epitaxy temperature reduction were not presented. Difficulties are encountered in constructing an adequate model of CdTe layered growth [25, 26]. There are no general recommendations for optimizing the growth regimes so that the epitaxy temperature can be reduced.

Components can be excited not only on the surface but also during vaporization, for example, by ionic or laser vaporization of sources. Ionic vaporization of CdTe and CdHgTe sources enabled epitaxy of CdTe up to 140°C; of CdHgTe, below 100°C [27]. Thus, the epitaxy temperature of CdTe was reduced. However, judging from the results presented, films grown at 240°C have the highest structural perfection. Films of CdHgTe grow epitaxially even at 70°C but up to only 2-μm thickness. The structure then degrades. The films become polycrystalline at a thickness greater than 10 μm. Inclusions of Te, initially causing twinning and then polycrystalline growth, are formed in the CdHgTe films. An important merit of ion sputtering is the ability to change sharply the ion beam current and therefore the sputtering rate, whereas thermal heating has a high inertia. Laser vaporization also has a low inertia. This is used to grow structures with a given profile of composition change with thickness [28]. A computer-controlled pulsed laser changes the vaporization rate of the material. The composition of the growing layer changes accordingly. An important element of the laser vaporization system is the laser scanning apparatus.

If there is no scanning, the target is heated locally and the vaporized material is sputtered. The composition distribution with depth coincides very precisely with the actual distribution obtained by laser vaporization and measured by Auger profiling. The laser method is the most reproducible of the various methods for modulating the flux intensity of the deposited substance. The treatment time is extremely short.

One of the unsolved problems of MBE of CMT SL is the lack of a reliable method for controlling thicknesses during epitaxy. Thus, according to intensity oscillations of RHEED patterns, thicknesses of the individual layers of the CdTe—HgTe SL are difficult to control precisely during growth due to the high Hg vapor pressure. Therefore, use of ellipsometry for these purposes is very interesting [29]. The dependence of the ellipsometric parameters ψ and Δ on the number of HgTe and CdTe layers deposited is plotted in Fig. 3. Estimates show that the thicknesses of individual layers can be controlled to an accuracy of at least 0.03 nm by following the parameter Δ during growth.

Increasing the structural perfection of CMT layers and SL based on CMT can be a problem during deposition on foreign substrates. The need for CMT heteroepitaxy is due to a number of reasons. These include the difficulty of preparing high-quality CdTe substrate and the low mechanical strength of CdTe substrates. The convenience of fabricating optoelectronic devices and signal treatment circuits during CMT epitaxy on Si substrates is indisputable. The possibilities for heteroepitaxy using MBE surpass all other methods due to the low temperature; lack of aggressive medium; ability to achieve large supersaturation necessary, for example, where the heterojunction has large excess energy; use of powerful cleaning methods; and control of the substrate surface state. Silicon substrates are especially interesting for fabricating monolithic microphotoelectronic devices. Attempts are described in the literature to fabricate heterosubstrates CdTe/Si by MBE [30, 31] and MOCVD [32]. However, the epitaxial layers grown directly on Si have significantly poorer structural perfection than other heterosubstrates since the mismatch is maximal for this heteropair of all the combinations of elemental semiconductors with A_2B_6 compounds ($f = 19.3\%$). Therefore, progress for the system CdTe/Si is not expected to improve the crystalline perfection of CdTe layers using any preparation method. An analogous situation occurs for the pair CdTe/Ge [33]. Nevertheless, the attractiveness of Si is so great that investigators continue to search for ways of fabricating heterosubstrates based on Si. Recently two directions related to a rejection of the simplest two-layered CdTe/Si structure and a transition to a three-layered one in which a buffer layer that substantially improves the CdTe properties is inserted between the Si and CdTe have been noted. One of these directions is directly based on successes of recent years in heteroepitaxy of GaAs/Si structures [34]. The first steps along this path were taken in [35]. The possibility in principle of preparing CMT of device quality on CdTe/GaAs/Si substrates has already been demonstrated [36].

The other direction is just as interesting. It involves fabrication of CdTe—buffer—Si heterosubstrates and is based on use of epitaxial dielectric films of alkaline earth difluorides as the buffer [37, 38]. Thus, successful preparation of a CdTe (5-20 μm)/BaF$_2$ (2000 Å)/CaF$_2$ (70 Å)/Si(111) heterosubstrate was reported in [38]. The idea here is very simple. The large mismatch of lattice constants between the terminal members of this heterocomposition is distributed over the three heterojunctions CaF$_2$/Si ($f = 0.6\%$), BaF$_2$/CaF$_2$ ($f = 14\%$), and CdTe/BaF$_2$ ($f = 4.5\%$). Due to this, the structural perfection of the upper CdTe layer is high enough to recommend this heterosubstrate for further investigation.

It is noteworthy that IR photodetectors based on epitaxial films of lead and tin chalcogenides with spectral sensitivity ranges of 3-5 and 8-14 μm, respectively, have already been fabricated using buffer dielectric layers on Si [39-41]. The situation in general suggests that heterosubstrates based on Si are very promising. New results in this area should be expected in the very near future.

Of the alternate substrates for CdTe epitaxy, InSb has the best match of lattice constants ($f = 0.06\%$). High-quality CdTe films on InSb(100) [42] and on InSb(111) [43] were grown by MBE. The CdTe films do not have small-angle boundaries or twins. The halfwidth of the oscillation curve is less than 1'. The mechanical prop-

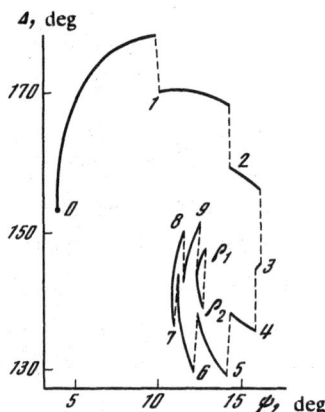

Fig. 3. Change of ellipsometric parameters ψ and Δ during growth of CdTe–HgTe SL on CdTe substrate. $d_{CdTe} = 4$ nm, $d_{HgTe} = 8$ nm. Solid line is HgTe growth; dashed line, CdTe. The numbers 1, 2,... are the period number. The quantities ρ_1 and ρ_2 are limiting values of the ellipsometric parameters of SL in the upper layer of HgTe and CdTe, respectively [29].

erties of the InSb substrates are not as good as desired. The substrates are not transparent to wavelengths less than 3 μm and cannot be used to prepare photodetectors illuminated from the substrate side. For this reason sapphire is not useful as substrate since it is opaque to wavelengths greater than 6-8 μm. However, epitaxy of CdTe on sapphire was successful [44]. The most probable alternative substrate for CdTe and CdHgTe epitaxy is considered to be GaAs. In spite of the large mismatch of lattice constants (13.6%), high-quality CdTe films on GaAs can be prepared [21, 45]. The orientational relations for epitaxy of CdTe on GaAs were theoretically examined in [46] on the basis of a coincidence-site lattice model. The approach is based on the concept that not the lattice constants but rather the translational symmetries of materials on both sides of the junction are examined. The calculations agree with experiment. In fact, CdTe film orientations (111) and (100) are observed during CdTe epitaxy on GaAs(100) depending on the substrate preparation conditions. The strongest bonds in the system Cd—Te—Ga—As are formed between Ga and Te atoms. Thus, it was observed experimentally that CdTe deposition begins with formation of a Te film [47, 48]. Adsorption of Cd and crystallization of CdTe occurs only after this. It is natural to assume that the Te atoms on the GaAs surface strive to form the maximal number of bonds to Ga atoms and the minimal number to the Cd atoms of the film. It was found unambiguously that the first layer of Te atoms during formation of the CdTe(111) film provides only one bond to the film. This causes polarity in the (111)B film [47]. Considering that Ga and Te form a series of compounds with different stoichiometry, Ga_3Te_2, $GaTe$, Ga_2Te_3, and $GaTe_2$, it is natural to assume that the initial ratio of Ga and Te activities have a decisive influence on the CdTe film orientation. This ratio determines the type of Te surface compound and the configuration of the Te bonds to the film. Certain elements of such an approach were used in [49]. However, formation of Te—Te bonds was presumed. The model predicts formation of CdTe(100) films on the GaAs(100) surface stabilized by Ga. Both of these concepts contradict experiment. X-ray photoelectron spectroscopy does not reveal Te—Te bonds. High-temperature annealing of GaAs(100) substrates that enriches the surface in Ga leads to growth of CdTe(111) [50]. It has been definitely established that oxygen on the GaAs(100) surface causes CdTe(100) films to grow [50]. The characteristic type of structural defects of CdTe(111)//GaAs(100) films are twinned lamellae. Crystals of CdTe are known to be prone to twinning. However, the density of twinned lamellae in the films is more than 10^5 times greater than the density of twins in the crystal bulk. In order to find the stages

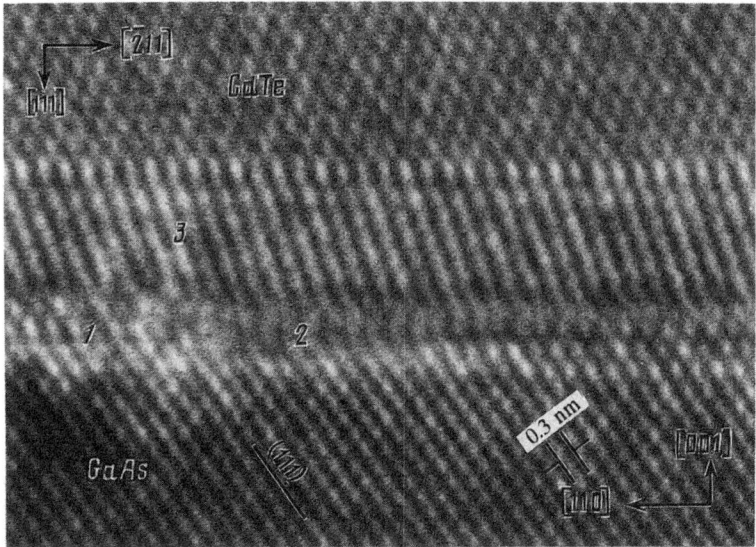

Fig. 4. Image of the (011) cross section of CdTe(111) film on the singular GaAs(100) surface obtained by high-resolution electron microscopy: 1, 2) Parts of the film at the boundary with the substrate that are located in the twinning position relative to each other. The photomicrograph was taken by V. Yu. Karasev of the Institute of Crystallography of the USSR Academy of Sciences.

of film growth responsible for the extensive formation of twinned lamellae, the structures of CdTe(111) films on GaAs(100) substrates were studied by high-resolution electron microscopy (HREM) of cross-sectional samples.

A multi-beam image of the $(0\bar{1}1)$ cross section of CdTe(111) film on the singular GaAs(100) surface is shown in Fig. 4. The slight mismatch of the film and substrate lattice constants ($f = 0.7\%$) along the heterojunction in the $[\bar{2}11]$ direction causes practically coherent joining of the GaAs(200) and CdTe(111) atomic planes. The parts of the film (1) and (2) at the junction with the substrate that are in the twinning position relative to each other are clearly seen from the slope of the {111} planes. The density of the parts of the film (1) and (2) along the heterojunction is about the same. This indicates an equal probability of forming CdTe nuclei in twinning positions during epitaxy on the singular GaAs(100) surface. Starting at a thickness greater than 3 nm, one of the two neighboring orientations outlives the other (region 3 in Fig. 4). According to *in situ* RHEED data, the film surface at thicknesses greater than 30 nm becomes smooth and a layered growth mechanism ensues. The twinned lamellae are observed over the whole thickness of the film (3 μm). However, their density with thickness is uneven. The number of twinning boundaries decreases sharply at a thickness greater than 30 nm. In a number of cases they are completely absent in separate parts of the film of thickness up to 0.5 μm.

The effect of substrate surface disorientation on twinning in the film was studied. The substrate GaAs disoriented relative to (100) by 6° was annealed at 580°C for 40 min. The surface became rough due to the extended annealing. Parts with an orientation close to the singular one and disoriented by up to 10° with a system of steps of the same sign were formed. A film practically without twins near the heterojunction is formed on the disoriented parts of the substrate according to HREM (Fig. 5). Extensive twinning in the thin film layer adjoining the substrate is observed in the parts oriented close to singular. It should be noted that in both cases inclined twinned lamellae growing into the film arise along with the twinning boundaries parallel to the growth surface (cf. Fig. 5). Thus, the twinning process in the initial epitaxy stage in the same sample, i.e., under identical growth conditions, proceeds differently in neighboring parts of the substrate with different orientations. The presence of a system of surface steps of the same sign greatly decreases the probability of twinning.

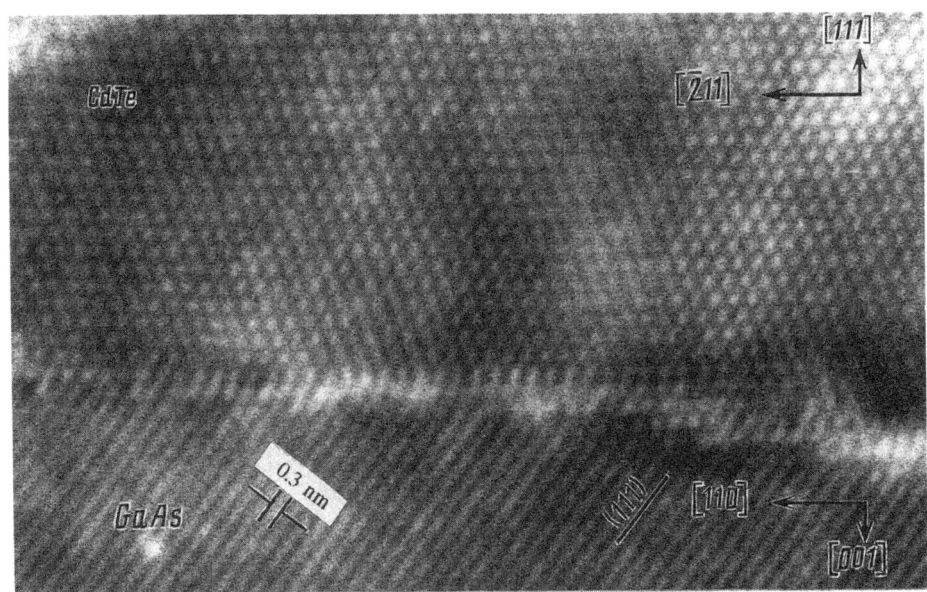

Fig. 5. Cross section of a CdTe(111) film on GaAs substrate disoriented relative to (001) by 6°. The photomicrograph was taken by V. Yu. Karasev of the Institute of Crystallography of the USSR Academy of Sciences.

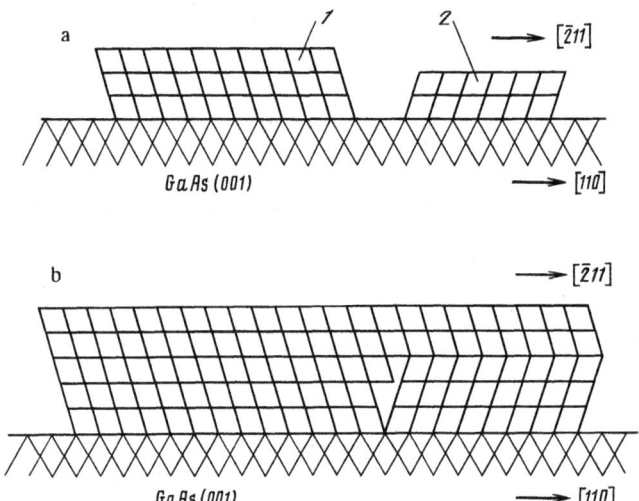

Fig. 6. Diagram of the formation of a twinning boundary in CdTe(111) film on GaAs(001) substrate during island growth: a) Isolated CdTe(111) islands 1 and 2 in twinning positions to each other; b) solid CdTe(111) film with a twinning boundary.

In agreement with the results obtained, the formation mechanisms of the twinning boundaries parallel to the heterojunction can be formulated as follows. Twinning on the singular GaAs(100) surface begins with nucleation of the film. There is an equal probability of forming nuclei rotated by 180° to each other around the axis perpendicular to the growth surface. This is shown schematically in Fig. 6a where *1* is the CdTe island formed earlier than the island *2* in the twinning position. Since the islands are not nucleated at the same moment they will have different heights. As a result the atomic planes of the higher island spread laterally over the surface of the neighboring islands (Fig. 6b). Therefore, firstly, the orientation of the higher island will survive and, secondly, a twinning boundary parallel to the substrate surface is formed. The density of twinning boundaries will be maximal up to the thickness at which the growth changes from insular to layered. This is observed using *in situ* RHEED. The film surface becomes smooth and layered growth occurs at a thickness of greater than 30 nm. According to HREM, the density of twinning boundaries is greatest at distances less than 30 nm from the heterojunction. It then decreases sharply. Nucleation is a statistical process. Therefore, oppositely oriented films can survive in different parts of the sample. As a result twinning can pervade large film thicknesses. However, the density of twinning boundaries should decrease steadily with distance from the heterojunction. The observed experimental increase in the number of twinned lamellae for certain samples indicates that other mechanisms leading to formation of twinning boundaries in the CdTe(111) films also exist.

If a regular system of nucleation centers with the same energy and same orientation is formed during epitaxy then extensive twinning in CdTe(111) films can be avoided. Surface steps of the same sign can act as such centers. Nuclei formed at an angle to the step have the preferred orientation due to interaction with the step. Apparently it is possible to ensure that nuclei of only one orientation form at a sufficient density of steps. Data obtained for the decrease of twinning boundary density during epitaxy on disoriented substrates prove that steps of the GaAs(100) surface can orient CdTe nucleation [51].

The empirical successes of MBE of CMT films are in general very impressive. After seven years this method has gone from the first CMT layers of small area (of the order of 1 cm^2) with anomalous electrical parameters on CdTe substrates to perfect films of 50 mm diameter on GaAs substrates. The homogeneous composition and electrical properties of these make them useful for fabricating multielement photodetectors.

REFERENCES

1. M. F. S. Tang and D. A. Stevenson, "Interdiffusion behavior of HgTe—CdTe junctions," *Appl. Phys. Lett.*, **50**, No. 18, 1272-1274 (1987).
2. S. J. C. Irvine, J. B. Mullin, H. Hill, et al., "Photostimulated II—VI crystal growth: A study of low temperature epitaxy," *J. Cryst. Growth*, **86**, No. 1/4, 188-197 (1988).
3. P.-Y. Lu, C.-H. Wang, L. M. Williams, et al., "Epitaxial Hg$_{1-x}$Cd$_x$Te growth by low-temperature metalorganic chemical vapor deposition," *Appl. Phys. Lett.*, **49**, No. 20, 1372-1374 (1986).
4. L. M. Williams, P.-Y. Lu, C.-H. Wang, et al., "Plasma enhanced chemical vapor deposition of epitaxial mercury telluride," *Appl. Phys. Lett.*, **51**, No. 21, 1738-1740 (1987).
5. L. S. Lichtman, L. D. Parsons, and E. H. Citrin, "Temperature-independent unassisted pyrolytic-MOCVD growth of cadmium telluride at 250°C using 2,5-dihydrotellurophene," *J. Cryst. Growth*, **86**, No. 1/4, 217-221 (1988).
6. P.-Y. Lu, L. M. Williams, C.-H. Wang, and S. N. G. Chu, "HgTe—CdTe superlattices and Hg$_{1-x}$Cd$_x$Te growth by low-temperature metalorganic chemical vapor deposition," *J. Vac. Sci. Technol., A*, **5**, No. 5, 3153-3156 (1987).
7. D. L. Smith, T. C. McGill, and J. N. Schulman, "Advantages of HgTe—CdTe superlattice as an infrared detector material," *Appl. Phys. Lett.*, **43**, No. 2, 180-182 (1983).
8. J. N. Schulman and T. C. McGill, "The CdTe/HgTe superlattice: Proposal for a new infrared material," *Appl. Phys. Lett.*, **34**, No. 10, 663-665 (1979).
9. J. P. Faurie, A. Million, and J. Piaguet, "CdTe—HgTe multilayers grown by molecular beam epitaxy," *Appl. Phys. Lett.*, **41**, No. 8, 713-715 (1982).
10. L. DiCioccio, A. Million, J. P. Gailliard, and M. Dupuy, "Observation of CdTe—HgTe superlattices by transmission electron microscopy," *Rev. Phys. Appl.*, **22**, No. 6, 465-468 (1987).
11. K. A. Harris, S. Hwang, Y. Lansari, et al., "Growth and properties of dilute magnetic semiconductor superlattices containing Hg$_{1-x}$Mn$_x$Te," *J. Vac. Sci. Technol., B*, **5**, No. 3, 699 (1987).
12. J. P. Faurie, A. Million, R. Boch, and J. T. Tissot, "Latest developments in the growth of Cd$_x$Hg$_{1-x}$Te and CdTe—HgTe superlattices by molecular beam epitaxy," *J. Vac. Sci. Technol., A*, **1**, No. 3, 1593-1597 (1983).
13. J. P. Faurie, X. Chu, S. Sivananthan, et al., "Type III—type I transition in Hg$_{1-x}$Cd$_x$Te—CdTe, Hg$_{1-x}$Mn$_x$Te—CdTe, and Hg$_{1-x}$Zn$_x$Te—CdTe superlattices," *J. Vac. Sci. Technol, B*, **5**, No. 3, 700 (1987).
14. R. Heckingbottom, G. J. Davies, and K. A. Prior, "Growth and doping of gallium arsenide using molecular beam epitaxy (MBE): Thermodynamic and kinetic aspects," *Surf. Sci.*, **132**, No. 1/3, 375-389 (1983).
15. J. P. Faurie, A. Million, and J. Piaguet, "Characterization of Cd$_x$Hg$_{1-x}$Te p-type layers grown by MBE," *J. Cryst. Growth*, **59**, No. 1/2, 10-14 (1982).

16. J. P. Gailliard, "A thermodynamic model of MBE, application to the growth of II–VI semiconductors," *Rev. Phys. Appl.*, **22**, No. 6, 457-463 (1987).

17. A. Koukitu, H. Nakai, T. Suzuki, and H. Seki, "Thermodynamic analysis of MBE of II–VI semiconductors," *J. Cryst. Growth*, **84**, No. 3, 425-430 (1987).

18. S. Sivananthan, X. Chu, and J. P. Faurie, "Dependence of the condensation coefficients of Hg on the orientation and stability of the Hg–Te bond for the growth of $Hg_{1-x}M_xTe$ (M = Cd, Mn, Zn)," *J. Vac. Sci. Technol., B*, **5**, No. 3, 694-698 (1987).

19. R. F. Brebrick and A. J. Strauss, "Partial pressures of Hg(g) and $Te_2(g)$ in Hg–Te system from optical densities," *J. Phys. Chem. Solids*, **26**, No. 6, 989-1002 (1965).

20. K. A. Harris, S. Hwang, D. K. Blanks, et al., "Growth of HgCdTe and other Hg-based films and multilayers by MBE," *J. Vac. Sci. Technol., A*, **4**, No. 4, 2061-2066 (1986).

21. H. A. Mar, K. T. Chee, and N. Salansky, "CdTe films on (001)GaAs:Cr by MBE," *Appl. Phys. Lett.*, **44**, No. 2, 237-239 (1984).

22. H. H. Farrell, J. P. Harbison, and L. D. Peterson, "MBE growth mechanism of GaAs(100) surfaces," *J. Vac. Sci. Technol., B*, **5**, No. 5, 1482-1489 (1987).

23. M. Pessa, O. Jylhä, P. Huttunen, and M. A. Herman, "Epitaxial growth and electronic structure of CdTe films," *J. Vac. Sci. Technol., A*, **2**, No. 2, 418-422 (1984).

24. M. Pessa, O. Julhä, and M. A. Herman, "Atomic layer epitaxy of CdTe on the polar (111)A and (111)B surfaces of CdTe substrates," *J. Cryst. Growth*, **67**, No. 2, 255-260 (1984).

25. M. A. Herman, O. Julhä, and M. Pessa, "Growth mechanism in atomic layer epitaxy. I. Re-evaporation of Cd and Te from CdTe(111) surfaces monitored by Auger electron spectroscopy," *Cryst. Res. Technol.*, **21**, No. 7, 841-851 (1986).

26. M. A. Herman, O. Julhä, and M. Pessa, "Growth mechanism in atomic layer epitaxy. II. A model of the growth process of CdTe on CdTe(111) substrates," *Cryst. Res. Technol.*, **21**, No. 8, 969-974 (1986).

27. S. V. Krishnaswamy, J. H. Rieger, N. G. Doyle, and M. H. Frankombe, "Ion beam sputter deposition and epitaxy of CdTe and HgTeCd films," *J. Vac. Sci. Technol., A*, **5**, No. 4, 2106-2110 (1987).

28. J. T. Cheung and J. Madden, "Growth of HgCdTe epilayers with any predesigned compositional profile by laser MBE," *J. Vac. Sci. Technol., B*, **5**, No. 3, 705-708 (1987).

29. A. V. Rzhanov, K. K. Svitashev, A. S. Mardezhov, and V. A. Shvets, "Control of superlattice parameters during their preparation by ellipsometry," *Dokl. Akad. Nauk SSSR*, **297**, No. 3, 604-607 (1987).

30. Y. Lo, R. N. Bicknell, T. H. Myers, et al., "Growth of CdTe films on silicon by molecular layer epitaxy," *J. Appl. Phys.*, **54**, No. 7, 4238-4240 (1983).

31. R. N. Bicknell, T. H. Myers, and J. F. Schetzina, "Growth of CdTe films on alternative substrates by molecular beam epitaxy," *J. Vac. Sci. Technol., A*, **2**, No. 2, 423-426 (1984).

32. R.-L. Chou, M.-S. Lin, and K.-S. Chou, "Characteristics of CdTe grown on Si by low pressure metalorganic chemical vapor deposition," *Appl. Phys. Lett.*, **48**, No. 8, 523-525 (1986).

33. N. Matsumura, T. Ohshima, J. Saraie, and Y. Yodogawa, "Preparation of CdTe thin films on Ge substrates by molecular beam epitaxy," *J. Cryst. Growth*, **71**, No. 2, 361-370 (1985).

34. R. Fischer, D. Neuman, H. Zabel, et al., "Dislocation reduction in epitaxial GaAs on Si(111)," *Appl. Phys. Lett.*, **48**, No. 18, 1223-1225 (1986).

35. R. C. Bean, K. R. Zanio, K. A. Hay, et al., "Epitaxial CdTe films on GaAs/Si and GaAs substrates," *J. Vac. Sci. Technol., A*, **4**, No. 4, 2153-2157 (1986).

36. R. Kay, R. Bean, K. Zanio, et al., "HgCdTe photovoltaic detectors on Si substrates," *Appl. Phys. Lett.*, **51**, No. 26, 2211-2213 (1987).

37. H. Zogg and S. Blunier, "Molecular beam epitaxial growth of high structural perfection CdTe on Si using a $(Ca, Ba)F_2$ buffer layer," *Appl. Phys. Lett.*, **49**, No. 22, 1531-1533 (1986).

38. H. Zogg, P. Maier, and H. Melchior, "Graded IIa fluoride buffer layers for heteroepitaxy of lead chalcogenides and CdTe on Si," *J. Cryst. Growth*, **80**, No. 2, 408-416 (1987).

39. H. Zogg and P. Norton, "Heteroepitaxial PbTe–Si and (Pb, Sn)Se–Si structures for monolithic 3-5 μm and 8-12 μm infrared sensors arrays," Int. Electron. Develop. Meet., IEEE Publ., New York (1985), pp. 121-124.

40. H. Zogg, W. Vogt, and H. Melchior, "Heteroepitaxial IV–VI infrared sensors on Si substrates with fluoride buffer layers," *Nucl. Instrum. Methods Phys. Res., Sect. A*, **A253**, No. 3, 418-422 (1987).

41. H. Zogg, W. Vogt, and H. Melchior, "Growth of heteroepitaxial lead chalcogenides infrared detector arrays on fluoride covered silicon substrates," *Mater. Res. Soc. Symp. Proc.*, No. 71, 87-95 (1986).

42. R. F. C. Farrow, G. R. Jones, G. M. Williams, and I. M. Young, "Molecular beam epitaxial growth of high structural perfection hetero-epitaxial CdTe films on InSb(100)," *Appl. Phys. Lett.*, **39**, No. 12, 954-956 (1981).

43. T. H. Myers, L. O. Yaucheng, J. F. Schetzina, and S. R. Jost, "Properties of CdTe/InSb heterostructures prepared by MBE," *J. Appl. Phys.*, **53**, No. 12, 9232-9234 (1982).

44. S. T. Edwards, A. F. Schreiner, T. H. Myers, and J. F. Schetzina, "Photoluminescence from CdTe/sapphire films prepared by MBE," *J. Appl. Phys.*, **54**, No. 11, 6785-6786 (1983).

45. R. N. Bicknell, R. W. Yanka, N. C. Giles, et al., "Growth of (100)CdTe on high structural perfection on (100)GaAs substrates by MBE," *Appl. Phys. Lett.*, **44**, No. 3, 313-315 (1984).

46. A. Zur and T. C. McGill, "Lattice match: An application to heteroepitaxy," *J. Appl. Phys.*, **55**, No. 2, 378-386 (1984).

47.   C. Hsu, S. Sivananthan, X. Chu, and J. P. Faurie, "Polarity determination of CdTe(111) orientation grown on GaAs(100) by MBE," *Appl. Phys. Lett.*, **48**, No. 14, 908-914 (1986).

48.   H. A. Mar, N. Salansky, and K. T. Chee, "Study of the initial stages of growth of CdTe on (001)GaAs," *Appl. Phys. Lett.*, **44**, No. 9, 898-900 (1984).

49.   G. Cohen-Solal, F. Bailly, and M. Barbe, "Model for heteroepitaxial growth of CdTe on (100) oriented GaAs substrates," *Appl. Phys. Lett.*, **49**, No. 22, 1519-1521 (1986).

50.   J. P. Faurie, C. Hsu, S. Sivananthan, and X. Chu, "CdTe—GaAs(100) interface: MBE growth, RHEED and XPS characterization," *Surf. Sci.*, **168**, No. 1, 477-482 (1986).

51.   S. A. Dvoretsky, A. K. Gutakovsky, V. Yu. Karasev, et al., "Twinning in CdTe(111) films on (100)GaAs substrates," *Inst. Phys. Conf. Ser.*, No. 93, Vol. 2, 407-408 (1988).

δ-STRUCTURES IN GALLIUM ARSENIDE

D. I. Lubyshev, V. P. Migal', V. N. Ovsyuk, B. R. Semyagin, and S. I. Stenin

A new class of semiconducting structures with δ-doped layers and a high concentration of layered quasi-two-dimensional charge carriers is currently of great interest. On one hand these layers have interesting physical properties, on the other they are promising for practical applications [1].

The term "δ-doping" refers to formation during molecular-beam epitaxy of a monoatomic layer doped until degenerate. The dopant distribution with depth in such structures is described by the Dirac δ-function, from which it gets its name. Screening of the ionized dopants by electrons gives rise to a V-shaped potential, as shown in Fig. 1. The characteristic sizes of the potential well are comparable to the electron wavelength. This leads to dimensioned quantification. Analysis of the Shubnikov—de Haas oscillations shown in Fig. 2 is consistent with the existence of at least three dimensioned quantum levels in the studied separate δ-layers at a concentration per unit surface $\Gamma_0 = 6 \cdot 10^{12}$ cm^{-2} [2].

Formation conditions of modulated structures based on GaAs with periodically placed δ-layers of Si are described in the present work. The influence of the structure period on the mobility and concentration of charge carriers is studied.

The δ-structures were grown in a modified PMA-12 molecular-beam epitaxy apparatus. A buffer layer of intentionally undoped GaAs was deposited on the atomically clean surface of semi-insulating (100)GaAs at a rate 1 μm/h. The buffer layer grown was p-type with a concentration $(2-5) \cdot 10^{14}$ cm^{-3} and mobility 5000-7800 cm^2/(V·sec). These values were measured at 77 K. The ratio of As and Ga molecular fluxes was chosen so that superstructure 3 × 6, corresponding to the stoichiometric surface composition, was observed at any epitaxy temperature in the RHEED pattern. The RHEED intensity oscillations indicated that growth occurred by a two-dimensional mechanism. They were used to determine the film growth rate as well as its thickness to an accuracy ±0.15 nm. The Ga flux was interrupted after the buffer obtained a thickness of 0.5 μm. The growth surface was smoothed by diffusion as monitored by the increased intensity of the specular reflection. The required amount of Si atoms was deposited on the smooth surface. Then undoped GaAs was again grown to 0.5 μm. The layers were grown at 530°C, i.e., below the temperature at which the δ-layer is broadened by diffusion [3]. Modulated structures were grown by repeating the cycle "δ-layer—undoped intermediate layer" five times and finishing with an undoped layer of thickness 0.5 μm. The amount of Si atoms that gave an electron concentration at room temperature in a single δ-layer of $3 \cdot 10^{12}$ cm^{-2} was deposited in each δ-layer. The intermediate layers of undoped GaAs within each modulated structure were identical.

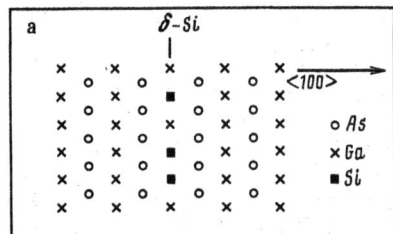 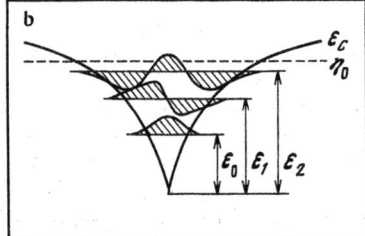

Fig. 1. The δ-layer: a) doping pattern; b) energy diagram. η_0 is the Fermi level; ϵ_c is the energy of the bottom of the conductivity band; ϵ_0, ϵ_1, ϵ_2 are the dimensioned quantum levels.

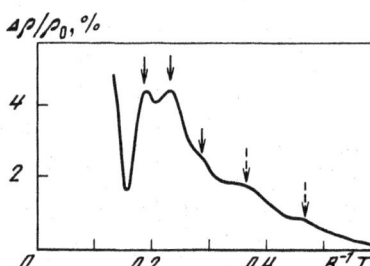

Fig. 2. Shubnikov—de Haas oscillations of the GaAs δ-layer. Solid arrows denote the first quantum level ϵ_1; dashed arrows, the second quantum level ϵ_2.

A series of modulated structures with 2, 8, 30, 60, 120, 150, 180, 210, and 240 nm between δ-layers was grown under identical conditions. Dumbbell-shaped samples were prepared on these structures by photolithography. The temperature dependences of the Hall and electric conductivity coefficients were measured. The surface concentrations and effective mobilities of electrons were calculated from measurements at 300, 77, and 4.2 K as functions of distance between δ-layers d, as shown in Fig. 3. On changing d from 30 to 240 nm, the charge carrier concentration does not change and has a value slightly less than the sum of the concentration in the five separate δ-layers. The effective mobility of the five-layered δ-structure depends on the distance between layers and reaches a maximum 7750 cm^2/(V·sec) near 77 K and d = 150 nm.

The temperature dependence of the effective mobility in structures with a single δ-layer (curve *1*) and with five δ-layers with different d (curves *2-6*) is shown in Fig. 4. In a single δ-layer with $\Gamma_0 = 3 \cdot 10^{12}$ cm^{-2} the electron mobility at 300 K was 3000 cm^2/(V·sec). It depended weakly on temperature throughout the measurement range. This is characteristic for a degenerate quasi-two-dimensional channel with scattering at contaminant centers [4]. In contrast with the single δ-structure, the nature of the mobility change in the periodic structure is similar to that in nondegenerate bulk GaAs which is determined mainly by scattering over lattice vibrations and dopant ions.

In order to explain these functions, we will examine the energy diagram of a semiconductor with δ-doping layers (Fig. 5). The dimensionless potential $y(z) > 0$ is calculated from the edge of the conductivity band ϵ_{c0} of the unperturbed semiconducting matrix. We will assume for simplicity that quasi-Boltzmann statistics are valid for electrons with energies above the Fermi level η_0. Then for $y(z) >> 1$ the path of the potential between $z \in (0, d_0/2)$ follows from the equation

$$\frac{dy}{dz} = \frac{1}{L} \sqrt{\frac{n_0}{N_A}} \sqrt{\exp y - \exp \bar{y} + \frac{N_A}{n_0}(y - \bar{y})}, \tag{1}$$

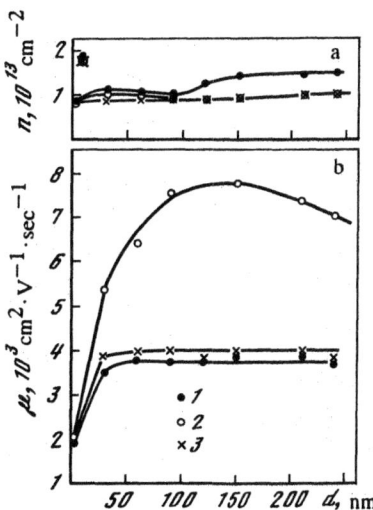

Fig. 3. Hall concentration (a) and mobility (b) in a periodic five-layered δ-structure as a function of distance between δ-layers. Measurements were made at 300 (1), 77 (2), and 4.2 K (3).

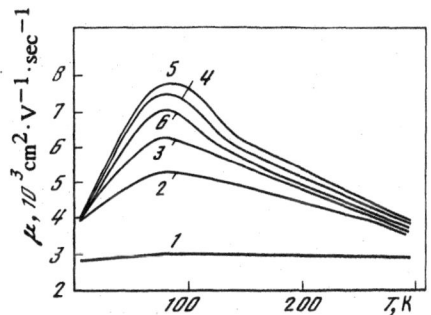

Fig. 4. Temperature dependence of Hall electron mobility in a periodic δ-structure: 1) Single δ-layer. $d = 30$ (2), 60 (3), 90 (4), 150 (5), 240 nm (6).

where $L = [\bar{\varepsilon}_0 \bar{\varepsilon}_s kT/(2q^2 N_A)]^{1/2}$, $\bar{\varepsilon}_0$ is the dielectric constant of a vacuum, $\bar{\varepsilon}_s$ is the relative dielectric constant of the semiconductor, k is Boltzmann's constant, t is the absolute temperature, q is the elementary positive charge, N_A is the concentration of acceptor dopant, n_0 is the free electron concentration far from the δ-doped layers, and $\bar{y}$ is the potential in the symmetry plane $z = d_0/2$ (here $d_0 = d - 2a$, $2a$ is the localized electron wave function in the quantum well at the Fermi level).

At sufficiently small values d_0 the inequality $\bar{n} = n_0 \exp \bar{y} >> N_A$ is fulfilled. Since this is valid, the linear terms under the radical in the right part of Eq. (1) can be neglected. This gives the relation

$$\arcsin \exp\left(-\frac{y-\bar{y}}{2}\right) - \arcsin \exp\left(-\frac{y_s-\bar{y}}{2}\right) = \frac{z}{2L}\sqrt{\frac{\bar{n}}{N_A}}, \qquad (2)$$

where y_s is the value of the potential at $z = 0$ (Fig. 5). We then obtain the dependence of electron concentration $\bar{n}$ in the symmetry plane on the distance d_0 between the edges of the quantum wells at the Fermi level

$$d_0 = 2L\sqrt{N_A/\bar{n}}(\pi - 2\arcsin\sqrt{\bar{n}/N_c}), \qquad (3a)$$

at $\bar{n} \lesssim N_c/3$, the approximation below is valid

$$\bar{n}/N_A = (2\pi L)^2 (d_0 + 4L\sqrt{N_A/N_c})^{-2}, \qquad (3b)$$

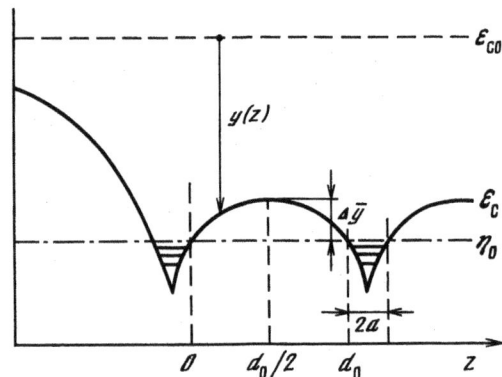

Fig. 5. Energy diagram for secondary charge carriers near the two terminal δ-layers. The quantity ε_{c0} is the edge of the conductivity band in the neutral bulk semiconductor and η_0 is the Fermi level.

where N_c is the effective density of states in the conductivity band. Using our approximation this is equal to the electron concentration $n_s = n_0 \exp y_s$ in the plane $z = 0$. Equations (3a) and (3b) are valid as long as $\bar{n} \gg N_A$.

The total excess of free electrons Γ_n in the half-layer $(0, d_0/2)$ is $\Gamma_n = 2L \, [N_A(N_c - \bar{n})]^{1/2}$. The height of the potential barrier between points $z = 0$ and $z = d_0/2$ is $\Delta y = y_s - \bar{y} = \ln(N_c/\bar{n})$.

Comparison of Fig. 4, curve 1, for a single δ-layer with the functions $\mu_H(d)$ in Figs. 3 and 4 suggests that "suprabarrier" electrons with energies greater than $kT\Delta\bar{y}$ (cf. Fig. 5) play the main role in increasing μ_H. We will set the layered concentration of these electrons equal to $\Gamma_{n2} \simeq \bar{n}d_0/2$ and approximate their effective drift mobility μ_2 by the phenomenological equation

$$\mu_2 = \mu_1(2z_0 + b_0 d)/(2z_0 + d),\qquad(4)$$

where μ_1 is the effective electron drift mobility in a single δ-layer, $b_0 = \mu_{n0}/\mu_1$, μ_{n0} is the bulk electron mobility in the unperturbed semiconducting matrix with doping level N_A, and z_0 is a certain characteristic distance. We will also assume that all remaining electrons with concentration $\Gamma_{n1} = \Gamma_0 - 2\Gamma_{n2}$ have a drift mobility μ_1. The measured Hall mobilities μ_H and the layered electron concentration Γ_{nH} in this case are

$$\mu_H = (\sum_j r_{Hj}\mu_j^2 K_j \Gamma_{nj})/(\sum_j \mu_j K_j \Gamma_{nj}),\qquad(5)$$

$$\Gamma_{nH} = (\sum_j \mu_j K_j \Gamma_{nj})^2 /(\sum_j r_{Hj}\mu_j^2 K_j \Gamma_{nj}),\qquad(6)$$

where K_j is the number of layers (along the z coordinate) with a single layered electron concentration of the given type Γ_{nj} and r_{Hj} is the corresponding Hall factor, in our case $K_1 = 5$ and $K_2 = 8$.

The calculated functions $\mu_H(d)$ and $\mu_H(T)$ obtained from Eq. (5) for $r_{H2} = r_{H1}$ and $2a = 20$ nm and $z_0 = 50$ nm are plotted in Fig. 6a and b. Since the temperature of the bulk electron mobility in the p-type material is unknown, we used the experimental function $\mu_{n0}(T)$ [5] obtained for epitaxial GaAs layers of the opposite conductivity type but with the same donor concentration $N_D \simeq 2 \cdot 10^{14}$ cm^{-3}. This is a bell-shaped function with $\mu_{n0} \simeq 6.8 \cdot 10^4$ cm^2/(V·sec) at the maximum at 50 K, $7 \cdot 10^3$ cm^2/(V·sec) at 300 K, and the extrapolated value of 10^4 cm^2/(V·sec) at helium temperature. The above value $2a = 20$ nm was chosen so that the mobility $\mu_H(d)$ dropped sharply from $3.9 \cdot 10^3$ to $2 \cdot 10^3$ cm^2/(V·sec) at $d < 30$ nm (cf. Fig. 3). This is apparently due to overlap of neighboring potential wells for quantized electrons. The value $z_0 = 50$ nm was selected so that the calculated and experimental values of the maximal mobility in the function $\mu_H(d)$ at 77 K agreed.

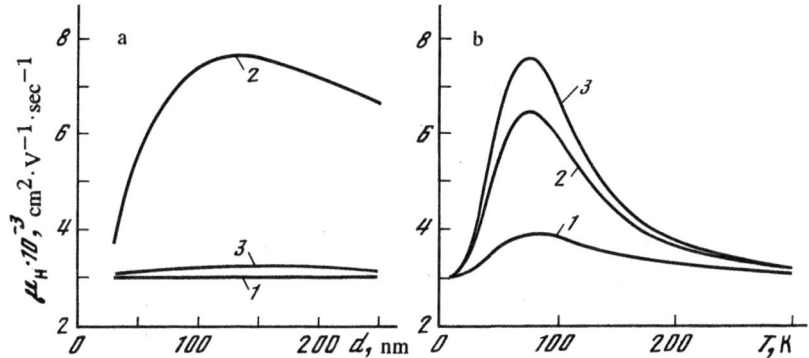

Fig. 6. Calculated Hall mobilities μ_H as functions of distance between δ-layers for various T (a) and of temperature for various d (b): 4.2 (1), 77 (2), 300 K (3) (a); 30 (1), 60 (2), 150 nm (3).

The qualitative shape of the theoretical functions $\mu_H(d)$ and $\mu_H(T)$ in Fig. 6a and b is similar to that measured experimentally in Figs. 3 and 4. This confirms the assumption that the suprabarrier electrons play a decisive role in determining the kinetic effects in samples with δ-doped planes separated by a sufficiently short period. At the temperatures of the high bulk electron mobility values, the function $\mu_H(d)$ has a maximum due to two competing mechanisms. At small d the effective drift mobility $\mu_2(4)$ decreases due to the increased frequency of collisions with scattering planes. At large d the barrier $\Delta \bar{y}$ increases and the concentration of suprabarrier electrons $\bar{n}$ decreases. The shape of the curves $\mu_H(T)$ (cf. Fig. 6b) describes mainly the temperature dependence of the bulk mobility $\mu_{n0}(T)$. Features related to a constant concentration of suprabarrier electrons appear as T decreases to the left of the maximum in $\mu_H(T)$. As a result, the maximum in the function $\mu_H(T)$ shifts toward higher temperatures relative to the maximum in $\mu_{n0}(T)$ by about 25 K. At $T \leq 10$ K, the mobility $\mu_H = \mu_{1H}$. The last result does not agree with experiment (Fig. 3, curve 3). For this reason, it is possible that the electron mobility μ_1 near the δ-doped layers is greater than for a single δ-layer due to the change of shape of the potential wells for localized electrons for sufficiently small d. An exact theory of the kinetic effects in multilayered δ-doped samples that takes into account the effects noted above must be developed for a more detailed analysis of the experimental data.

Thus, a dependence of Hall mobility on distance between δ-doped planes in GaAs is observed. This is due to the influence of suprabarrier electrons in determining the kinetic effects in periodically δ-doped structures.

REFERENCES

1. K. Ploog, "Delta-(δ)-doping in MBE-grown GaAs: concept and device application," *J. Cryst. Growth*, **81**, No. 1, 304-313 (1987).
2. G. M. Gusev, Z. D. Kvon, D. I. Lubyshev, et al., "Anisotropic negative magnetostrictive quasi-two-dimensional δ-doped layers of GaAs," *Fiz. Tverd. Tela*, **30**, No. 10, 3148-3150 (1988).
3. E. F. Schubert, T. H. Chiu, and B. Tell, "Diffusion of atomic silicon in gallium arsenide," *Appl. Phys. Lett.*, **53**, No. 4, 293-295 (1988).
4. A. V. Chaplik, "Energy spectrum and electron scattering processes in inversion layers," *Zh. Éksp. Teor. Fiz.*, **60**, No. 5, 1845-1852 (1971).
5. *Gallium Arsenide: Preparation, Properties, and Application* [in Russian], Nauka, Moscow (1973).

Part III

GROWTH OF CRYSTALS AND FILMS
FROM SOLUTIONS AND FLUXES

INFLUENCE OF IMPURITIES ON GROWTH KINETICS

AND MORPHOLOGY OF PRISMATIC FACES

OF ADP AND KDP CRYSTALS

L. N. Rashkovich and B. Yu. Shekunov

The influence of impurities on crystal growth has been the subject of numerous studies. The topic has been reviewed thoroughly [1, 2]. The principal concepts of the mechanism of action of impurities are defined by essentially three phenomena. The impurity decreases the number of sites for adding building blocks by entering kinks. It stops the step from turning by adsorbing to the surface. This increases the curvature of the step and reduces its velocity. The rate of impurity adsorption should be comparable with the exposure time of the surface before the following growth layer appears. The quantitative relations obtained on the basis of these phenomena often cannot be confirmed experimentally. This is due to both their incompleteness and the deficiency of the method that in most cases enables operation on only one property, the normal growth rate. It is even more difficult to find the influence of impurities on the morphological stability of an echelon of elementary steps since few microscopic studies of surface relief have been performed *in situ* and the sensitivity of the method is often insufficient.

Impurities in actual solutions, which are usually not monitored, decisively influence the kinetics by changing the dependence of the tangential step velocity on the supersaturation. As a result, the shape of the dislocation spiral and the normal growth rate of the face change [3]. In the present work, we attempted to use all the capabilities of the interference method of investigating *in situ* crystallization in solution to study in detail the influence of impurities on growth of ADP and KDP crystals, typical crystals growing by a dislocation mechanism.

METHOD

The interference method [4] was used. It enabled the growing face of the crystal to be observed and the normal growth rate R and the curvature of the vicinal hillock steepness p to be determined independently. The tangential rate of step advance V was calculated from the equation $R = pV$ or was measured directly from the advance of the interference bands. The photographs presented below are images of crystal faces and the interference pattern from it, a topographical picture of the face. The distance between neighboring horizontals according to the height of the relief was $\Delta d = \lambda'/(2n) = 0.23\ \mu\text{m}$, where $\lambda' = 0.6328\ \mu\text{m}$ is the wavelength of the light source and $n \simeq 1.37$ is the index of refraction of a solution saturated at 35°C. The uncertainty in R was at worst 3%; that of p, 5-7%.

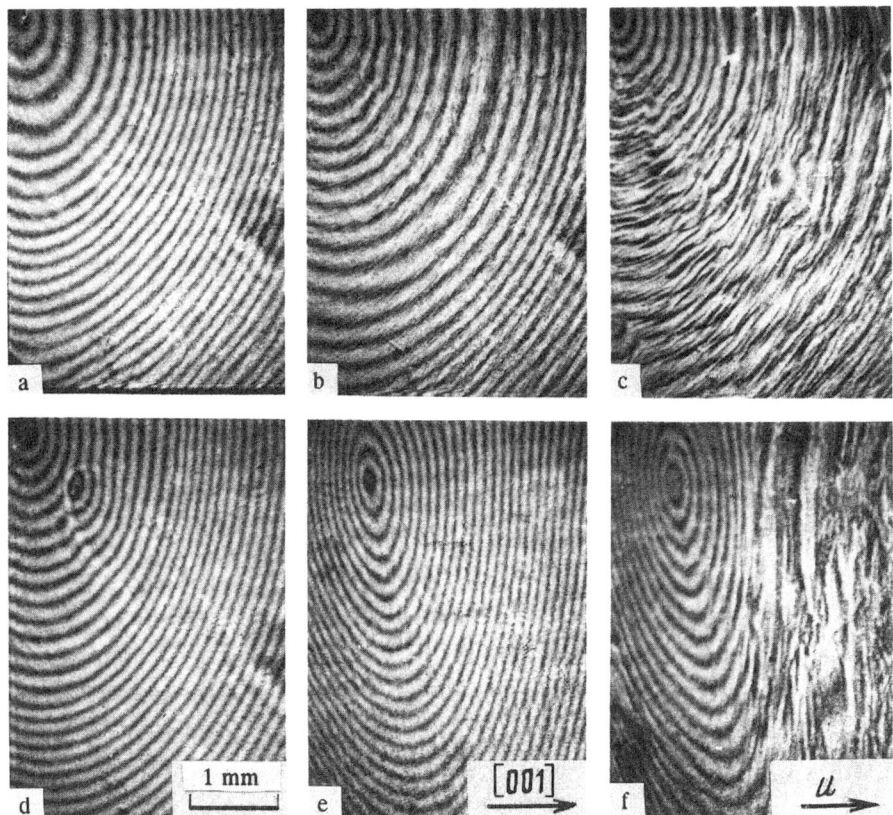

Fig. 1. Change of shape of hillocks and formation of macrosteps as σ increases for an ADP prismatic face. The arrow u denotes the flow direction. σ, 10^{-2}: 0.83 (a), 1.24 (b), 1.49 (c), 2.01 (d), 2.73 (e), 3.29 (f).

The solution was prepared using doubly distilled water and salt of at least 99.97% purity. Spectral analysis found no detectable impurities in the doubly distilled water. The analyzed impurities for KDP were (mass %): Si, 10^{-3}; Al, Fe, $2 \cdot 10^{-4}$; Mg, Pb, Zn, $2 \cdot 10^{-5}$; and Cr, V, Mo, Co, $2 \cdot 10^{-6}$. The values for ADP were Si, 10^{-3}; Al, Fe, $8 \cdot 10^{-5}$; Mg, $5 \cdot 10^{-5}$; Zn, $2 \cdot 10^{-5}$; Pb, $5 \cdot 10^{-6}$; Cr, V, Mo, Co, $2 \cdot 10^{-6}$. The solution pH values were not adjusted.

During the experiment a measured amount of $CrCl_3 \cdot 6H_2O$ impurity was added.

Solution (1.5 liters) was pumped from a thermostatted container through a cuvette with a seed crystal. The seeds were placed in the cuvette so that the studied face (5 × 5 mm) was parallel to the solution flow. The flow rate was 30-80 cm/sec. This ensured that the face growth rate was independent of the flow rate. The temperature in the cuvette was held constant within ±0.005°C. The supersaturation σ created by lowering the temperature T was calculated as $\sigma = \ln(c/c_0)$, where c is the concentration of the starting solution determined from its saturation temperature to an accuracy ±0.01°C and c_0 is the equilibrium concentration calculated from the formulas $c_0 = 16.73 + 0.484\,T \pm 0.2$ mass % (ADP) and $c_0 = 10.68 + 0.3616\,T \pm 0.04$ mass % (KDP).

1. SURFACE MORPHOLOGY CHANGE WITH INCREASING σ

The change of surface relief of an ADP prismatic face as σ increases is shown in Fig. 1. We note three characteristic features of this process. The surface ceases to be smooth in a comparatively narrow range of small σ values at a certain distance from the apex of the hillock. At moderate values the equidistant elementary steps (Fig. 1a) coalesce into an irregular macrostep (Fig. 1b and c). As σ increases the macrosteps disappear (Fig. 1d). The

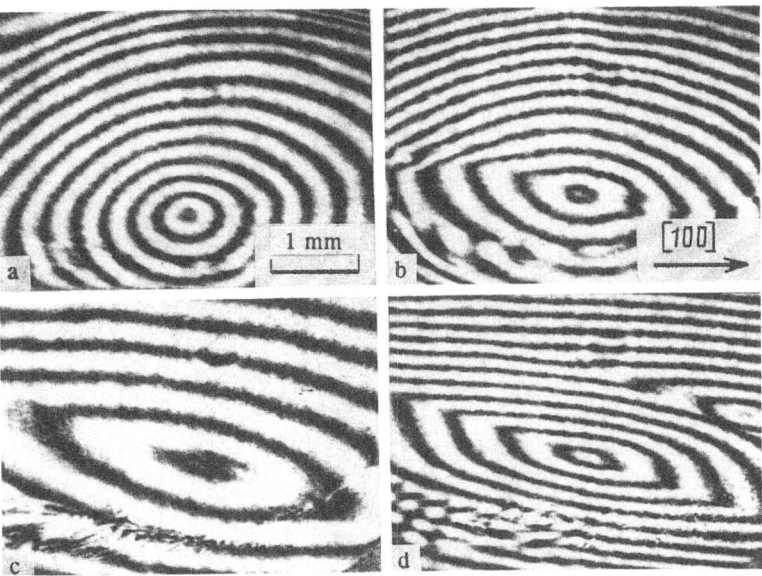

Fig. 2. Change of dislocation hillock shape on a KDP prismatic face as supersaturation changes. σ, 10^{-2}: 2.73 (a), 3.61 (b), 4.51 (c), 6.02 (d).

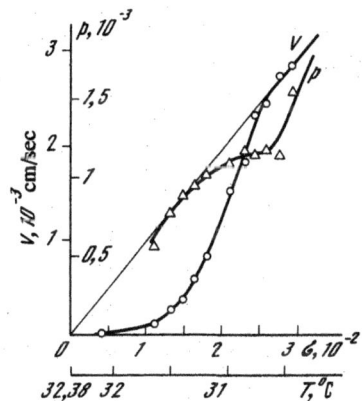

Fig. 3. Dependence of step velocity and hillock curvature in the [001] direction on supersaturation for an ADP prismatic face.

hillock then changes its eccentricity and rotates by 10-15°. Its curvature decreases in spite of the increased supersaturation (Fig. 1e). Next macrosteps again arise on the leeward side (right) relative to the flow (Fig. 1f).

The generation of macrosteps at large σ values is well known [1] although the mechanism of their formation is still not clear. They do not disappear as σ increases further. Rather they expand over the whole surface. iN [5-7] macrosteps are described whose formation and extent are related to the magnitude and direction of the solution flowing past the crystal and are explained by a variable supersaturation over the surface.

The appearance and subsequent disappearance of the macrosteps at low supersaturations are apparently observed for the first time. The solution hydrodynamics does not influence their formation. A minimum in the function $p(\sigma)$ was found in [3, 8]. It is due to $V(\sigma)$ being distinctly nonlinear in this supersaturation range. However, the change of the hillock eccentricity and its rotation have not yet been explained.

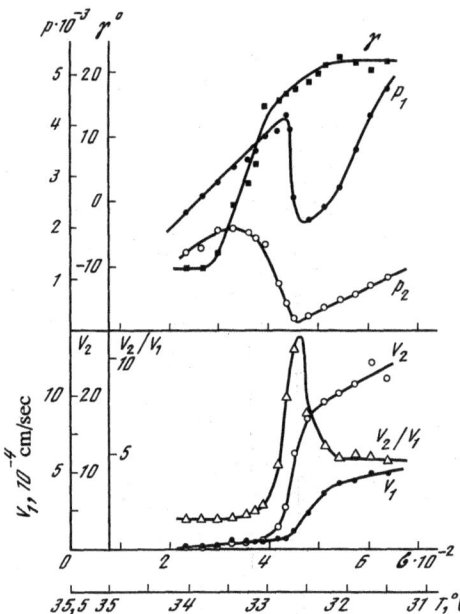

Fig. 4. Influence of supersaturation on V and p of a hillock for the directions of minimal (1) and maximal (2) step advance rates on a KDP prismatic face.

A new dislocation hillock appears in Fig. 1d. As σ increases further, this grows rather than the original. This effect is also known. Competition of growth sources is due to the difference in their structure caused by intersection of the corresponding functions $p(\sigma)$ [4].

The same processes develop on the prismatic face of a KDP crystal as on that of ADP. Nevertheless, they appear more variegated externally. The change of hillock shape is illustrated in Fig. 2. The position of the minimum on the function $p(\sigma)$ corresponds to Fig. 2c. The macrosteps appear and disappear in the range of supersaturations corresponding to Fig. 2b and c and can be seen as breaks in the interference patterns in the lower part (Fig. 2b). (In Fig. 2 the macrosteps are relatively weak due to the small hillock curvature. However, numerous macrosteps make the shape of the dislocation hillock difficult to observe.) The process of appearance and subsequent disappearance of macrosteps will be illustrated in more detail below.

2. GROWTH KINETICS OF FACES

2.1. The Functions $V(\sigma)$ and $p(\sigma)$

The influence of supersaturation on the rate of step advance and hillock curvature in the [001] direction on the ADP prismatic face is plotted in Fig. 3. At $\sigma = \sigma_* = 0.015$-$0.017$, V increases sharply. It reaches a plateau passing through the origin at $\sigma \approx 0.025 \, V(\sigma)$. The slow kinetics at $\sigma < \sigma_*$ are explained in [3, 8] by inhibition by relatively mobile impurity stoppers adsorbed to the surface. At $\sigma = \sigma_*$, the steps penetrate the palisade of stoppers. Inhibition by the impurities becomes unimportant on the linear part of $V(\sigma)$ at large σ. The minimum in the curve $p(\sigma)$ is shallow apparently due to the structure of the dislocation source (cf. Sec. 2.3).

The macrosteps described in Sec. 1 arise in the range $\sigma \approx 0.01$-0.02, which includes σ_*. The rotation and change of hillock shape occur near the most curved part of $V(\sigma)$. Macrosteps of hydrodynamic origin usually appear under these same conditions.

The behavior of V and p for KDP is analogous to that of ADP (Fig. 4, curves V_1 and p_1). The sharp change of p and hillock shape is noteworthy. A dependence of V on p could not be found in experiments with a kinetic growth regime. The value σ_* was determined only by the purity of the solution and the temperature.

The data of Figs. 3 and 4 enable the fundamental characteristics of dislocational growth, the kinetic coefficient β, and the free energy of the surface-step end α, to be determined. For the linear function V, $\beta = V/(\sigma c_0^* \rho^{-1})$, where c_0^* is the bulk solution concentration and ρ is the crystal density. From the data presented for steps parallel to the x axis we have $\partial V/\partial \sigma$ equal to 0.1 and 0.0085 cm/sec for ADP and KDP, respectively. Hence we find that at 31°C for ADP $\beta_x \approx 0.5$ cm/sec; for KDP, $\beta_x = 0.08$ cm/sec. The value $\beta_x \approx 0.005$ cm/sec was obtained for ADP in [4]. It is now clear to us that this value refers to the initial part of $V(\sigma)$ where $\sigma << \sigma_*$.

The value α_x for KDP can be determined from the initial part of the curve $p(\sigma)$ assuming that the extended linear part ($dp/d\sigma = 0.095$) corresponds to the single dislocation generating this hillock. For ADP, $\alpha_x = 16.1$ ergs/cm^2 [4]; for KDP, $\alpha_x = 16.7$ ergs/cm^2.

2.2. Influence of Surface Energy Anisotropy

The influence of σ on the change of anisotropy in the shape and on the rotation of a hillock on a KDP prismatic face is plotted in Fig. 4. The values V_1 and V_2 characterize the minimal and maximal rates of step advance. The curves p_1 and p_2 define the corresponding hillock steepness. The value γ is the angle between V_2 and [100]. The hillock anisotropy is characterized by the ratio V_2/V_1, equal to p_1/p_2 since $R = pV$ does not depend on direction.

The principal result derived from Fig. 4 is that σ_* for the maximal and minimal rates are different ($\sigma_{*1} > \sigma_{*2}$). This same shift causes a sharp maximum in the function $(V_2/V_1)/(\sigma)$. If all directions σ_* were constant, then V_2/V_1 would change smoothly from one constant (at $\sigma << \sigma_*$) to another (at $\sigma >> \sigma_*$).

In our opinion, this effect is explained readily by the following hypothesis. Evenly distributed surface impurity stoppers (Cabrera–Vermilli palisades) initially grow in the direction where the critical nucleus radius of curvature r_c is lower, i.e., the step-end surface energy is lower at the given supersaturation. In fact, the average distance between impurity stoppers in this same direction initially becomes greater than $2r_c$. Thus, the step may pass between them. For other directions where α is greater, a larger σ is required in order to decrease $2r_c$ to the same value.

The anisotropy a can be determined on this basis. The distance between neighboring impurity stoppers becomes greater than $2r_c$ at a normal face growth rate $R > R_* = j\pi r_c^2/h$, where j is the frequency at which impurity adsorbs at an arbitrary surface site and h is step height [6]. Since $r_c \approx \alpha/\sigma$, $(\alpha_1/\alpha_2)^2 = (R_{*1}/R_{2*})(\sigma_{*1}/\sigma_{*2})^2$, where $R_* = pV$ at $\sigma = \sigma_*$. Using the data of Fig. 4, we find $\alpha_1/\alpha_2 = 1.4$-1.5. It should be noted that α_1/α_2 is about the same as the hillock anisotropy at small σ (cf. curve V_2/V_1 in Fig. 4). Apparently this means that the anisotropy V in this region is determined by the anisotropy α.

The fact that σ_* for V_2 is lower is consistent with a smaller α for this direction. Thus, β and α are inversely proportional (of course, if these values are related at all). The proportionality can be explained by a decrease of α with increasing number of breaks on the step, i.e., increasing β. (Vul'f's rule presumes that β and α are directly proportional.)

The hillock shape at $\sigma > \sigma_*$ reflects the kinetic coefficient anisotropy. At $\sigma < \sigma_*$ the shape is determined by V that is now related not only to the anisotropy β but also to the anisotropy α. We assume that the hillock shape turns and changes at $\sigma \approx \sigma_*$ when the impurities and, thus, the influence of α on the anisotropy in the tangential rate of step advance no longer have an effect. The function $\gamma(\sigma)$ is the most sensitive characteristic of these processes. The quantity γ begins to change when changes in hillock shape can no longer be observed (cf. Fig. 4).

2.3 Depth of the Minimum on $p(\sigma)$

It can be seen from Fig. 4 that the depth of the minimum p is different for different directions of the first hillock. The quantity p_2 decreases many times whereas p_1 decreases only by a factor of two. At the minimum, p for

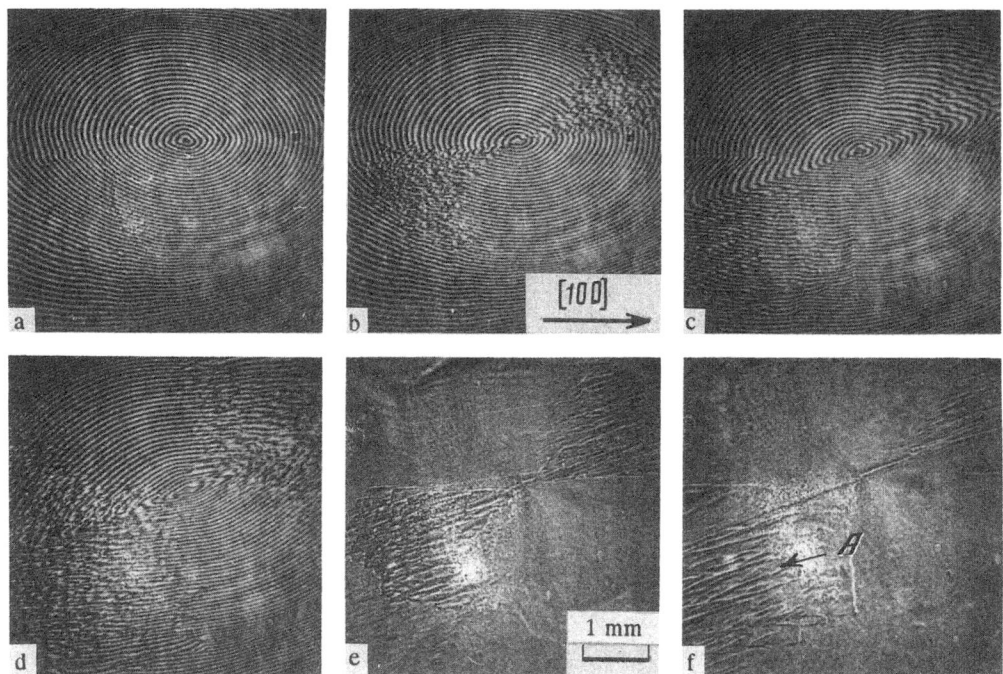

Fig. 5. Development of impurity macrosteps on a KDP prismatic face: Interferograms (a-d), surface photographs in one beam (e-f), b and c correspond to e and f. σ, 10^{-2}: 4.52 (a), 4.82 (b, c), 5.13 (d). Time after attaining supersaturation: 0.5 (b), 16 min (c).

many hillocks in the direction of maximal step rate is almost zero. The quantity p for a single direction but different vicinal hillocks can be substantially different for a single function $V(\sigma)$. These facts can be explained using the concepts developed in [4, 5, 7] for a dislocation with one Burgers vector and an isotropic (or square) spiral. It follows from these works that p is defined for nonlinear kinetics by $\sigma/\sigma_* > 1$ [for linear $\sigma < \sigma_* p(\sigma)$]. The decrease of p at $\sigma = \sigma_*$ is equal to the ratio of kinetic coefficients for $V(\sigma)$ at $\sigma < \sigma_*$ and $\sigma > \sigma_*$. A simple analysis shows that the last statement is valid for different directions of the first hillock in the anisotropic case. It can be seen from Fig. 4 that the ratio of slopes of the linear portions of $V(\sigma)$ for V_2 in these ranges of σ is much greater than for V_1. Thus, p_2 decreases more than p_1.

The conclusions of [3, 8, 9] should be modified for different hillocks generated by complex (not single) dislocation sources. The hillock steepness from an extended source with linear step kinetics is known to be defined by $p = mh/(19r_c + 2L)$, where mh is the total Burgers vector of the source from m distinct dislocations and $2L$ is the source perimeter. The quantity p is almost constant at $L \gg r_c$ (greater than σ) as σ increases since the curvature of the spiral decreases at the start of the first turn and the Gibbs—Thomson effect becomes inoperative. The function $p(\sigma)$ begins to smooth out at supersaturation $\sigma > \sigma_1$ determined from the relation $19r_c = 2L$. For this reason the minimum on $p(\sigma)$ depends on the ratio of σ_1 and σ_*. The minimum becomes less evident as σ_*/σ_1 increases and disappears at $\sigma_* \gg \sigma_1$. These features can easily be expressed quantitatively for a square spiral surrounding a square source from side L using the scheme of [8].

If $V = b_1\sigma$ at $\sigma < \sigma_*$ and $V = b_2\sigma$ at $\sigma > \sigma_*$, then at $1 \ll \sigma/(\sigma - \sigma_*) \ll b_2/b_1$, we obtain roughly (at $\alpha \approx 20$ erg/cm^2)

$$p/p_0 = (1 + \sigma L/10^{-7})/(\sigma/(\sigma - \sigma_*) + \sigma L/10^{-7}),$$

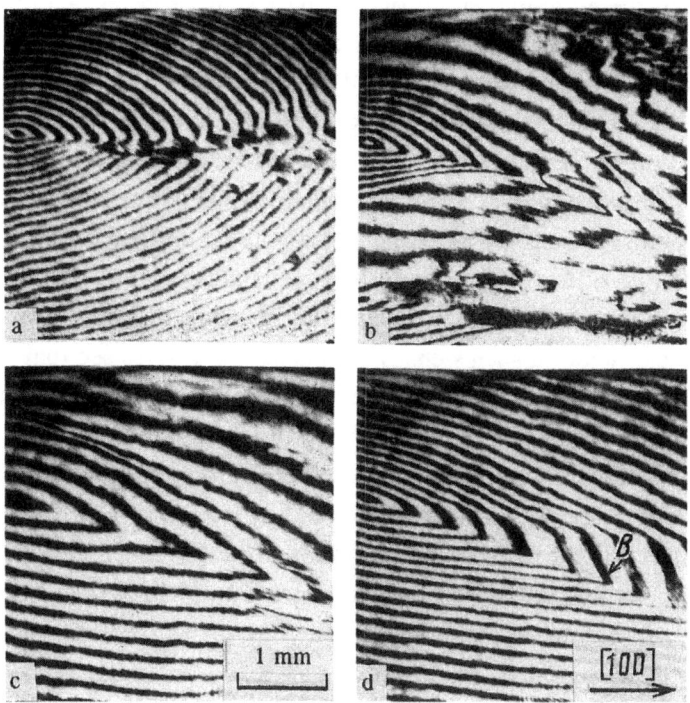

Fig. 6. Morphology of a KDP prismatic face at the time impurity macrosteps appear and disappear. σ, 10^{-2}: 4.06 (*a*, *b*), 4.51 (*c*), 5.11 (*d*). Time after attaining σ: 0.5 (*a*), 17 min (*b*).

where L is in cm and p_0 is the hillock steepness for $\sigma_* = 0$. It is clear from the above formula that p/p_0 increases as L increases. For example, $p/p_0 \approx 0.1$ at $\sigma = 1.1$, $\sigma_* = 0.04$, and $L \to 0$ and $p/p_0 \approx 0.8$ at $L = 10^{-4}$ cm.

3. IMPURITY MECHANISM OF MACROSTEP FORMATION

The fact that morphological instability exists within a certain distance of point σ_* is important to macrostep formation in a comparatively narrow range of small supersaturations.

Let us examine the experimental data in more detail. Characteristic features of macrostep formation on the surface of a rather steep hillock as σ and time change are visible in Fig. 5. Firstly, macrosteps begin to develop in the direction of maximal V (shown in Fig. 5 by the arrow A). The region captured by macrosteps forms sectors with the apex in the hillock center. The angle of the sectors increase with increasing σ (Fig. 5*b-d*). Secondly, the number of macrosteps initially increases and then decreases (Fig. 5*b* and *c*) after a constant supersaturation is attained. If σ changes slightly, the number of macrosteps again changes with time (Fig. 5*c* and *d*). Thirdly, the macrosteps themselves are mostly oriented with the direction of maximal rate (clearly seen in Fig. 5*e* and *f*), i.e., almost perpendicular to the orientation of the elementary steps.

Yet another feature of impurity macrostep formation is seen in Fig. 6. Here, like in Fig. 5, coagulation starts in the direction of maximal rate (Fig. 6*a*). However, the average steepness of the hillock periphery decreases compared with the center (Fig. 6*b*). The surface clearly becomes devoid of macrosteps at first as σ increases (Fig. 6*c*). Then, macrosteps of a completely different type appear and become more distinct as σ increases. These are indicated in Fig. 6*d* by arrow B.

Although all details of the process cannot yet be explained, the general picture is as follows. Elementary steps penetrate stopper palisades at supersaturation near σ_*. The step rate increases by an order of magnitude or more. The step breaking through the stopper palisades at any site leaves behind a surface relatively free of impurities, initiating penetration of the next step, etc. If one of these steps stops at random, the following one catches up and forms a higher step. Steps of different height should advance at different rates [1, 9]. This forms macrosteps tens and hundreds of times higher than an elementary step.

Anisotropy α has the effect of initially directing the step penetrations along the minimum α and along other directions with increasing σ. Thus, macrosteps will capture increasingly larger surfaces of the neighboring hillock. Elementary steps penetrating into a certain portion are extended in the direction of maximal rate (cf. Fig. 5c and f). The distance between steps is much greater in this direction than along the perpendicular. Since the initial step is not rectilinear, the sides of neighboring "fingers" coalesce. However, the end of the penetrated step has a stepped shape. The portions extended along the step-advance direction alternate with portions oriented at an acute angle to the first. If the echelon of such steps is examined, it is easy to see that the sides extended along the maximal rate direction will be much steeper. These sides are the macrosteps portrayed in Fig. 5f.

As noted above, macrosteps are more numerous on dislocation hillocks with greater steepness p. This is evidently due to the fact that a larger number of elementary steps are joined into macrosteps on steep sides. Thus, the macrostep will be higher and the distance between them less than for a gently sloped hillock.

Macrosteps formed during the initial supersaturation arise on the surface as σ changes. The direction of elementary layer advance on such a surface does not correspond to V for the new σ. This causes morphological instability. Eventually the surface is formed by steps generated by a source corresponding to the new anisotropy V and this factor is no longer effective. This explains the presence of a maximum in the number of macrosteps with time at $\sigma = $ const. At $\sigma > \sigma_*$, the surface again becomes stable since new macrosteps are not generated and those present earlier vacate the face by advancing to the edge.

The last photograph in Fig. 6 illustrates macrosteps of another type. These arise at sufficiently high supersaturations. They are oriented in the same direction as the elementary steps. Macrosteps appear at the same time as the hillock becomes polygonal. As σ increases, they gradually encompass a larger part of the surface. They do not appear due to impurities although the reason is not purely hydrodynamical [5, 7] since macrosteps also arise on those sides of the hillock where elementary steps advance against the current.

4. EXPERIMENTS WITH Cr IMPURITY

4.1. Change of Face Morphology

Adding Cr impurity to the solution causes R and V to decrease at all supersaturations. The hillock steepness changes likewise except for the region of σ where $p(\sigma)$ has a minimum and the p curves at different impurity concentrations c_i can intersect. The quantity p decreases in all directions. This cannot be explained by the change of surface free energy. Impurity adsorption should decrease α and r_c and increase p.

The change of surface relief with increasing σ in a solution with $c_i = 4.5$ ppm is illustrated in Fig. 7. The impurity is added at such a σ so that $V \approx 0$ at the given c_i. Therefore, Fig. 7a almost reflects the relief of the starting surface. A new very flat surface (Fig. 7b) arises as σ increases. The steepness gradually increases. The minimum in $p(\sigma)$ is very broad. The number of impurity macrosteps is also small. This last condition is apparently due to the fact that the hillock is gently sloped.

4.2. Change of Growth Kinetics

The kinetic curves obtained with Cr added are plotted in Fig. 8. Similar curves were studied earlier [6, 8]. Here only new facts are noted.

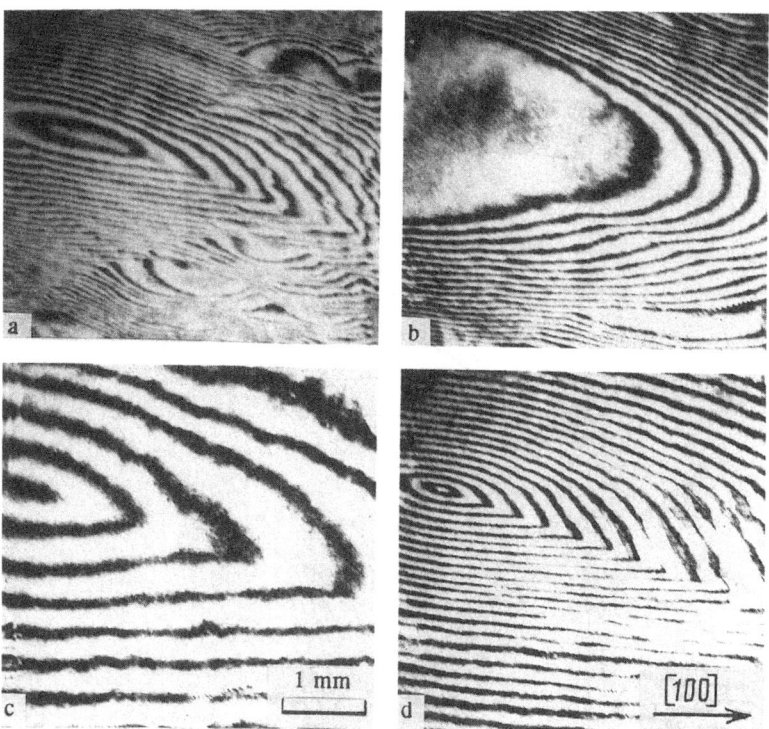

Fig. 7. Influence of supersaturation on KDP prismatic face relief in a solution with Cr impurity. $c_i =$ 4.5 ppm. σ, 10^{-2}: 4.51 (a), 4.66 (b), 5.41 (c), 6.32 (d). The hillock is the same one as in Fig. 6.

It can be seen from Fig. 8 that at $c_i \neq 0$ the function $p(\sigma)$ at $\sigma < \sigma_*$ does not approach the origin. Extrapolation to $p \to 0$ gives the same intercepts on the abscissa as extrapolation of $V(\sigma)$ to $V \to 0$. Our attempts to determine the growth of faces during days at supersaturations corresponding to the first points on curves 3-5 ($c_i > 1.5$ ppm) were unsuccessful. Thus, a "dead" zone of supersaturations $0 \leq \sigma \leq \sigma_d < \sigma_*$ (σ_d corresponds to the start of growth) exists in solutions with Cr where growth is negligible ($V < 10^{-7}$ cm/sec). This was claimed earlier in [10], where analysis of the functions $R(\sigma)$ in the presence of Fe and Al revealed points analogous to σ_d and σ_*.

Naturally, a dislocation spiral is generated only at $\sigma = \sigma_d$ and the hillock steepness at this supersaturation should be zero (cf. also the corresponding analysis in [11]) if a dead zone is present. The increase of σ_d with increasing c_i also explains the decrease of hillock steepness on adding Cr. The curves $p(\sigma)$ shift to the right together with σ_d.

Analysis of the data in Fig. 8 indicates that the gap between σ_d and σ_* decreases, the rise of $V(\sigma)$ at $\sigma \gtrsim \sigma_*$ becomes steeper, and the minimum of p becomes less and less distinct with increasing c_i. The points σ_d and σ_* almost coincide and the minimum in $p(\sigma)$ disappears at $c_i \gtrsim 5 \cdot 10^{-6}$ moles $CrCl_3 \cdot 6H_2O$ per mole KDP.

A possible explanation is that impurities that can be divided into two groups always exist in solution. One group contains impurities with a relatively short lifetime on the surface. These enable steps to advance at $\sigma < \sigma_*$. The other group contains impurities with a lifetime (compared with λ/V, λ is the distance between steps) sufficiently long so that they can be considered permanently fixed. The Cr impurity (probably like Al and Fe) belongs to the second group. The concentration of long-lived impurities determines σ_d. The total concentration of surface stoppers determines σ_*.

The Cr concentration in the starting solutions is much less than 1 ppm. However, Fe and Al are of the order of 4 ppm in KDP and 1.6 ppm in ADP. Thus, judging from Figs. 3 and 8, $\sigma_d \lesssim 0.008$ (KDP) and 0.003 (ADP) for them.

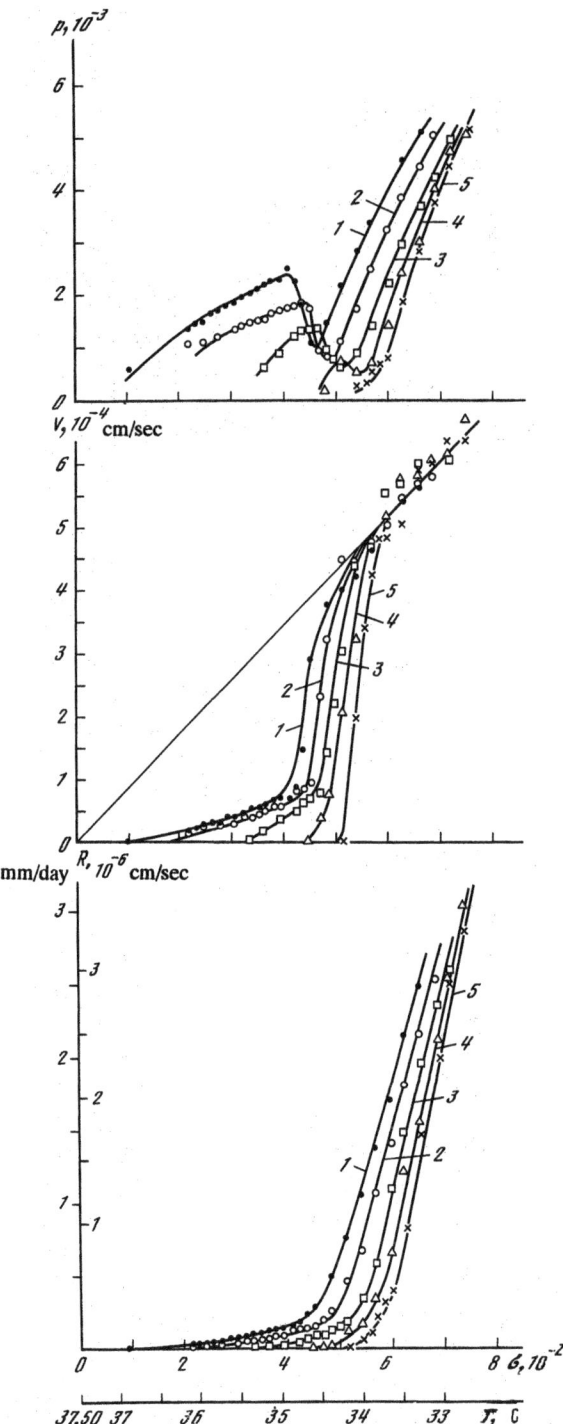

Fig. 8. Steepness p, tangential rate V, and normal growth rate R in solutions with various Cr concentrations. KDP prismatic face. c_i, ppm: 0 (1), 1.5 (2), 3 (3), 4.5 (4), 6 (5).

The Cabrera–Vermilli model of step penetration through palisades of fixed stoppers presumes that σ_d and σ_* are inversely proportional to the average distance between stoppers, which in turn should be inversely proportional to the square root of the bulk impurity concentration. Thus, σ_d^2 and σ_*^2 should be proportional to the impurity concentration in solution. The corresponding functions plotted in Fig. 9 are in fact close to linear. In spite of this contradiction, it is difficult to propose another explanation for the existence of points σ_d and σ_*.

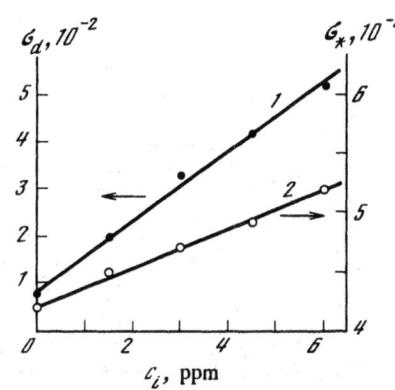

Fig. 9. Dependences of σ_d (1) and σ_* (2) on Cr impurity
concentration c_i. KDP prismatic face.

The values σ_d in Fig. 9 were taken from the $R(\sigma)$ curves in Fig. 8. The values σ_* are from $V(\sigma)$ and $p(\sigma)$ curves [where $p(\sigma)$ starts to decrease]. Extrapolation of the lines in Fig. 9 to the intercept with the abscissa gives a long-lived impurity content in a nominally pure solution of the order of 1 ppm and a total content of impurity stopper-steps of 25 ppm ($3 \cdot 10^{-3}$ mass %).

4.3. Face Growth Features at $c_i \gtrsim 5$ ppm

As already noted, the supersaturations σ_* and σ_d determined from the path of $V(\sigma)$ and $p(\sigma)$ practically coincide at Cr impurity concentrations $c_i \gtrsim 5$ ppm. Apparently certain face growth features in solutions with high Cr content are related to these.

The change with time of the KDP prismatic face relief at $\sigma = 0.0777$ in a solution with $c_i = 9$ ppm is shown in Fig. 10. The crystal was previously kept in the dead zone at $\sigma < \sigma_d = 0.074$ for 12 h. Only one hillock (lower in Fig. 10) was seen on the face at this supersaturation. A second upper hillock appeared during the change of σ at a certain $\sigma > \sigma_d$. The first hillock was stationary during the whole experiment. This is clear from its configuration in Fig. 10a-e. However, a smooth surface that was gradually filled with steps generated by the second hillock appeared on the periphery of this hillock beginning from the [001] edge of the face. The latter hillock (the steeper) eventually occupied the whole face (Fig. 10f).

Experiment shows that a hillock poisoned with an impurity can start to grow again if (as most often happens) the planar surface reaches its vertex or the vertex reaches the side of an unpoisoned flatter hillock arising on a previously formed planar surface.

The surface self-purification effect described above can be explained as follows. The supersaturation is usually slightly greater near the edge than over the face. However, the impurity concentrations are less since their distribution coefficients are less than unity and the impurity accumulates over the face. Therefore, the elementary step nearest to the edge passes more easily through the stopper palisade and reaches the edge, leaving behind a surface without Cr impurity. This initiates advance of the following steps. The surface free of impurity stoppers (with $p \approx 0$) will expand from the edges to the center of the hillock. Thus, the delay of crystal face growth is determined by the time required for the unpoisoned surface to reach the center of the leading growth-dislocation hillock.

The following experiment was performed to explain how a hillock poisoned with impurity behaves at sufficiently large supersaturations far from the dead zone. The dislocation hillock formed at large σ has no time to flatten out if σ is quickly lowered to values less than σ_d (Fig. 11). Then the supersaturation is increased at a rate of the order of 0.01 min^{-1}. This causes planar portions and macrosteps to form on the hillock surface, initially in

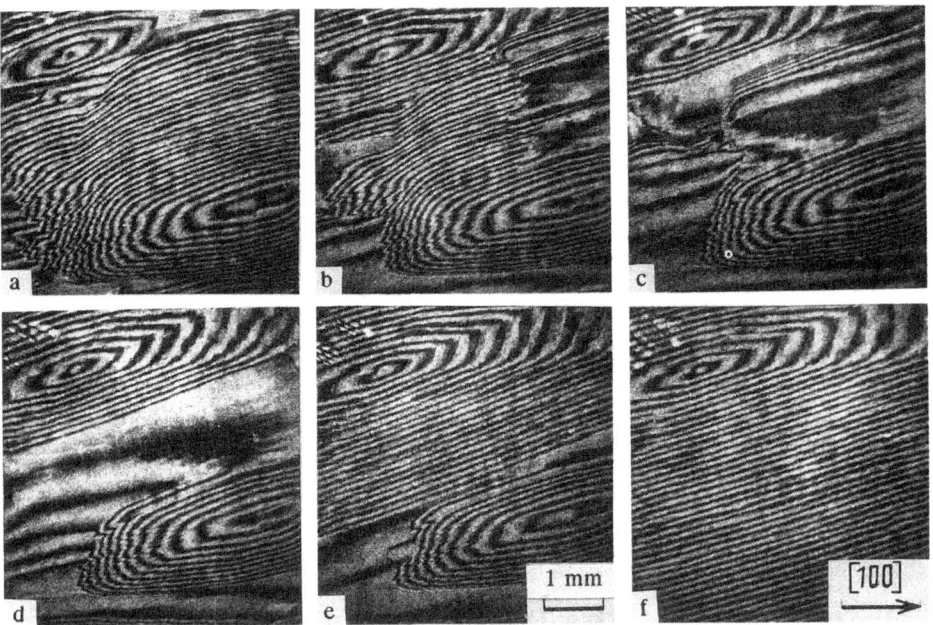

Fig. 10. Purification of a poisoned KDP prismatic face. $c_i = 9$ ppm, $\sigma = 0.0777$. Time after attaining supersaturation, min: 1 (a), 6 (b), 10 (c), 19 (d), 27 (e), 41 (f).

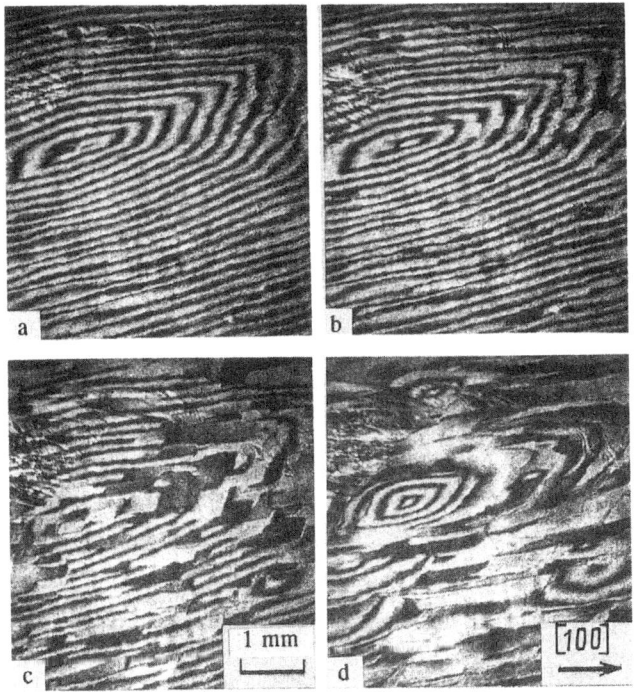

Fig. 11. Change of KDP prismatic face surface relief with rapid change of supersaturation. $c_i = 9$ ppm, $\sigma_d = 0.074$. σ, 10^{-2}: 8.23 (a), 8.87 (b), 9.19 (c), 9.73 (d).

the direction of maximal tangential rate (Fig. 11*b*) and then everywhere (Fig. 11*c*). When one of these portions reaches the hillock vertex (Fig. 11*d*), the hillock begins to grow again. The macrosteps formed earlier were the impurity type and are absent on the surface of the growing hillock.

In conclusion, it should be noted that the "exposure" effect of the surface during growth apparently plays an important role in the phenomena described above. To a certain degree, results of [12] are indicative of this. Nevertheless, our data are sufficiently contradictory to require further investigations.

The authors thank A. A. Chernov for his interest in the work and useful discussions.

REFERENCES

1.   A. A. Chernov, E. I. Givargizov, Kh. S. Bagdasarov, et al., *Modern Crystallography*, Vol. 3, *Crystal Formation* [in Russian], Nauka, Moscow (1980).

2.   E. B. Treivus, *Kinetics of Growth and Dissolution of Crystals* [in Russian], Leningrad State Univ. (1979).

3.   A. A. Chernov and L. N. Rashkovich, "Spiral growth of crystals with nonlinear dependence of step rate on supersaturation: (100) faces of KH_2PO_4 crystals in solution," *Dokl. Akad. Nauk SSSR*, **295**, No. 1, 106-109 (1987).

4.   L. N. Rashkovich, A. A. Mkrtchyan, and A. A. Chernov, "Optical interference investigation of growth morphology and kinetics of (100) face of ADP from aqueous solution," *Kristallografiya*, **30**, No. 2, 380-387 (1985).

5.   A. A. Chernov, Yu. G. Kuznetsov, I. L. Smol'skii, and V. N. Rozhanskii, "Hydrodynamic effects during growth of ADP crystals from aqueous solutions in the kinetic regime," *Kristallografiya*, **31**, No. 6, 1193-1200 (1986).

6.   A. A. Chernov, L. N. Rashkovich, I. A. Smol'skii, et al., "Crystal growth processes from aqueous solutions (KDP group)," in: *Growth of Crystals*, Vol. 15, Consultants Bureau, New York (1988).

7.   L. N. Rashkovich, B. Yu. Shekunov, and Yu. G. Kuznetsov, "Relation of solution hydrodynamics to morphological stability. The KDP prismatic face," in: Extended Abstracts of the Seventh All-Union Conf. on Crystal Growth, Vol. 2 [in Russian], Moscow (1988), pp. 12-13.

8.   A. A. Chernov, L. N. Rashkovich, and A. A. Mkrtchyan, "Interference-optical investigation of KDP, DKDP, and ADP crystal surface growth processes," *Kristallografiya*, **32**, No. 3, 737-754 (1987).

9.   J. P. Van der Eerden and H. Müller-Krumhar, "Dynamic coarsening of crystal surface by formation of macrosteps," *Phys. Rev. Lett.*, **57**, No. 19, 2431-2433 (1986).

10.   V. I. Bredikhin, V. P. Ershov, V. V. Korolikhin, and V. N. Lizyakin, "Influence of impurities on the growth kinetics of a KDP crystal," *Kristallografiya*, **32**, No. 1, 214-219 (1987).

11.   A. S. Mikhailov, L. N. Rashkovich, V. V. Rzhevskii, and A. A. Chernov, "Isotropic dislocation spiral with nonlinear dependence of step growth rate on supersaturation," *Kristallografiya*, **34**, No. 2, 439-445 (1989).

12.   A. A. Chernov and A. I. Malkin, "Kinetics of impurity stopper formation preventing the growth of (101) ADP faces in aqueous solution," *Kristallografiya*, **33**, No. 6, 1487-1491 (1988).

CONTROLLED FLUX GROWTH OF COMPLEX OXIDE SINGLE CRYSTALS

G. A. Emel'chenko, V. M. Masalov, and V. A. Tatarchenko

Crystallization from fluxes was developed in general to grow complex oxide single crystals [1-4]. Many synthetic single crystals of ferroelectrics, ferrites, laser materials [1-3], and, starting in 1987, the rediscovered high-temperature superconductors [5] are prepared by crystallization from fluxes. Spontaneous crystallization is the most widely used method at present. This is due to its use as the first step in searching for new crystals of complex multicomponent systems, incongruently-melting compounds, and high-melting materials as well as its technical simplicity [4].

The first attempts to control crystallization involved localization of nucleation centers by heat conduction away from the crucible bottom [6-8]. Combination of heat conduction and intermittent accelerated crucible rotation was demonstrated to be the most effective method of control [9, 10].

The next development in control methods was growth on a seed [9, 11-21]. This had several variants, including slow cooling and vaporization of solution, crystallization with gradient mass transport (temperature gradient method), the modified Czochralski method, and zone melting with a solvent.

Further development of crystallization control methods requires a deeper understanding of the physico-chemical characteristics of the system and crystallization parameters and the construction of reliable devices for controlling the process [4].

With respect to apparatus design and control possibilities, the modified Czochralski method based on the usual melt method has been most widely used for nonvolatile systems [21-23]. Experimental data for systems used in flux crystallization indicated that lead compounds are the most widely used fluxes. These have high dissolving capabilities but are also highly volatile [4]. The modified Czochralski method has limited application in such systems.

Tolksdorf [9] and Bennet [17] simultaneously proposed a method for volatile systems in which a seed is placed in a saturated flux and the crystal grown is separated from the flux by rotating a closed crucible inside a furnace at a given temperature. Special crucibles were then constructed [24]. The composition and growth rate of ferrogarnet crystals could be effectively controlled [25, 26].

One of the principal problems in preparing single crystals of solid solutions is the maintenance of conditions that produce homogeneous composition and properties throughout the whole growing crystal. Thus, one of the Tolksdorf crystallizers [25] uses a nutrient crucible enabling growth at constant temperature in a temperature gradient that is intensified and localized using cooling. The flux is effectively stirred using intermittent accelerated crucible rotation (ACRT, accelerated crucible rotation technique) around its vertical axis [10]. The most homogeneous $Y_3Fe_{5-x}Ga_xO_{12}$ single crystals of any known [25] were prepared using this method.

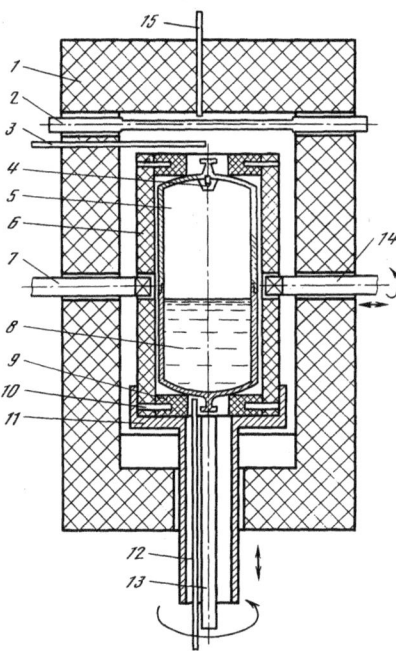

Fig. 1. Diagram of the apparatus for growing single crystals from a flux on a seed using a rotating crucible: heating chamber (1), SiC heater (2), control thermocouples (3, 12), seed (4), platinum crucible (5), corundum tube (6), rotation axes (7, 14), flux (8), lock washer (9), lock bolt (10), support (11), cooler (13), regulating thermocouple (15).

Unfortunately, this method has not been widely used due to shortcomings in construction of the platinum crystallizer. This requires welding and cutting after each experiment and severely complicates its use.

Several modifications of bulk crystallizers without these shortcomings were developed in the Laboratory of Directed Crystallization of the Institute of Solid-State Physics at the Academy of Sciences of the USSR. These were used in the present work to investigate crystallization of complex ferrogarnet oxide [27-34] and high-temperature superconductor single crystals [35, 36]. As a rule, these compounds melt incongruently and are prepared at present practically only by the flux method [35-41]. Data for the hydrodynamics in model and actual experiments as applied to the rotating crucible method are also presented.

This method enables the physicochemical state of the system to be monitored in situ even in scouting experiments since the crucible with the reaction products can be freed of flux at any stage. Crystallization itself can be conducted by slow cooling and gradient mass transfer under isothermal conditions.

ROTATING CRUCIBLE METHOD

The first rotating crystallizer was developed by us to grow yttrium-iron garnet (YIG) single crystals on a seed by slow cooling (Fig. 1) [27]. A separable platinum crucible is inserted into a ceramic casing and placed in a furnace so that it can be rotated by 180° around the horizontal axes 7 and 14. A seed 4 is fixed to the upper half of the crucible in the starting position. After the flux is homogeneous, it is cooled to the starting crystallization temperature. The crucible is rotated around the horizontal axes to immerse the seed into the flux. A special program cools it slowly. The crystal grown is taken out of the flux by the opposite rotation after the crystallization

Fig. 2. Crystal of $Y_3Fe_5O_{12}$ grown on a seed.

temperature range is passed. The flux is cooled further to room temperature. The crucible is opened. The crystal is extracted. The flux is stirred by the ACRT during crystallization [10]. Single crystals of YIG of mass up to 90 g with a habit close to that of the seed are prepared on this apparatus from a flux of 1 kg mass (Fig. 2).

The following rotating crystallizer was developed to grow single crystals of solid solutions by the temperature gradient method with continuous feed of crystallizing material to the flux (Fig. 3). As demonstrated by Linares [42] and Tolksdorf [25], this is one of the most effective isothermal methods for preparing homogeneous crystals of solid solutions.

The feeder is set up in the crystallizer developed earlier (cf. Fig. 1). A charge of the desired crystal composition is placed in the crystallizer. The charge is prepared as sintered ceramic pellets or spontaneously grown fine crystals. The feeder is covered with a perforated lid and has overflow outlets in the bottom. The flux level in the starting position is below the bottom of the feeder. In the working position, it is above the edge of the feeder covered with the lid. The apparatus works as follows. In the starting position, crystallizer 1 is heated. Mother liquor 6 is melted and homogenized. The crystallizer is rotated around the vertical axis using the ACRT with periodic acceleration to assist dissolution and stirring. The rotational axes protrude from the ceramic casing 10 during this period. After the mother liquor is homogeneous, it is cooled to the saturation temperature. A seed is immersed into the flux by rotating the crystallizer around the horizontal axes 13 and 14.

After the crystallizer is rotated the flux enters the feeder 3 through the perforated lid and bathes the charge 7. A constant temperature gradient is established in the flux. The upper flux layers are hotter, the lower ones are cooler. The temperature gradient is controlled by heat transfer from the bottom of the crucible using the cooler 16. After rotation, growth is conducted at constant temperature in the solution stirred by rotating the crystallizer around the vertical axis (ACRT). At the end of the growth period, the growing crystal is removed from the hot flux by a second rotation.

Further developments using the rotating crucible method involve modifications of the two main apparatuses described above to actual problems and systems.

The rotating crucible method was used in [29] to obtain data on phase equilibrium in the Bi—Ca—V—Fe—Pb—O system. Instead of one large crucible, three 30-cm³ separable crystallizers in a ceramic casing were used. The composition of the liquidus was calculated analogously to [43].

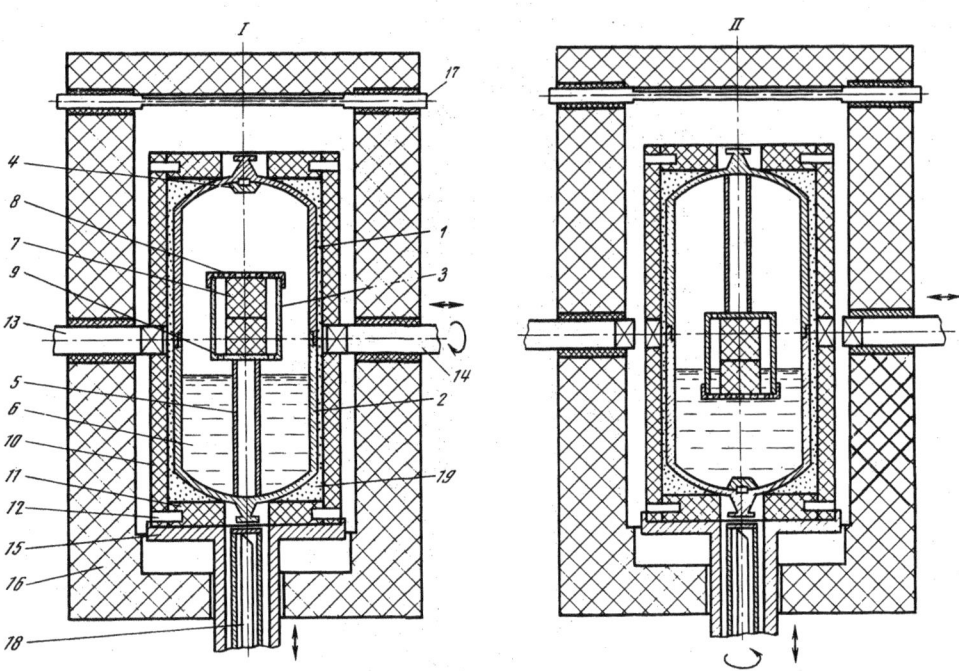

Fig. 3. Apparatus for growing single crystals of solid solutions. Starting and final positions (*I*), working position (*II*). Upper and lower halves (1, 2), feeder (3), seed (4), holder (5), flux (6), charge (7), cover (8), overflow outlets (9), casing (10), lock washer (11), bolt (12), rotational axes (13, 14), support (15), heating chamber (16), SiC heater (17), cooler (18), insulation (19).

Improved hemispherical crucibles with a removable nutrient crucible have been constructed [44].

MODELLING SOLUTION HYDRODYNAMICS WITH ACCELERATED CRUCIBLE ROTATION*

During growth of crystals from a flux, hydrodynamics determine the mass transfer and homogenization in the liquid phase. They also affect the shape of the crystallization front and temperature fluctuations at the interface, as well as the growth rate, structure, and composition of the crystals.

Recently, attempts have been substantially intensified to physically model hydrodynamics in order to find a quantitative dependence between the hydrodynamic parameters and the physicochemical properties of the liquid phase [45-47].

In the present work, hydrodynamics of model liquids are studied with respect to growth of crystals from fluxes using a rotating crucible [27, 28]. The solution was stirred using the ACRT [10]. Aqueous solutions of glycerine (50 vol. %) were used as the model solutions. A glass crucible of 70-mm diameter similar to the actual shape was used [28, 32]. The feeder diameter and its position in the crucible were varied in the experiments. Mass flow was visualized using fine graphite powder. The rotation parameters varied as follows: maximal rotation rate 10-60 rpm, acceleration (deceleration) duration 5-60 sec, rotation time 0-60 sec.

Experiments with a crucible without a feeder revealed pulsating mass flows in the shape of a torus. These have been theoretically described earlier [40]. During acceleration, solution in the central part begins to descend along the crucible axis and is driven outward by Eckman currents on the bottom that dissipate against the walls, ascending along them. The currents reverse during deceleration. The velocity of the currents and the solution height from which the liquid begins to descend depend mainly on the acceleration and to a lesser extent on the maximal crucible rotation rate.

---

*V. Nikolov (Institute of General and Inorganic Chemistry, Academy of Sciences of Bulgaria) participated in this work.

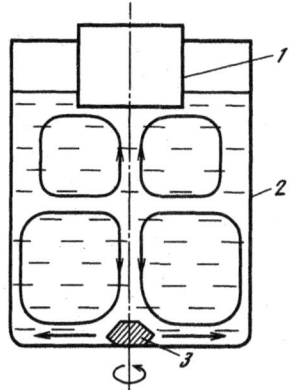

Fig. 4. Shape of mass flows during acceleration (in the center cross-section): feeder (1), crucible (2), crystal (3).

Introduction of the feeder in the surface layer causes two toroidal current vortexes of opposite direction to form (Fig. 4). The lower vortex is analogous to that described above. However, it is shorter. The upper vortex is related to the horizontal level of the feeder. Its size depends on the feeder diameter. The vortexes are separated by a thin stationary layer. The position of this layer depends on the ratio of the feeder and crucible diameters, the flux height, the geometric shape of the crucible bottom, and the temperature gradient. The vortexes form synchronously with a period corresponding to that of the accelerated–decelerated crucible rotation.

A preliminary analysis of the data demonstrated that the amplitude of temperature fluctuations at the crystallization front is the most important hydrodynamic characteristic with respect to physical applications of modelling. In real situations, this value can be derived from the growth of crystals (ensuring that results obtained can be compared). It is directly connected with the supercooling at the growth front.

The distribution of temperature oscillations along the whole solution height (Fig. 5, curve 3) is consistent with the presence of a stationary solution layer near the center. The maximal oscillations are observed at 10- and 37-mm height. This corresponds to toroidal regions with the maximal mass flow rates. As the temperature gradient is varied from 0 to 8°C/cm, the amplitude of oscillations (δT) increases with increasing gradient. As n_{max} increases at constant acceleration time τ_a, deceleration time τ_d, and rotation time τ_r, δT also increases. Simultaneous observations of the average temperature gradient with varying n_{max} showed that the gradient falls with increasing rotation rate.

The limiting conditions of parameter variation must be defined before the optimal solution stirring regimes can be chosen. The upper limit for the oscillation amplitude is the supercooling. Therefore, optimization of the rotation parameters should minimize the temperature oscillations at the growth front to a level less than the supercooling. For ferrogarnet fluxes, this value is of the order of 1°C [26].

The upper limit of maximal rotation rate at the given times ($\tau_a = \tau_d = 6$ sec, $\tau_r = 12$ sec) and temperature gradient (less than 7.5°C/cm) is 25 rpm.

HYDRODYNAMICS OF ACTUAL FLUXES

The influence of stirring regimes on mass transfer and crystallization was studied on the apparatus shown in Fig. 3. The temperature distribution was measured on a solution of composition analogous to that used for growing single crystals of $Y_3Fe_{5-x}Ga_xO_{12}$ (mole %): 37.02 PbF_2 + 27.93 PbO + 22.01 Fe_2O_3 + 3.29 Ga_2O_3 + 10.45 Y_2O_3. The measurement method is given in [28].

Temperature measurements in the flux without rotation are shown in Fig. 6. The average temperature gradient in the static position was of the order of 7.5°C/cm. Stirring the flux decreases the gradient to 5.5-6°C/cm.

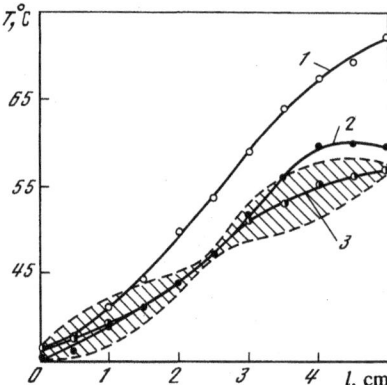

Fig. 5. Vertical temperature distribution in the crucible along the side wall without rotation (1), along the central axis without rotation (2), along the central axis with rotation $n_{max} = 35$ rpm, $\tau_a = \tau_d = 6$ sec, $\tau_r = 12$ sec (3). The quantity l is the flux height. The hatched region shows T-oscillations.

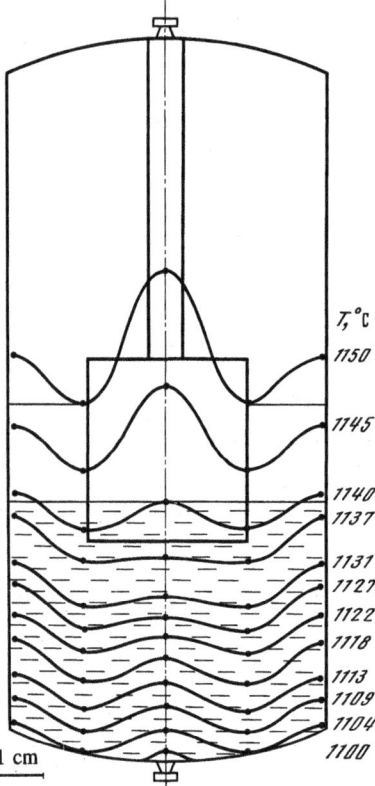

Fig. 6. Temperature distribution in the flux.

The gradient is dependent on the stirring regimes. Temperature oscillations up to $+2.5°C$ are observed in the flux synchronously with crucible rotation.

The size of the oscillations depends on the rotation parameters τ and n_{max} and the observation site in the flux (Fig. 7a and b). The maximal oscillations are observed on the crucible bottom at $n_{max} = 60$ rpm (Fig. 7a). As the distance from the bottom increases, the oscillations decrease and disappear at 1/3 of the distance from the feeder. Above this point the oscillations increase. However, they decrease as the surface is approached.

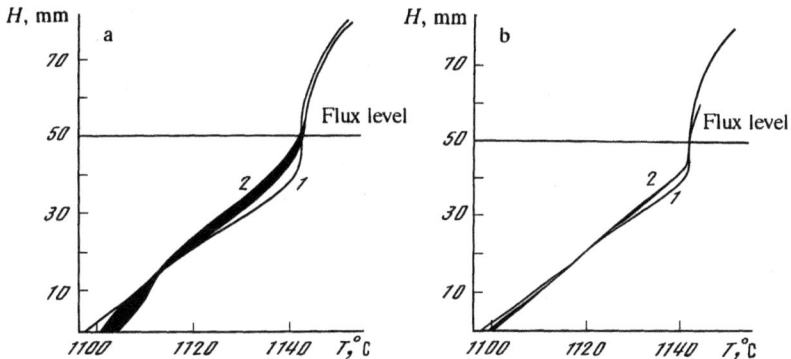

Fig. 7. Vertical axial temperature distribution in the crucible without rotation (1) and with rotation (2) at $n_{max} = 60$ (a) and 20 rpm (b). $\tau_a = \tau_d = 6$ sec, $\tau_r = 12$ sec.

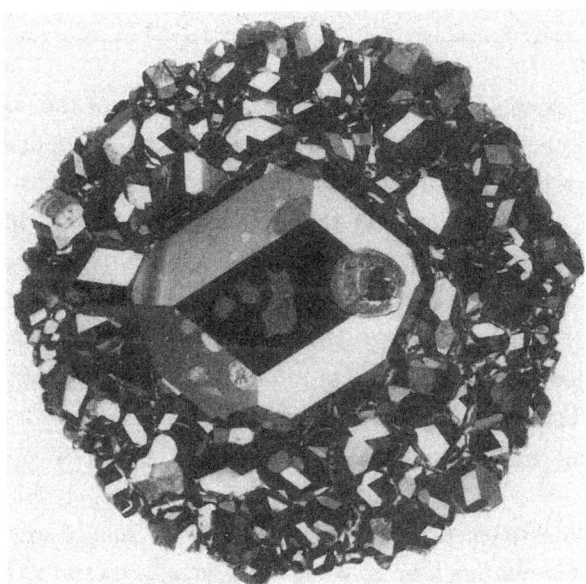

Fig. 8. Crystals of YIG prepared at high rotation rates (n_{max}).

As n_{max} is varied, the oscillation amplitude increases with increasing rotation rate. As τ_a, τ_d, and τ_r decrease, the oscillation amplitude also decreases. At $n_{max} = 20$ rpm (Fig. 7b), a wide band with minimal temperature oscillations exists in the flux. This band is located between 10 and 30 mm from the crucible bottom.

Analysis of the temperature distribution in the flux indicates that crystal growth will be initiated in the center of the crucible bottom, where the seed is placed. However, such a temperature distribution (hotter above) causes concentrational layering of the flux and makes the role of free convection insignificant in mass transfer. As a result, segregation effects and formation of spontaneous crystals at various sites can be observed. Forced convection in the axial direction is necessary to suppress these effects and to ensure transfer of charge to the crystallization zone. This is done using the ACRT [10].

Concentrational layering of the solution is confirmed by experiment. Without rotation, the seed almost does not grow and spontaneous crystallization is observed on the crucible walls mainly near the surface.

Mass transfer to the crucible bottom also is not observed at a stationary rotation up to 40 rpm. Spontaneous crystals form on the wall in a narrow band of 10- to 20-mm width in the upper half of the flux. High mass

Table 1. Amplitude of Temperature Oscillations (°C) in a Flux for Various Stirring Regimes

n_{max}, rpm	Flux height, mm						
	0	10	15	20	30	40	50
				$\tau_a = \tau_d = \tau_r = 15$			
20	3.6	2.0	1.0	2.1	3.3	1.0	0.7
30	3.1	1.4	1.2	1.4	4.1	1.7	2.9
40	5.0	1.7	2.2	2.5	4.2	1.8	1.7
60	5.1	1.5	2.7	4.9	4.1	2.1	1.7
				$\tau_a = \tau_d = \tau_r = 10$			
20	3.0	2.6	0.9	0.7	2.7	0.6	–
30	2.4	1.8	0.6	1.3	2.3	0.7	0.3
40	2.6	1.1	1.2	2.5	1.7	0.3	1.0
60	3.1	1.3	1.3	2.9	2.2	2.2	0.4
				$\tau_a = \tau_d = \tau_r = 12$			
20	1.1	0.8	0.4	0.3	2.7	1.7	0.4
30	2.6	2.1	0.7	0.7	2.9	1.7	0.5
40	2.6	0.8	0.8	1.7	1.3	1.3	0.3
60	2.8	0.6	1.5	2.0	2.0	0.8	0.5

transfer to the seed is achieved using periodic acceleration-deceleration of the crystallizer around its vertical axis. Use of accelerated rotation ensures simultaneous stirring in the axial and radial directions due to formation of Eckman currents and spiral currents in the flux [48]. Mass transfer to the crystallization zone increases with in-increasing n_{max}. This increases the crystal growth rate. However, competing spontaneous crystals form on the crucible bottom as the seed grows. These reduce the yield of large specimens (Fig. 8).

The rotation parameters were optimized by studying the amplitude of temperature oscillations in the flux as a function of stirring regimes. The optimal rotation parameters should give the minimal oscillations in the crystallization zone with sufficient mass transfer of nutrient from the dissolution zone. The oscillation amplitude in the growth zone should be below the supercooling (of the order of 1°C according to [26]). Measured oscillations with flux height beginning from the crucible bottom ($H = 0$) for several rotation regimes are given in Table 1.

Variations of n_{max} demonstrated that decreasing it below 20 rpm resulted in insufficient mass transfer to the crystallization zone whereas increasing it up to 40 rpm and greater increased temperature oscillations throughout the whole flux above the supercooling.

The ratio of τ_a (τ_d) to τ_r is important in reducing the oscillations. During acceleration, Eckman currents conduct part of the flux away from the lower cold part of the crucible and up the walls [48]. These currents slow as the rotation slows and can even reverse direction when the rotation rate of the upper layers is greater than the lower ones. A steady rotation time was added to the rotation program to eliminate this effect. Oscillations measured at τ_a (τ_d) = 15 sec and different τ_r did not reveal zones with oscillation amplitude less than 1°C in the flux. At τ_a (τ_d) = 10 sec, such a thin zone does appear for n_{max} = 20-30 rpm. The widest zone, practically from the bottom to 25 mm into the flux, is found for $\tau_a = \tau_d = 6$ sec and $\tau_r = 12$ sec ($n_{max} = 20$ rpm) (cf. Table 1).

By changing the temperature gradient, it was demonstrated that these rotation parameters for a 7-8°C/cm gradient provide mass transfer of the nutrient to the seed zone without forming competing spontaneous crystals on the bottom. Decreasing the gradient defocuses the growth zone and forms spontaneous crystals on the walls. Increasing the gradient is deleterious since the temperature oscillations increase in the crystallization zone and competing crystals form spontaneously on the bottom (cf. Fig. 8).

Thus, analysis of the influence of stirring regimes and temperature gradient on mass transfer and crystallization suggests that the following rotation parameters are closest to optimal (for a 7-8°C/cm temperature gradient): n_{max} = 20-25 rpm, $\tau_a = \tau_d = 6$ sec, $\tau_r = 12$ sec. This agrees with model experiments.

GROWTH OF $Y_3Fe_{5-x}Ga_xO_{12}$ SINGLE CRYSTALS

A charge of the composition indicated above was used to grow $Y_3Fe_{5-x}Ga_xO_{12}$ ($x \approx 1.0$) single crystals. The nutrient of polycrystalline pellets or spontaneous crystals of $Y_3Fe_4GaO_{12}$ was placed in the feeder and covered with the perforated lid.

Spontaneous single crystals of YIG (2-5 g mass) were used as seeds. The crucible with charge and seed was set in the ceramic casing, heated to 1230-1250°C, and held for 20-25 h at this temperature with constant stirring. The solution was then cooled at 60°C/h to 1120-1130°C and held there for up to 25 h. The cooler was set in the working position when the crucible started to cool. The crystallizer was rotated by 180° around the horizontal axes to immerse the seed in the saturated flux (Fig. 3, position II). After rotation, crystals were grown at constant temperature, stirring the solution by rotation around the vertical axis. At the end of the growth period, the growing crystal was extracted from the hot flux by a second rotation and cooled to room temperature. The crystal obtained had a mass up to 85 g.

The magnetization saturation $4\pi M_s$, the ferromagnetic resonance (FMR) linewidth ΔH_f at 700 MHz, and the excitation intensity of magnetostatic types of vibrations (MSTV) Δ_{200} [49] were measured for 0.6-mm diameter spheres cut from the crystal obtained. They indicate that the crystal is highly homogeneous in magnetization saturation with a homogeneous distribution of Ga dopant throughout. More than 50% of the spheres measured have magnetization ±5 G. About 80% are in the range ±10 G. According to electron microprobe analysis, the Ga concentration within uncertainty limits (~1%) is constant across the crystal cross section. These results are typical [25].

Analysis of the FMR linewidth ΔH_f confirms the high quality of the crystals. Thus, 70% of the measured spheres have $\Delta H_f < 1$ Oe. As a rule, all spheres with a narrow resonance line either have no 200 MSTV lines or exhibit an excitation intensity Δ_{200} close to zero. Tests of crystals grown under conditions different from these regimes reveal a large scatter of magnetization saturation and a broader FMR line. Thus, the scatter of $4\pi M_s$ values was ~100 G and ΔH_f was from units to tenths of an Oe if crystals were grown at $n_{max} = 60$ rpm, $\tau_a = \tau_d = 15$ sec, and $\tau_r = 15$ sec.

Thus, experimental results (both model and actual systems) for the choice of optimal growth conditions for single crystals of $Y_3Fe_{5-x}Ga_xO_{12}$ solid solutions and tests of samples obtained under these conditions suggest that the conditions found are close to optimal. The crystals prepared, like crystals prepared in [25], are very homogeneous with respect to magnetization saturation.

GROWTH OF Ca—Bi—V FERROGARNETS

The isothermal rotating crucible method is very effective for preparing homogeneous crystals of solid solutions in the multicomponent system Bi_2O_3—CaO—Fe_2O_3—V_2O_5—In_2O_3—PbO.

Preparation of $Bi_{3-2x}Ca_{2x}Fe_{5-x}V_xO_{12}$ (CBVG) crystals doped with In and Nb by a spontaneous method is difficult due to simultaneous crystallization of several phases, a highly inhomogeneous bulk distribution of the elements forming the crystals, and inhomogeneous magnetic properties as a result.

The phase diagram of the pseudoternary system PbO—CaO/Bi_2O_3—Fe_2O_3/V_2O_5/In_2O_3 in the region where CBVG crystallizes with an average magnetization saturation of 150 G [29, 33] contains a very narrow region of ferrogarnet primary crystallization. Phases such as magnetoplumbite $PbFe_{12}O_{19}$, calcium ferrite $CaFe_2O_4$, and calcium vanadate $Ca_3V_2O_8$ cocrystallize if the charge deviates slightly (by about 2 mole %) from the given composition (Fig. 9).

In such a situation, the temperature gradient method using a rotating crucible seemed most applicable for preparing homogeneous crystals of solid solutions with the given composition. The given solution composition is adjusted automatically and maintained during growth using the following procedure. If the starting composition

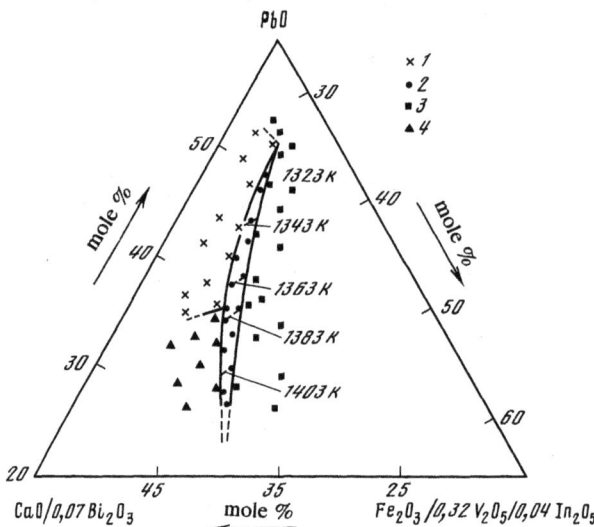

Fig. 9. Projection of the CBVG primary crystallization field in the pseudoternary system PbO—CaO/0.07Bi$_2$O$_3$—Fe$_2$O$_3$/0.32V$_2$O$_5$/0.04In$_2$O$_3$: CaFe$_2$O$_4$ (1), CBVG (2), PbFe$_{12}$O$_{19}$ (3), Ca$_3$V$_2$O$_8$ (4).

deviates from the garnet toward the magnetoplumbite region (cf. Fig. 9) during preparation of the saturated solution, excess PbFe$_{12}$O$_{19}$ cocrystallizes. This occurs from the moment that the seed is immersed into the flux. The composition shifts toward garnet crystallization as a result. After the first crucible rotation, the excess phases are withdrawn from the solution (cf. Fig. 3) (they remain on the bottom and walls of the upper half). A continuous feed compensates for a further shift of the flux composition during crystallization of garnet, i.e., the system will be kept at the same point of the phase diagram during growth.

As an illustration, let us examine results for growth of CBVG crystals from a flux containing (mole %) 30 PbO + 2.2 Bi$_2$O$_3$ + 32 CaO + 27.2 Fe$_2$O$_3$ + 8.6 V$_2$O$_5$. Dopant amounts of In and Nb are added to the starting charge to obtain the required magnetization saturation. This composition lies in the primary crystallization field of magnetoplumbite (Fig. 9). Crystals of magnetoplumbite, calcium ferrite, and calcium vanadate in addition to ferrogarnet form during growth by slow cooling between 1160-1060°C. The CBVG crystals prepared this way are highly homogeneous. This gives rise to a magnetization saturation between 100-200 G throughout the crystal bulk.

Polycrystalline CBVG as pellets with $x \approx 1.38$ containing In and Nb dopant were used as nutrient during temperature gradient growth with a rotating crystallizer [34].

The charge was heated to 1200°C in the crucible (Fig. 3, position I), held there for several hours while accelerating and decelerating the crucible, and cooled to 1080°C. The seed was immersed into the flux after one hour (Fig. 3, position II). A crystal was then grown at constant temperature for 10 days, after which it was withdrawn from the flux by a second rotation. Thus, CBVG single crystals (Fig. 10) of mass up to 40 g with magnetization saturation 150 ± 10 G were prepared. Accompanying phases were not observed.

GROWTH OF SINGLE CRYSTALS OF HIGH-TEMPERATURE SUPERCONDUCTORS

A new class of compounds, high-temperature superconductors (HTSC), was discovered recently in the systems Ln—Ba(Sr)—Cu—O, where Ln = REE and Y [52, 53]. A distinguishing feature of these compounds is their thermal instability. Superconductors of composition La$_{2-x}$Sr$_x$CuO$_4$ melt with decomposition near 1370°C [50]. The phases LnBa$_2$Cu$_3$O$_{7-\delta}$, depending on the REE, melt incongruently at 980-1060°C [51].

Fig. 10. A CBVG crystal of 25-g mass prepared on a seed
by temperature gradient with nutrient.

The first attempts to grow single crystals of these compounds used slow cooling from a nonstoichiometric flux [37] and a modified Czochralski method [38]. Single crystals of HTSC were also grown using the rotating crucible method [52, 53].

The composition of the starting charge was chosen and the temperature range of La_2CuO_4 crystallization was determined by studying the liquidus in the system La_2O_3–CuO_x [36]. The charge was heated to 1250°C, held there for several hours, and quickly cooled to a temperature greater than the liquidus by 10-20°C. After this, it was cooled at a slow rate (0.5-1.5°C/h) until the end of the crystallization range was reached. The crucible was then rotated around the horizontal axes (cf. Fig. 1), extracting the crystals grown from the flux. The flux was stirred during the growth period by periodic acceleration–deceleration of the crucible around the vertical axis. A general view of the La_2CuO_4 crystals in the crucible after the experiment is shown in Fig. 11.

The single crystals prepared are pyramidal with a cross section at the base of 8 × 6 mm and a length along the c axis up to 15-20 mm. The crystals are bounded by the slower growing faces of the rhombic dipyramid (111). The apex of the truncated pyramid is bounded by the pinacoid plane (001), which is sometimes completely hollowed out. The orthorhombic cell constants are $a = 5.36$, $b = 5.40$, and $c = 13.14$ Å.

The specific resistivity of the crystals as grown has a semiconducting temperature dependence. Heat treatment of single crystals of thickness 0.5 mm at 900°C for 10 h in an oxygen atmosphere ($P_{O_2} = 10$ atm) did not produce a superconducting state down to 4.2 K.

Single crystals of $La_{1.85}Sr_{0.15}CuO_{4-\delta}$ prepared in the same way as La_2CuO_4 are platelike with maximal size 10 × 10 × 1 mm and are bounded by base planes (001). A superconducting transition is not observed in crystals doped with Sr even with analogous heat treatment ($T = 900$°C, $P_{O_2} = 10$ atm, 10 h).

It was reported [54] that superconducting crystals $(La, Sr)_2CuO_4$ were prepared under conditions further from equilibrium. Nonequilibrium growth conditions are modelled in our work. For this, the starting flux containing the same Sr concentration was cooled at a rate 20 times greater than in previous experiments. The crystals prepared were thin rectilinear plates (0.2-0.5 mm thick) of composition $La_{1.86}Sr_{0.14}CuO_{4-\delta}$, i.e., the Sr distribution coefficient changed little. However, the crystals had a superconducting transition at $T_c \approx 11$ K and $\Delta T_c = 5$ K. The transition begins at 35 K according to resistivity measurements.

The composition of the starting charge and the crystallization temperature range for growing $YBa_2Cu_3O_{7-\delta}$ single crystals from a homogeneous flux were chosen by studying the polythermal cross section $YBa_2Cu_3O_{7-\delta}$–BaO/CuO [55]. Platinum crucibles with a protective coating and corundum crucibles were used to grow the crystals. The crucible (cf. Fig. 1) was heated until the charge was completely melted, held there for several hours, and quickly cooled to a temperature slightly greater than the liquidus. After this, it was cooled at

Fig. 11. General view of La_2CuO_4 single crystals in the crucible.

0.5-1.5°C/h until the end of the crystallization range was reached. The flux was stirred and the temperature gradient was maintained by local heat transfer from the bottom center of the crucible. The crucible was rotated around the horizontal axes to remove the crystals grown from the flux after the process was finished.

The crystals were square plates of dimensions up to $3 \times 3 \times 1$ mm fixed to the bottom of the crucible by the center of one edge and elongated in the direction of the temperature gradient. The crystal surface on both sides was mirrorlike. A broad superconducting transition starting at 60-70 K was found by measuring the magnetic susceptibility. Heat treatment at 450°C and elevated oxygen pressure (6-7 atm) raised T_c to 75 K and narrowed the transition to 3-4 K.

CONCLUSIONS

1. Features of the rotating crucible method are examined for directed crystallization from fluxes in multicomponent oxide systems, including ferrogarnets and high-temperature superconductors.

2. The liquid phase hydrodynamics are studied by physical modelling for the accelerated crucible rotation technique (ACRT). The sizes and changes of temperature oscillations at certain characteristic points of the flux are determined as a function of the crucible rotation and temperature gradient. The shape and direction of mass flows in crucibles of various geometry are determined.

3. The influence of stirring mode and temperature gradient on mass transfer and crystallization in an actual flux is studied. The temperature distribution and amplitude of temperature oscillations are measured. The dependence of the oscillation size on rotation parameters and observation site in the flux is found.

4. Single crystals of the ferrogarnets $Y_3Fe_5O_{12}$, $Y_3Fe_{5-x}Ga_xO_{12}$, $Bi_{3-2x}Ca_{2x}Fe_{5-x}V_xO_{12}(In, Nb)$, and the high-temperature superconductors La_2CuO_4, $(La, Sr)_2CuO_4$, $YBa_2Cu_3O_{7-\delta}$ are prepared by the rotating crucible method.

REFERENCES

1. R. A. Laudise, "Precipitation from molten salt solutions," in: *Art and Science of Growing Crystals*, Wiley, New York (1963), pp. 112-118.

2. B. M. Wanklyn, "Practical aspects of flux growth by spontaneous nucleation," in: *Crystal Growth*, Vol. 1, Pergamon Press, Oxford (1974), pp. 93-262.

3. D. Elwell and H. J. Scheel, *Crystal Growth from High-Temperature Solutions*, Academic Press, London, etc. (1975).

4. V. A. Timofeeva, *Growth of Crystals from Fluxes* [in Russian], Nauka, Moscow (1978).

5. H. Hasegawa, U. Kawabe, T. Aita, and T. Ishiba, "Single crystal growth of layered perovskite metal oxides," *Jpn. J. Appl. Phys.*, **26**, No. 5, L673 (1987).

6. J. W. Nielsen, "Improved method for the growth of yttrium-iron and yttrium—gallium garnets," *J. Appl. Phys.*, **31**, 518-525 (1960).

7. A. B. Chase and J. A. Osmer, "Localized cooling in flux crystal growth," *J. Am. Ceram. Soc.*, **50**, No. 6, 325 (1967).

8. L. I. Averin, B. I. Birman, O. M. Konovalov, and T. R. Mnatsakanova, "Growth of single crystals of iron—yttrium garnet by spontaneous crystallization with a fixed amount of crystallization centers," in: *Single Crystals and Technology*, Vol. 1 [in Russian], All-Union Scientific Research Institute of Single Crystals, Khar'kov (1970), pp. 3-11.

9. W. Tolksdorf, "Growth of yttrium iron garnet single crystals," *J. Cryst. Growth*, **3/4**, 463-466 (1968).

10. H. J. Scheel and F. O. Schulz-Dubois, "Flux growth of large crystals by accelerated crucible rotation technique," *J. Cryst. Growth*, **8**, 304-307 (1971).

11. R. A. Laudise, R. C. Linares, and E. F. Dearborn, "Growth of yttrium iron garnet on seed from molten salt solution," *J. Appl. Phys.*, **35**, Suppl. 3, 1362-1363 (1962).

12. R. C. Linares, A. A. Ballman, and L. G. Van Uitert, "Growth of beryl single crystals for microwave application," *J. Appl. Phys.*, **33**, No. 11, 3209-3210 (1962).

13. R. C. Linares, "Growth of single crystal garnets by a modified pulling technique," *J. Appl. Phys.*, **35**, No. 2, 433-434 (1964).

14. V. A. Timofeeva and I. Kvapil, "On the solubility and crystallization of $Y_3Al_5O_{12}$ from $PbO—B_2O_3$ and $PbO—B_2O_3—PbF_2$ fluxes," *Kristallografiya*, **11**, No. 2, 289-294 (1966).

15. M. Kestigian, "Yttrium—iron garnet single crystal growth by the combined Czochralski—molten salt solvent technique," *J. Am. Ceram. Soc.*, **50**, 65-66 (1967).

16. V. I. Voronkova, V. K. Yanovskii, and V. A. Koptsik, "Growth of single crystals of corundum from tungstate fluxes," *Izv. Akad. Nauk SSSR, Neorg. Mater.*, **4**, No. 10, 1727-1731 (1968).

17. G. A. Bennet, "Seeded growth of garnet from molten salts," *J. Cryst. Growth*, **3**, No. 4, 458-462 (1968).

18. L. N. Bezmaternykh, G. I. Shvartsman, D. V. Tsynchik, et al., "Study of YIG single-crystal growth conditions during growth from high-temperature fluxes on seeds," *Izv. Akad. Nauk SSSR, Ser. Fiz.*, **34**, No. 6, 1246-1249 (1970).

19. O. M. Konovalov, E. N. Sablin, and V. I. Salo, "Growth of YIG single crystals from a flux on a seed," in: Proceedings of a Conference on Electronics Technology [in Russian], Central Scientific Research Institute of Electronics, Moscow (1970), No. 9(25), pp. 154-155.

20. S. H. Smith and D. Elwell, "Growth of nickel ferrite crystals from barium borate by a pulling method," *J. Cryst. Growth*, **3/4**, 471-474 (1968).

21. P. V. Klevtsov, L. P. Kozeeva, and A. A. Pavlyuk, "Polymorphism and crystallization of potassium-rare earth molybdates $KLn(MoO_4)_2$ (Ln = La, Ce, Pr, Nd)," *Kristallografiya*, **20**, No. 6, 1216-1220 (1975).

22. P. V. Klevtsov and L. P. Kozeeva, "Synthesis and polymorphism of crystals of binary lithium tungstates of rare earths and yttrium, *Kristallografiya*, **15**, No. 1, 57-61 (1970).

23. P. V. Klevtsov and L. P. Kozeeva, "Synthesis, x-ray, and thermographic study of potassium-rare earth tungstates $KLn(WO_4)$, Ln = RE," *Dokl. Akad. Nauk SSSR*, **185**, No. 3, 571-573 (1969).

24. W. Tolksdorf and F. Welz, "Improved crucible molds for forming yttrium-iron garnet single crystals from melt solutions," *J. Cryst. Growth*, **35**, No. 3, 285-296 (1976).

25. W. Tolksdorf and F. Welz, "Growth of gallium-substituted yttrium iron garnet single crystals from a molten solution at constant temperature," *J. Cryst. Growth*, **20**, No. 1, 47-52 (1973).

26. W. Tolksdorf and F. Welz, "Crystal growth of magnetic garnets from high-temperature solution," in: *Crystals for Magnetic Application*, Vol. 1, Springer, Berlin, etc. (1978), pp. 3-52.

27. G. A. Emel'chenko and V. T. Ushakovskii, "Growth of $Y_3Fe_5O_{12}$ single crystals from a flux on a seed," *Izv. Akad. Nauk SSSR, Neorg. Mater.*, **18**, No. 2, 334-336 (1982).

28. G. A. Emel'chenko, V. V. Masalova, G. F. Zakharyugina, and V. V. Petrov, "Growth of homogeneous single crystals of solid solutions based on YIG from a flux," *Izv. Akad. Nauk SSSR, Neorg. Mater.*, **23**, No. 5, 837-840 (1987).

29. V. M. Masalov, G. A. Emel'chenko, and V. A. Tatarchenko, "The crystallization of substituted CBVG and CVG in a lead oxide melt," *Izv. Akad. Nauk SSSR, Neorg. Mater.*, **25**, No. 3, 451-457 (1989).

30. G. A. Emel'chenko and V. V. Masalova, "Growth of Ga-substituted YIG single crystals from a flux in a temperature gradient field with accelerated crucible rotation," in: Abstracts of Papers of the Third All-Union Seminar on Hydromechanics and Thermal Mass Exchange in Zero Gravity [in Russian], Chernogolovka (1984), pp. 213-215.

31. G. A. Emel'chenko and V. V. Masalova, "Growth of $Y_3Fe_{5-x}Ga_xO_{12}$ single crystals by the temperature gradient method on a seed and study of the influence of temperature oscillations in solution on their homogeneity," in: Abstracts of Papers of the Sixth All-

Union Conf. on Crystal Growth, Vol. 2 [in Russian], Tsakhkadzor, Armenian SSR (1985), pp. 122-123.

32. G. A. Emel'chenko, V. Nikolov, V. M. Masalov, and V. A. Tatarchenko, "Physical modelling of hydrodynamics during growth of crystals from high-temperature solutions with accelerated—decelerated crucible rotation," in: Abstracts of Papers of the Second All-Union Conf. on Modelling Crystal Growth [in Russian], Riga (1987), pp. 378-380.

33. V. M. Masalov and G. A. Emel'chenko, "Crystallization of substituted CBVG and CVG single crystals in a lead oxide flux," in: Abstracts of Papers of the Seventh All-Union Conf. on Crystal Growth, Vol. 2 [in Russian], Moscow (1988), pp. 227-229.

34. G. F. Zakharyugina, G. A. Emel'chenko, V. V. Maslova, et al., "Use of constant gradient method for growing CBVG single crystals from a flux," *Elektron. Sverkhvysok. Chastot*, No. 1/2, 247-248 (1987).

35. G. A. Emel'chenko, N. V. Abrosimov, V. M. Masalov, et al., "Growth of single crystals of high-temperature superconductors in the systems Ln_2O_3—$BaO(SrO)$—CuO (Ln = RE and Y) and their characteristics," in: Abstracts of Papers of the Seventh All-Union Conf. on Crystal Growth, Vol. 2 [in Russian], Moscow (1988), pp. 400-402.

36. V. A. Tatarchenko, G. A. Emelchenko, N. V. Abrosimov, et al., "Single crystal growth of high temperature superconductors and investigation of their physical properties," *Int. J. Mod. Phys. B*, **3**, No. 1, 71-83 (1989).

37. S. Takekawa and N. Iyi, "Single crystal preparation of $Ba_2YCu_3O_x$ from non-stoichiometric melts," *Jpn. J. Appl. Phys.*, **26**, No. 5, L851-L853 (1987).

38. K. Oka and H. Unoki, "Phase diagram of the La_2O_3—CuO system and crystal growth of $(LaBa)_2CuO_4$," *Jpn. J. Appl. Phys.*, **26**, No. 10, L1590-L1592 (1987).

39. G. A. Emel'chenko, M. V. Kartsovnik, P. A. Kononovich, et al., "Bulk nature of the superconductivity of $YBa_2Cu_3O_x$ single crystals," *Pis'ma Zh. Eksp. Teor. Fiz.*, **46**, No. 4, 162-164 (1987).

40. A. B. Bykov, L. N. Dem'yanets, N. D. Zakharov, et al., "Superconductivity and crystal structure of $(La_{1-x}Sr_x)CuO_{4-y}$ single crystal," *Pis'ma Zh. Eksp. Teor. Fiz.*, **46**, Suppl., 19-22 (1987).

41. L. F. Schneemeyer, J. V. Waszczak, T. Siegrist, et al., "Superconductivity in $YBa_2Cu_3O_7$ single crystals," *Nature*, **328**, 601-605 (1987).

42. R. C. Linares, "Substitution of aluminum and gallium in single crystal yttrium iron garnets," *J. Am. Ceram. Soc.*, **48**, No. 2, 68-80 (1965).

43. D. Jonker, "Investigation of phase diagram of the system PbO—B_2O_3—Fe_2O_3—Y_2O_3 for growth of single crystals of $Y_2Fe_5O_{12}$," *J. Cryst. Growth*, **28**, No. 2, 231-239 (1975).

44. Inventor's Certificate No. 1,345,681 (USSR), "Apparatus for growing single crystals of solid solutions," *Byull. Izobret.*, No. 38 (1987).

45. V. Nikolov, K. Iliev, and P. Peshev, "Relationship between the hydrodynamics in the melt and the shape of the crystal/melt interface during Czochralski growth of oxide single crystals," *J. Cryst. Growth*, **89**, No. 2/3, 313-330 (1988).

46. C. D. Brandle, "Simulation of fluid flow in $Gd_3Ga_5O_{12}$ melts," *J. Cryst. Growth*, **42**, 400-404 (1977).

47. C. M. Lawrence and D. Elwell, "Mass transport limitation of top-seeded solution growth rate," *J. Cryst. Growth*, **32**, No. 3, 287-292 (1976).

48. E. O. Schulz-Dubois, "Accelerated crucible rotation: Hydrodynamics and stirring effect," *J. Cryst. Growth*, **12**, 81-87 (1972).

49. G. M. Galaktionova, S. Sh. Gendelev, and Yu. R. Shil'nikov, "Ferromagnetic resonance in zonar iron garnet crystals," *Kristallografiya*, **29**, No. 6, 1109-1113 (1984).

50. J. M. Tarascon, L. H. Greene, W. R. Mckinnon, et al., "Superconductivity at 40 K in the oxygen-defect perovskites $La_{2-x}Sr_xCuO_{4-y}$," *Science (Washington, D. C., 1883-)*, **235**, No. 4794, 1373-1376 (1987).

51. A. Katsui, Y. Hidaka, and H. Ohtsuka, "Single crystal growth of $Ba_2NdCu_3O_{7-\delta}$ from $BaCO_3$—Nd_2O_3—CuO solution," *Jpn. J. Appl. Phys.*, **26**, No. 9, L1521-L1523 (1987).

52. J. G. Bednorz and K. A. Muller, "Possible high T_c superconductivity in the Ba—La—Cu—O system," *Z. Phys. B: Condens. Matter*, **64**, 189-193 (1986).

53. M. K. Wu, J. R. Ashburn, C. J. Torng, et al., "Superconductivity at 93 K in a new mixed-phase Y—Ba—Cu—O compound system at ambient pressure," *Phys. Rev. Lett.*, **58**, No. 9, 908-910 (1987).

54. Y. Hidaka, Y. Enomoto, M. Suzuki, et al., "Single crystal growth of $(La_{1-x}A_x)_2CuO_4$ (A = Ba or Sr) and $Ba_2YCu_3O_{7-y}$," *J. Cryst. Growth*, **85**, 581-584 (1987).

55. A. A. Zhokhov, G. A. Emel'chenko, N. V. Abrosimov, et al., "Phase formation in the system Y_2O_3—BaO—CuO and synthesis of $YBa_2Cu_3O_{7-\delta}$ single crystals," in: Abstracts of Papers of the First All-Union Conf. on the Physics, Chemistry, and Technology of High-Temperature Superconductors [in Russian], Moscow (1988), pp. 51-52.

MECHANISM OF RELAXATION OF THE NONEQUILIBRIUM
LIQUID–SOLID INTERFACE BEFORE LIQUID-PHASE
HETEROEPITAXY OF III–V COMPOUNDS

Yu. B. Bolkhovityanov

Any contact of the liquid and solid phases before heteroepitaxy is known to be nonequilibrium due to the different amount or ratio of components in the contacting phases. The importance of understanding processes occurring at the nonequilibrium interface is obvious. Such contact is used both in preparation for epitaxial growth from the liquid and in production of the heterostructures themselves. In light of the tendency in all technologies to grow thinner and thinner (quantum-well) films, nonequilibrium processes occurring at the liquid–solid interface become an important part of the whole process of epitaxial growth. They determine the initial stages of formation and the structural and electrophysical properties of such a heterostructure.

Various final results of such contact are described in the literature. Nonequilibrium contact of a multicomponent liquid phase–binary substrate (or other ratio of components in the liquid and solid phases) has been studied. Either unexpected etching of the substrate or precipitation on the substrate in spite of the undersaturation of the liquid phase occur.

We demonstrated that quasiequilibrium is possible in a liquid–solid system (for example, In–Ga–As–GaAs, Ga–As–Sb–GaAs, etc.) in which the liquid phase is supersaturated and the supersaturation depends on the misfit of the cell constants a_s of the solid used for contact with the liquid (substrate) and a_0 of the solid in equilibrium with this liquid. The subsurface interlayer formed by diffusion and the important role of strains in this layer have been examined in such systems [1, 2]. The different character of the relaxation of the nonequilibrium liquid–solid interface as a function of the size and sign of the misfit of substrate cell constant (a_s) and that of the solid intended to be grown (a_0) was illustrated for a large number of examples in [3]. It can be characterized by the relationship $(a_0 - a_s)/a_s = \Delta a/a$.

In the present article, we will attempt to demonstrate that the different manifestations of the liquid–solid interface seen in various systems are stages of a single relaxation mechanism that are exhibited and occur with different rates in one case or another.

THEORY

It can be demonstrated [2] that a multicomponent liquid–multicomponent solid system on becoming nonequilibrium seeks a new equilibrium through further etching of the crystal if the solid composition deviates

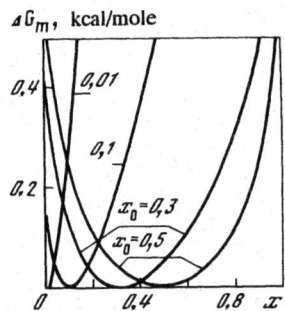

Fig. 1. Change of thermodynamic potential of In—Ga—As—
In$_x$Ga$_{1-x}$As calculated for isothermal deviation of solid
composition x from equilibrium x_0 for various initial x_0.
Equilibrium temperature $T_0 = 800°$C.

from the equilibrium value. Thus, the change of thermodynamic potential $\Delta G_m = G_S - G_L$ with an isothermal
deviation of the composition of the three-component solid solution x from its equilibrium value x_0 is presented in
Fig. 1 for the solid solution In$_x$Ga$_{1-x}$As. Here ΔG_m is that part of the total change of thermodynamic potential
ΔG due to the deviation of the system from equilibrium as a result of changing the composition of one phase (in
this case, the solid). It is clear from Fig. 1 that ΔG_m is positive regardless of the sign of the deviation x. This is
consistent with the need to dissolve more of the solid. Therefore, the first step in moving a liquid—solid system of
independent composition toward a new equilibrium should begin with etching of solid. A diffusion layer close to
the equilibrium composition or in equilibrium with the multicomponent liquid should form in the subsurface layer
of the dissolving solid [2]. Such a layer was found experimentally.

A relatively long-lived quasiequilibrium of the liquid—equilibrium diffusion interlayer (skin)—substrate
system is possible at this stage since the limiting stage of system relaxation is solid-state diffusion that thickens the
skin. At certain low temperatures characteristic of a real system, this process can be so slow that the quasiequilibrium will be prolonged. We will call this quasiequilibrium of the first type.

If the skin in the subsurface layer has a composition such that its cell constant differs from that of the
solid, then this layer should be strained. Thus, the thermodynamic potential of the solid should increase to a value
equal to the energy of these strains. The saturated liquid should also dissolve the solid, becoming supersaturated.
This effective supercooling of the liquid ΔT_ϵ has been called "equilibrium" supercooling initiated by strains [1].
The liquid in quasiequilibrium contact with the solid through such an interlayer will be supersaturated as long as
the strains remain. Thus, the quasiequilibrium supersaturated liquid—strained skin is a particular case of the first
type of quasiequilibrium.

The calculated dependences of ΔT_ϵ for two widely used three-component systems liquid—binary substrate,
Ga—As—P—GaP (a) and In—Ga—P—GaP (b), are shown in Fig. 2. The composition of the corresponding solid in
equilibrium with the changing liquid is plotted along the horizontal axis. It can be seen that a third component in
the liquid [i.e., the solid in equilibrium with it becomes the three-component GaAs$_{x_0}$P$_{1-x_0}$ (a) or In$_{x_0}$Ga$_{1-x_0}$P (b),
which differs more in composition as x_0 increases from the substrate GaP, for which $x_0 = 0$] gives rise to supercooling ΔT_ϵ and its increase to very large calculated values. Kuznetsov et al. [4] expanded the calculation of the
equilibrium supercooling initiated by strains to four-component systems.

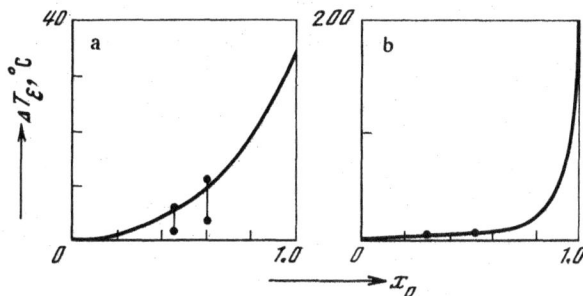

Fig. 2. Equilibrium supercooling increase initiated by strains calculated for changing liquid composition of equilibrium solid $GaAs_{x_0}P_{1-x_0}$ (a) or $In_{x_0}Ga_{1-x_0}P$ (b). GaP substrate. Equilibrium temperature $T_0 = 800°C$. Experimental points are from [3].

Thus, according to the thermodynamic arguments presented above, the multicomponent saturated liquid—binary solid systems (and other systems with a nonequilibrium ratio of components) should have a quasiequilibrium state resulting from formation of a skin on the crystal surface, quasiequilibrium of the first type. If this layer is strained, then additional solid must dissolve to attain this quasiequilibrium state. This creates supersaturation in the liquid that compensates for the strain energy in the skin.

However, there are two fundamental reasons interfering not only with the persistence of such a quasiequilibrium state but also with its very existence in certain cases. These are a) the microscopic inhomogeneity in the thermodynamic properties of the solid subsurface adjoining the homogeneous liquid (dislocations, point defect pile-ups, and other imperfections of the crystal structure) that leads to formation of nuclei of a new phase on the solid surface and to initiation of the next stages of nonequilibrium interface relaxation and b) the limiting supersaturations in the liquid (the case $\Delta a/a \neq 0$) which when exceeded can give rise to nucleation of new phases that relax the supersaturation of the liquid and change its composition. These reasons will be discussed in more detail below.

INTERFACE METASTABILITY

ISOTHERMAL FORMATION OF NEW-PHASE NUCLEI

The liquid becomes supersaturated and capable of forming new-phase nuclei as substrate dissolves. However, additional conditions are necessary for their generation. Figure 3 diagrams schematically the new-phase nucleus of composition $A_xB_{1-x}C$ (dashed) on the surface of substrate of composition BC. A chemical and geometric misfit generally exists between these two solids. As a result, the solid—solid interface energy σ_{ss} is elevated. If the skin in the substrate is thicker than δ, where δ is the chemical bond length, due to diffusion during liquid—solid contact, then the change of chemical and geometric properties between phases $A_xB_{1-x}C$ and BC is smeared out over a greater substrate depth and σ_{ss}, an additional barrier to new-phase nucleation, should decrease.

Thus, the penetration depth of the element diffusing into substrate on skin formation determines the probability of forming new-phase nuclei on the solid surface during isothermal contact of the multicomponent liquid—binary substrate since it affects the surface energy σ_{ss}. If this probability is small, then quasiequilibrium is possible for a certain time in the saturated or supersaturated multicomponent liquid—solid system. Such a state was observed experimentally in a number of III—V systems. These are Ga—As—Sb—GaAs [1], In—Ga—P—GaAs, In—Ga—As—P—GaAs [5, 6], and In—Ga—As—InP [7]. This quasiequilibrium was especially evident in the system used to grow strained garnets [8]. Quasiequilibrium of the solid and liquid supercooled up to 30 K was seen.

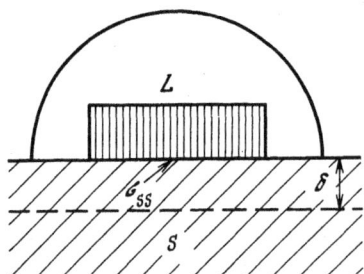

Fig. 3. Diagram illustrating a portion of the diffusion layer at the new-phase nucleation center. Liquid (L) and substrate (S).

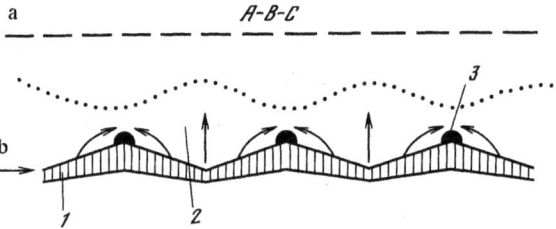

Fig. 4. Diagram of the new-phase nucleation model taking into account the skin: initial liquid–solid interface position (a), interface position after etching of part of the substrate (b). Skin (1), local supersaturation in the liquid (2), new-phase nucleus (3). Arrows indicate the direction of dissolved particle movement.

Experience shows that substrate dissolution into saturated and weakly undersaturated multicomponent liquids is selective. As a result, the solid surface becomes corrugated and covered with vicinal faces [9]. Thus, the skin thickness will be different due to the different solid etching rate. The striped band in Fig. 4 shows schematically the thickness of such a skin. On the other hand, the supersaturation is also uneven due to irregular etching of solid into the liquid layer adjoining the crystal. It is maximal over those parts of the solid that dissolve at the greatest rate. These local supersaturation clusters expand due to liquid diffusion processes. The most probable sites for new-phase nucleation are clearly positive step outcroppings into which elements absent in the solid penetrate to a great depth and decrease σ_{ss}. Diffusion in the liquid from surrounding parts with accelerated etching creates supersaturation over these parts.

Thus, inhomogeneous substrate etching should eventually lead to irregular new-phase nucleation. The inability to maintain a quasiequilibrium state of the first type will be determined by the generation and development of new-phase nuclei, i.e., kinetic factors. Slow etching of the crystal and/or formation of an ultrathin skin (small solid diffusion coefficients from the liquid to the crystal subsurface) hinder formation of new-phase nuclei. As will be demonstrated by experimental examples in the following section, the rates of these two processes depend not only on temperature and the actual liquid–solid system but also on the size and sign of $\Delta a/a$.

EXPERIMENTAL

Detailed experimental studies of the relaxation kinetics of the nonequilibrium interface multicomponent liquid–solid suggested the presence of a certain regularity in the relaxation. A thin skin was found on the dissolving crystal surface. Nucleation of a new phase was then observed. The nuclei grew, coalesced, and formed a continuous epitaxial interlayer separating the crystal from the liquid not in equilibrium with it. We arbitrarily divided the whole relaxation process into three stages.

The first stage is partial etching of substrate in the multicomponent liquid and formation of the skin.

The second stage includes formation of new-phase nuclei and their growth simultaneously with continued substrate etching. This is the most characteristic stage in relaxation of a nonequilibrium interface, the so-called etch-back—regrowth mechanism. Its distinguishing feature is the concurrent etching and regrowth localized at different parts of the substrate.

The third stage involves formation of a continuous epitaxial interlayer (EIL) preventing direct contact of the liquid with the substrate and generation of a long-lived quasiequilibrium state in the system consisting of one liquid and two successive solids. We will call such a state quasiequilibrium of the second type.

One of the obvious parameters affecting the relaxation process and its rate is the temperature. Other important parameters are the size and sign of the difference of substrate and new-phase cell constants ($\Delta a/a$). Therefore, the following discussion of experimental data is divided into three parts differing in this parameter, $\Delta a/a > 0$, $\Delta a/a \approx 0$, and $\Delta a/a < 0$.

$\Delta a/a > 0$. This group includes such systems as the liquids of Ga—As, Ga—As—P, In—Ga—P, and Al—Ga—As in contact with GaP, In—Ga—As and Ga—As—Sb in contact with GaAs, In—Ga—Sb and In—As—Sb in contact with InAs, etc.

The calculated supercooling ΔT_ϵ necessary for quasiequilibrium of the first type is greater than 35 K in Ga—As—GaP due to the large $\Delta a/a$ (cf. Fig. 2). Therefore, local equilibrium of this type in such a system is impossible due to the fact that the liquid cannot withstand such large supercoolings. The substrate etching that begins at the initial moment of contact is accompanied by formation of new-phase nuclei on the GaP surface. These coalesce and form a continuous epitaxial layer. The formation time of the EIL decreases with increasing temperature. It is hours at 700°C and seconds at 900°C. Therefore, experimental temperature conditions can be found at which all phases of this process can be studied.

The GaP(111)B surface five minutes after contact with liquid Ga—As at 700°C is shown in Fig. 5. Large nuclei of the new phase GaAsP are visible. The dissolved GaP surface lies between them.

Auger spectra of a control GaP substrate surface that did not contact liquid Ga—As (*a*), the surface of a $GaAs_{1-x}P_x$ nucleus analogous to that shown in Fig. 5 (*b*), and a portion of the GaP substrate between new-phase nuclei (*c*) are presented in Fig. 6. In the last case, the As peak at 1228 eV is very weak whereas the P peak is nevertheless decreased by a factor of two. Moreover, if the intensity of the low-energy As peak at 34 eV in Fig. 6*c* is 1.5 times less than for Fig. 6*b*, then the same ratio for the high-energy peak is about 10. These facts indicate that As on the GaP surface without the new-phase nuclei and occupying a P site is located in a layer thinner than the depth of the As 1228 eV Auger electrons (~2 nm) but comparable with the depth of the P Auger electrons (~0.5 nm). Thus, a diffusion layer of the solid solution GaAsP does exist on the dissolving GaP surface although it is very thin. Its thickness can be estimated as about one cell constant of the GaP crystal.

Isothermal contact In—Ga—As—GaAs for 10-30 min at 800°C did not lead to formation of new-phase nuclei on GaAs after 1-4 h. Auger analysis of the surface of such a sample also indicated that the skin is very thin, 0.5-3 nm [11].

$\Delta a/a \approx 0$. In spite of the large chemical misfit in the AlAs—GaAs system, an EIL AlGaAs is formed on GaAs even if the liquid Al—Ga—As is undersaturated. It was assumed earlier that this AlGaAs layer with an estimated thickness of 40-200 nm is the result of anomalous Al diffusion into the dissolving GaAs crystal [12]. However, it was proven recently that the AlGaAs layer is of epitaxial origin [13]. In particular, inhomogeneous AlGaAs deposition (Fig. 7a) was revealed using anodic oxidation of such surfaces. Light and dark portions are observed corresponding to AlGaAs and GaAs regions. The anodic oxide deposited on the sample is thinner on AlGaAs. This condition provides noticeable contrast appearing on the black-and-white images as light and dark portions. We also note that the formation of the AlGaAs layer observed experimentally in the GaAs subsurface region during continuous heating of the system, considered by Small et al. [12] as final proof of the diffusion

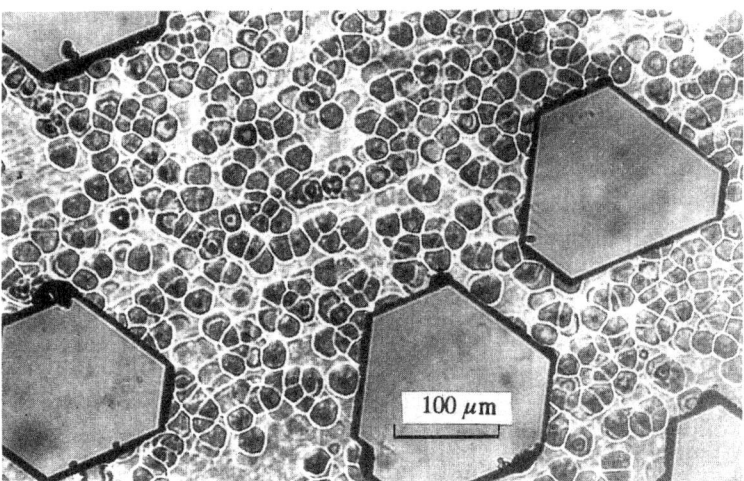

Fig. 5. Nuclei of the new phase GaAsP on GaP substrate.

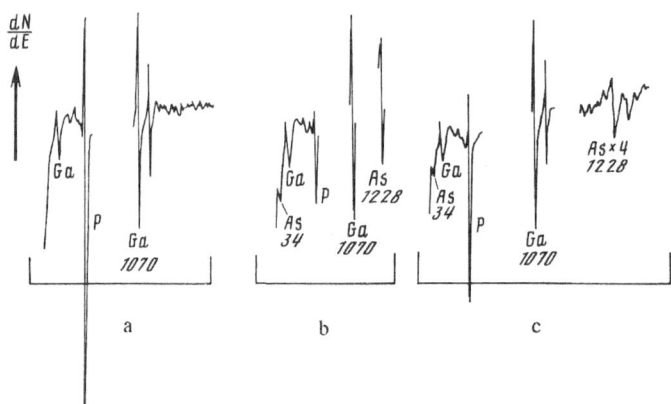

Fig. 6. Auger spectra of the GaP surface (*a*) from the control GaP substrate, (*b*) after Ga–As–GaP contact at a new-phase nucleus, (*c*) after Ga–As–GaP contact on the pure substrate surface.

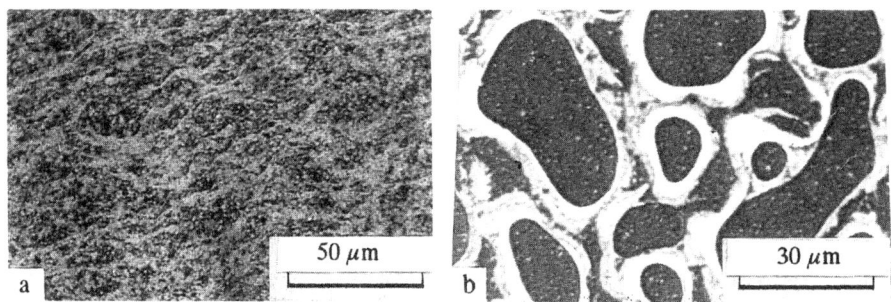

Fig. 7. GaAs(111)B substrate surface after contact with undersaturated liquid Al–Ga–As with subsequent anodic oxidation. $T_0 = 800°C$. Isothermal contact with liquid for 3 min at 802°C (*a*), continuous heating at 1°C/min after contact at 802°C (*b*).

nature for formation of a thick AlGaAs layer, is also due to substrate surface irregularity arising from continuous etching and inhomogeneous deposition of this layer. The surface of such a sample after anodic oxidation is shown in Fig. 7*b*. This picture changes unusually depending on the heating rate and the substrate orientation.

Fig. 8. InAs substrate after isothermal contact with Ga—Sb superheated by 2 K
(followed by sample anodization).

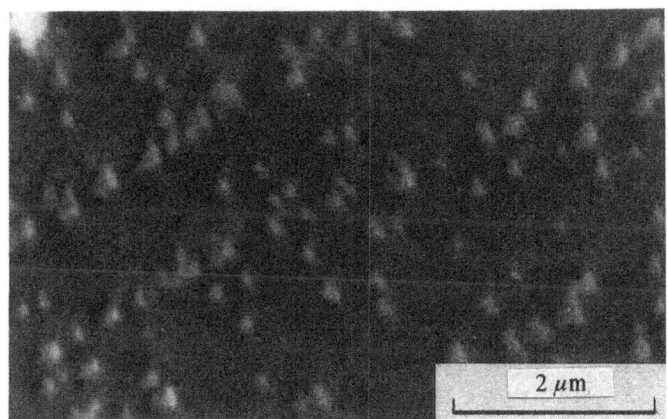

Fig. 9. Surface analogous to that shown in Fig. 8 in a BS-350 microscope (Tesla).

An analogous epitaxial layer was found in the system Ga—Sb—InAs [14]. Its inhomogeneity can also be observed using anodic surface oxidation (Fig. 8). The dark bands are the anodic oxide on GaSb (the anodic oxide on GaSb is thicker than on InAs for oxidation under identical conditions). At high resolution in a scanning electron microscope, these bands look like chains of InGaSbAs new-phase nuclei (Fig. 9).

Thus, it can be assumed that the skin in the Al—Ga—As—GaAs and Ga—Sb—InAs systems during contact of the undersaturated liquid—substrate is thick enough that σ_{ss} is insignificant and the new phase forms through a barrierless mechanism.

However, a study of the In—Ga—P—GaAs system (the equilibrium solid $In_{0.5}Ga_{0.5}P$ is commensurate with the substrate cell constant) found that there is no epitaxial layer of the new phase on the GaAs substrate after isothermal liquid—solid contact. This can be seen in Fig. 10, which presents the Auger profiles of P, In, and O with substrate depth. It is evident that In and P lie within the natural oxide of less than 2-nm thickness according to ellipsometric measurements. This singular difference of the Al—Ga—As—GaAs and In—Ga—P—GaAs systems may provide an explanation for the fact that the film—substrate transition layers (TL) are quite different. If in the first system the TL is 10-30 nm thick [15], then in the $In_xGa_{1-x}P$—GaAs system the TL is much less, of the order of 3 nm [16].

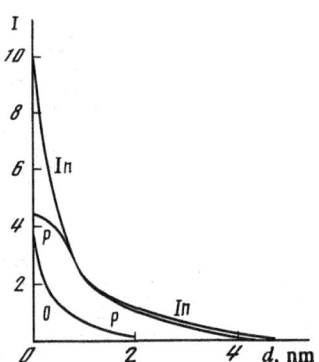

Fig. 10. Auger profiles of In, P, and O peaks with depth of GaAs substrate after 30 min isothermal contact with ᴵⁿ—Ga—As—P—GaAs(111)A at 800°C.

Fig. 11. Nucleus of the new phase InGaAs on InAs after contact with the undersaturated liquid In—Ga—As.

$\Delta a/a < 0$. This group includes such systems as Ga—As—P—GaAs, In—Ga—As—InAs, In—Ga—P—InP, and others. These systems also behave uniquely. Isothermal contact of a saturated or undersaturated liquid with the substrate causes the latter to dissolve partially and become covered with a solid solution layer of imperfect structure, in particular, with trapped solvent drops. Such an imperfect start of heteroepitaxial growth for this case is the reason studies on this system are practically nonexistent.

We will demonstrate using contact of the liquid In—Ga—As—InAs(111)A (the solid in equilibrium with the liquid, $In_{1-z}Ga_zAs$, had $z = 0$-0.2) as an example that the relaxation mechanism as before consists of three stages but has its own peculiarities. In the first stage, contact of InAs with the liquid differing from In—As by addition of a small amount of Ga produced separate round new-phase nuclei (Fig. 11) on the surface of the dissolving substrate. Figure 12 presents the Auger spectra of the principal elements on the surface without new-phase nuclei (Fig. 12a) and at the nucleus (Fig. 12b). Gallium is observed not only at the nucleus of the new phase but also on the substrate surface without nuclei. However, the Ga signal in the latter case is much weaker. Thus, a thin skin is also observed on the solid surface without epitaxial nuclei.

A distinguishing feature of the system is the growth rate of three-dimensional nuclei that is anomalously high for liquid-phase epitaxy. It increases substantially with increasingly negative $\Delta a/a$. This is demonstrated in Fig. 13, in which the dependences of the density of new-phase nuclei (Fig. 13a) and their tangential growth rate (Fig. 13b) are plotted as functions of solid solution composition z intended to be grown. If the function $v(z)$ is extrapolated to $z = 0.2$, the new-phase growth rate is greater than 10^{-2} cm/sec. At such a growth rate and a density of nuclei of the order of 200 cm^{-2}, a continuous new-phase film should be formed in less than one second. Such continuous imperfect layers were usually observed for isothermal contact of a multicomponent liquid and binary solid and for other systems with $\Delta a/a < 0$ [3].

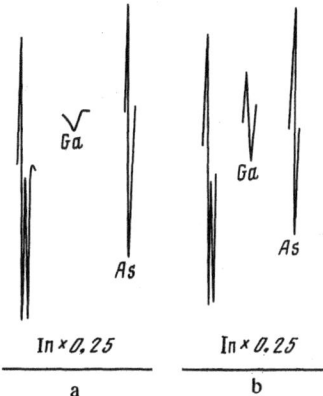

Fig. 12. Auger spectra of principal elements on the InAs surface without nuclei (*a*) and at the InGaAs nucleus (*b*).

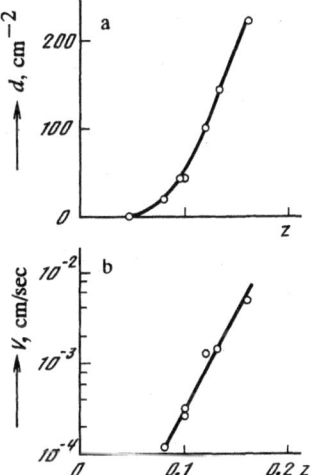

Fig. 13. Dependences of new-phase nucleus density (*a*) and their tangential rate (*b*) on $In_{1-z}Ga_zAs$ solid solution composition intended for growth.

DISCUSSION

Experimental studies of the relaxation kinetics of a nonequilibrium interface multicomponent liquid—solid suggests the presence of a certain commonality in the relaxation process. A thin skin is formed on the surface of the dissolving crystal. A new phase then nucleates, grows, coalesces, and finally forms a continuous EIL that separates the crystal from the liquid not in equilibrium with it.

However, the formation rate of the new phase is substantially different for the cases $\Delta a/a < 0$ and $\Delta a/a > 0$. The following reason for the difference is possible. For $\Delta a/a > 0$, etching of substrate surface covered with a strained skin is slow and limits the entire solid dissolution into the nonequilibrium multicomponent liquid. This is confirmed by the fact the etching pits due to various structural imperfections are always observed on the substrate surfaces as they dissolve (for example, the surface without nuclei in Fig. 5 is covered with etching pits). Selective etching of the solid surface is a morphological sign of the limiting influence of surface etching reactions. For the opposite inequality, $\Delta a/a < 0$, substrate etching is apparently limited by diffusion of dissolving components whereas the kinetics of surface etching reactions are fast. This mechanism is indirectly confirmed by the smooth parts of the InAs substrate after contact with liquid In—Ga—As. (Such a surface can be seen in Fig. 11 around the nucleus.)

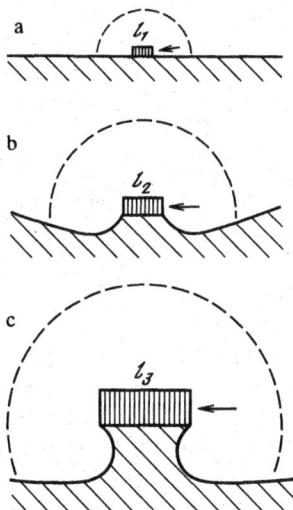

Fig. 14. Generation of an unstable liquid—solid interface as a result of new phase nucleation. The new phase nucleus is shown by an arrow.

A higher supersaturation than for $\Delta a/a > 0$ results from the rapid substrate etching for $\Delta a/a < 0$. This is accompanied by nucleus growth rates that are very high for liquid-phase epitaxy, of the order of 10^{-2} cm/sec. They are apparently even greater for greater differences between a_0 and a_s.

If such a mechanism for the different substrate etching rate in the nonequilibrium liquid is valid, then strains in the subsurface resulting from formation of the skin should contribute greatly to the different rates. For $\Delta a/a < 0$, this layer experiences distension strain; for $\Delta a/a > 0$, compression strain.

CATASTROPHIC SUBSTRATE ETCHING

Four-component systems have an additional degree of freedom. For a fixed cell constant, the composition of the film can vary over wide limits. The liquid composition in equilibrium with it will also be different. Thus, the second relaxation stage, etch-back—regrowth, becomes catastrophic solid etching in certain cases. The generation of a new-phase nucleus on the substrate surface changes the composition of the liquid surrounding it so much (Fig. 14a, region l_1) that the liquid becomes undersaturated relative to the substrate and dissolves it at an increasing rate (Fig. 14b and c). Finally, a significant part of the solid is dissolved including practically all of those parts in which nuclei arise. The new phase precipitates outside the substrate. The rapid relaxation processes end due to either a substantial change of the liquid composition toward the substrate composition or isolation of the bulk of the liquid from the substrate by a continuous mono- or polycrystalline solid layer.

As was demonstrated in [17], such catastrophic substrate etching occurs as a result of an array of factors making the initial liquid—solid system unstable. In systems with a variable solid cell constant such as In—Ga—As—P—GaAs, In—P—$In_{0.53}Ga_{0.47}As$, and In—As—GaSb, the principal driving force leading to catastrophic etching is the decrease of an already negative $\Delta a/a$ as the solid dissolves. In quaternary systems based on liquid Al—Ga—As, but differing in that the Ga solvent is replaced by others (Bi, In [13] and Sn [18]), the Al segregation coefficient increases sharply at small concentrations. Aluminum is known to decrease substantially the solubility of As in the liquid. As the nucleation of a new phase containing mainly AlAs fluctuates, the depletion of Al from the liquid region l (Fig. 14) substantially increases the As solubility. In this case, this is the driving force for catastrophic substrate etching.

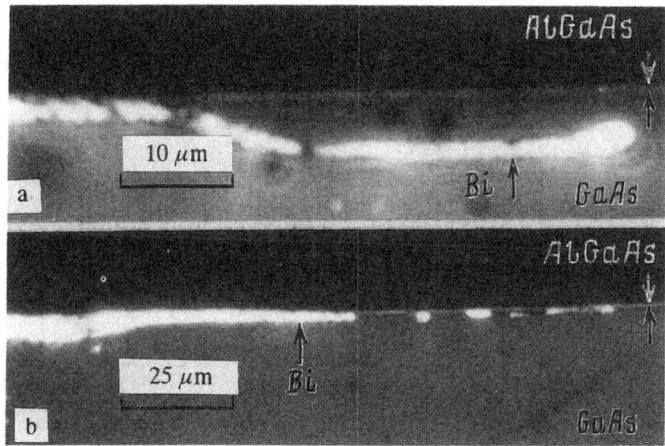

Fig. 15. Slices of various parts (*a* and *b*) of the structure generated after isothermal contact of liquid Bi + Al (0.1% Al atomic fraction) with GaAs(100) substrate at 700°C demonstrating trapping of Bi by the growing AlGaAs film.

The final result of such relaxation can be interpreted as a limiting case of drop entrapment by the developing crystal—film in which the drop can become practically continuous along the whole substrate surface. This separates the new phase from the substrate by a metallic adlayer. Slices of such structures are shown in Fig. 15.

CONCLUSION

Thus, the entire relaxation process of a nonequilibrium interface undersaturated (or saturated) multicomponent liquid—substrate can be divided into several stages.

1. Substrate etching with formation on its surface of a skin several angstroms thick consisting of a solid solution of all elements present in the liquid and solid. In certain cases, such multicomponent liquid contact [the multicomponent liquid becomes saturated or supersaturated ($\Delta a/a > 0$) due to partial substrate etching] converts the skin to a quasiequilibrium state in which no visible changes on the solid surface occur for tens of minutes.

2. Formation of new-phase nuclei and their growth simultaneously with continuing substrate etching. This is the most characteristic stage of the relaxation process, the so-called etch-back—regrowth mechanism. A distinguishing feature is the synchronicity of etching and regrowth localized at different parts of the substrate. Thus, this stage is mixed (or transitional). The liquid and solid are separated as parts of the diffusional and epitaxial layers. The shape of the new-phase nuclei and their density vary widely depending on the actual conditions. A whole spectrum of experimental results for contact of the nonequilibrium liquid and solid is determined by the rate of etching and regrowth processes that depend on contact temperature, actual liquid composition, and the difference of the new-solid and substrate cell constants. The rate of new-phase nucleation increases especially where $\Delta a/a$ is negative. The rate of coverage of substrate by new-phase nuclei is so great that stage 2 remains practically unnoticed whereas stage 3 is observed.

3. Formation of a continuous EIL separating the liquid from direct contact with substrate and generation of a prolonged quasiequilibrium state consisting of one liquid and two successively positioned solids.

REFERENCES

1. Yu. B. Bolkhovityanov, "The peculiarities of isothermal contact of liquid and solid phase during the LPE of A^3B^5 compounds," *J. Cryst. Growth*, **55**, No. 3, 591-598 (1981).

2. Yu. B. Bolkhovityanov, "The contact phenomena between the liquid phase and the substrate during LPE of A^3_B5 compounds," *J. Cryst. Growth*, **57**, No. 1, 84-90 (1982).

3. Yu. B. Bolkhovityanov, "Isothermal contact of III-V saturated solutions with different III-V substrates: Two modes of behavior," *Cryst. Res. Technol.*, **18**, No. 5, 679-686 (1983).

4. V. V. Kuznetsov, P. P. Moskvin, and V. S. Sorokin, "Coherent phase diagram of four-component systems based on $A^{III}B^V$ compounds," *Zh. Fiz. Khim.*, **60**, No. 6, 1376-1381 (1986).

5. Yu. B. Bolkhovityanov, R. I. Bolkhovityanova, and S. I. Chikichev, "Experimental examination of GaAs etching in In—P melt," *J. Electron. Mater.*, **12**, No. 3, 525-550 (1983).

6. Yu. B. Bolkhovityanov, "Study of $In_xGa_{1-x}P$/GaAs films formation on the basis of In—Ga—P liquidus precised investigation," *Cryst. Res. Technol.*, **17**, No. 12, 1483-1489 (1982).

7. K. Nakajima, T. Tanahashi, K. Akita, et al., "Determination of In—Ga—As phase diagram at 650°C," *J. Appl. Phys.*, **50**, No. 7, 4975-4981 (1979).

8. V. F. Dorfman, S. A. Petrushinina, and M. L. Shupegin, "Displacement of the solution—melt liquidus point during the epitaxy of stressed garnets," *Dokl. Akad. Nauk SSSR*, **246**, No. 5, 1159-1162 (1979).

9. Yu. B. Bolkhovityanov, "Experimental study of isothermal contact of a saturated liquid $A^{III}B^V$ phase with various binary substrates $A^{III}B^V$," *Élektron. Tekh., Ser. 6. Mater.*, No. 6 (167), 49-52 (1982).

10. Yu. B. Bolkhovityanov, Yu. D. Vaulin, B. Z. Ol'shanetskii, and S. I. Stenin, "Auger analysis of the GaP(111)B surface after isothermal contact with a saturated Ga—As liquid phase," *Poverkhnost*, No. 12, 68-71 (1984).

11. Yu. B. Bolkhovityanov, Yu. D. Vaulin, B. Z. Ol'shanetskii, and S. I. Stenin, "Auger analysis of the solid surface after isothermal contact of a multicomponent liquid and binary substrate in the $A^{III}B^V$ system," *Poverkhnost*, No. 9, 47-53 (1985).

12. M. B. Small, R. Ghez, R. M. Potemski, and J. M. Woodall, "The formation of $Ga_{1-x}Al_xAs$ layers on the surface of GaAs during continuous etching into Ga—Al—As solutions," *Appl. Phys. Lett.*, **35**, No. 3, 209-210 (1979).

13. Yu. B. Bolkhovityanov, R. I. Bolkhovityanova, Yu. D. Vaulin, et al., "The formation of $Al_xGa_{1-x}As$ layers by regrowth on the surface of GaAs during its contact with the undersaturated liquid containing Al," *J. Cryst. Growth*, **78**, No. 3, 335-341 (1986).

14. Yu. B. Bolkhovityanov, R. I. Bolkhovityanova, Yu. D. Vaulin, et al., "Auger study of initial stages of heterojunction formation during heteroepitaxy from a liquid," in: Abstracts of Papers of the Tenth All-Union Conf. on Semiconductor Physics [in Russian], Part 1, Sept. 17-19, 1985, Institute of Solid-State Physics and Semiconductors, Academy of Sciences of the Belorussian SSR, Minsk (1985), pp. 174-175.

15. T. Hayakawa, N. Miyauchi, T. Suyama, et al., "Influences of thin active layer in (GaAl)As double-heterostructure lasers grown by LPE," *J. Appl. Phys.*, **56**, No. 11, 3088-3095 (1984).

16. Yu. B. Bolkhovityanov, R. I. Bolkhovityanova, Yu. D. Vaulin, et al., "Transition layers in InGaAsP/GaAs heterostructures," in: Abstracts of Papers of the First All-Union Conf. Physical and Physicochemical Principles of Microelectronics [in Russian], Nauka, Moscow (1987), pp. 297-298.

17. Yu. B. Bolkhovityanov and S. I. Chikichev, "Instability of 'slow' solid—liquid interface relaxation before the hetero-LPE of III—V compounds," *Cryst. Res. Technol.*, **18**, No. 7, 847-857 (1983).

18. S. G. Zhilenis and V. U. Stankevich, "Relaxation of the 'melt—crystal' phase boundary on contact of GaAs with Al—Ga—Sn—As flux," in: Expanded Abstracts of the Seventh All-Union Conf. on Crystal Growth [in Russian], Vol. 2, Moscow (1988), pp. 309-311.

MODELLING AND CONTROL OF HEAT AND MASS TRANSFER
DURING LIQUID EPITAXY

N. A. Verezub and V. I. Polezhaev

Liquid-phase epitaxy (LPE) in which a layer is grown from dilute solution is one of the principal methods of preparing semiconducting structures for immediate use in devices. These devices are usually multilayered heterocomposites based on binary, ternary, or quaternary solid solutions.

The thickness and structure of the epitaxial layer depends on the physicochemical properties of the flux, the growth temperature range, the cooling rate, the configuration of the LPE region, and the orientation of the substrate relative to the gravitational force.

The present article examines hydrodynamic processes in a flux during epitaxial growth using mathematical modelling based on dynamic Navier—Stokes equations and investigates processes in a heating or cooling melt layer in a gravitational field in order to reveal convective heat- or mass-transfer effects during LPE and to control them.

The epitaxial growth rate depends on the rate with which the dissolving substance is separated from the flux and deposits on the growing surface. Since epitaxial growth is a nonequilibrium process [1], the rate of transfer of the dissolved substance cannot be determined directly from the temperature—composition diagram of the system. This requires a knowledge of transfer processes in the liquid.

The approximate growth rate and epitaxial layer thickness are found by solving the one-dimensional diffusion equation taking into account the experimental conditions. This model is used to analyze binary and ternary systems with and without considering kinetic processes on the growing surface, to calculate impurity profiles for doped layers during growth, and to investigate electroepitaxy of $A^{III}B^V$ systems [2-9].

However, the formation in some experiments of rough surfaces with steps and growth waves, the dependence of layer thickness on substrate orientation and melt thickness, and the variable composition over the substrate cannot be explained within the framework of the above model. These effects are due to kinetic processes at the crystallization front, inclusion of solvent in the layer due to constitutional supercooling, and convection in the melt [10-15].

The presence of convection in LPE experiments can explain many of the above effects, as noted previously [16-19]. Convection can arise due to a nonuniform flux density, surface forces on a free surface due to nonuniform concentration and/or temperature, and the presence of nuclei of a microphase in the melt. Concentrational convection dominates over thermal for a number of semiconducting systems. This enables the examination to be limited to a single motion mechanism.

The following problems arise in analyzing processes in the melt during epitaxial growth:

(1) determination of the critical conditions generating motion;
(2) investigation of the motion structure for actual experimental parameters;
(3) study of motion modes as a function of changing process parameters.

From a mathematical viewpoint, processes in a flux during LPE can be studied analogously to those in a heating or cooling liquid layer in massive force fields.

The equilibrium liquid state in a planar horizontal layer under certain conditions becomes unstable and develops Rayleigh—Bènard convection [20]. Subcritical convection during heating has a downward structure that depends on the Rayleigh and Prandtl numbers, the spatial uniformity of physical parameters, the presence of side boundaries, initial and limiting conditions, and the steadiness of heating. As the liquid layer is heated above and below, it separates into two LPE regions [21]: the equilibrium is unstable in the lower one (relative to the middle of the layer). Convection that gradually expands into the upper stable region develops. This is the so-called penetrating convection. A similar type of motion is seen where epitaxial structures are prepared from a limited volume of saturated flux onto two substrates.

Equilibrium is impossible in the vertical limiting liquid layer during heating from the side [20]. For a temperature drop as small as desired between the boundaries, motion arises with an intensity that increases with an increasing temperature difference.

Transitional motion regimes exist between the two limiting cases, Rayleigh—Bènard convection in the horizontal layer and hydrodynamic instability of opposing currents in the vertical layer. The instability and the appearance of irregular structures with increasing Rayleigh number depend on the slope of the layer [22-24].

Comparative criteria can be analyzed to evaluate the effect of process conditions on epitaxial layer quality. This is especially important for experiments in space.

Liquid-phase processes can be analyzed for fluxes of $A^{III}B^V$ systems using Prandtl (Pr), Schmidt (Sc), and Rayleigh (Ra) numbers; the concentrational Rayleigh number (Ra_c), the Marangoni number (Mn), and the concentrational Marangoni number (Mn_c). It is noteworthy that experimental data for the dependence of saturated solution parameters on temperature have not been reported, with the exception of density data for the Ga—GaAs system [17] and the surface tension coefficient of GaSb [25]. The values $\rho(T, c)$ and other parameters were calculated from values for Ga, In, and $A^{III}B^V$ taking into account phase diagrams [26].

The physicochemical parameters used to calculate comparative criteria of LPE for terrestrial conditions ($g_0 = 9.8$ m/sec^2) are given in Table 1.

The systems examined are dominated by concentration-gradient convection rather than thermal convection. This enables the epitaxial growth process to be studied under isothermal conditions, neglecting thermal convection.

A mathematical model of LPE was given in detail in [26, 27]. Therefore, we will provide only a short discourse on the subject. Growth of layers from a limited saturated flux volume onto two substrates under isothermal conditions is examined. The modelling is based on two-dimensional dynamic Navier—Stokes equations in the Boussinesq approximation with variables rotation ω, flux ψ, and concentration C. The temperature—composition diagram is accounted for by setting the boundary conditions for concentration at the boundaries of the melt with the substrates. The model was tested [26] before modelling the process numerically.

The principal characteristic is the function Q, which describes the dimensionless mass of crystallized component depositing on a unit length of substrate since the start of the experiment and determining the layer configuration.

Table 1. Comparative Criteria for A^{III}–$A^{III}B^{V}$ Systems

System	Pr	Sc	$\dfrac{Ra}{H^3 \Delta T}$, $cm^{-3} \cdot deg^{-1}$	$\dfrac{Ra_c}{H^3}$, cm^{-3}	$\dfrac{Mn}{H \Delta T}$, $cm^{-1} \cdot deg^{-1}$	$\dfrac{Mn_c}{H}$, cm^{-1}
In–InSb	0.007	118	$8.4 \cdot 10^2$	$2.36 \cdot 10^9$	20	$2.3 \cdot 10^8$
Ga–GaSb	0.0076	32	$3 \cdot 10^2$	$1.44 \cdot 10^8$	52	$3.4 \cdot 10^6$
In–InAs	0.0022	55	$1.1 \cdot 10^3$	$3.9 \cdot 10^9$	47.6	$2.7 \cdot 10^8$
In–InP	0.0021	4	$7.4 \cdot 10^2$	$1.4 \cdot 10^8$	51.4	$9 \cdot 10^7$
Ga–GaAs	0.0056	96	$4 \cdot 10^2$	$2.9 \cdot 10^{10}$	52	$1.1 \cdot 10^8$
Ga–GaP	0.0047	21	$5.6 \cdot 10^2$	$3.8 \cdot 10^8$	55	$3.5 \cdot 10^7$

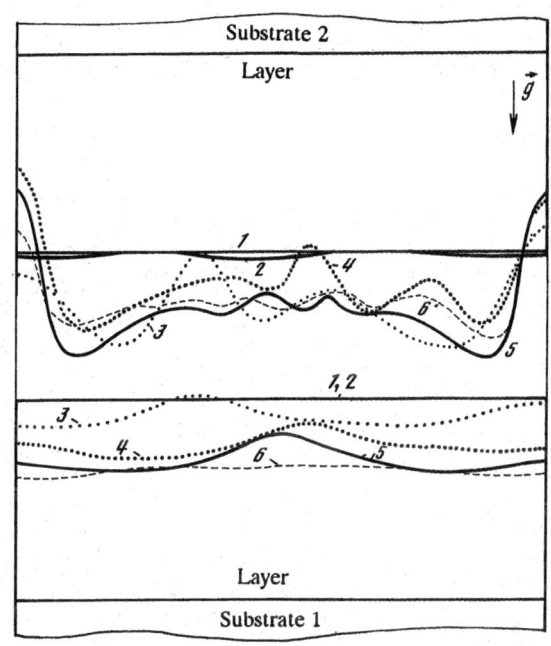

Fig. 1. Configuration of layers at the end of LPE. $V_{cool} = 1°C/min$. $Ra_c = 10^5$ (1), $5 \cdot 10^5$ (2), 10^6 (3), $5 \cdot 10^6$ (4), 10^7 (5), $2 \cdot 10^7$ (6).

The geometric nonuniformity of the layer is characterized by the quantity

$$Q_i' = \frac{Q_{max} - Q_{min}}{Q_{max} + Q_{min}} \cdot 100\% \ (i = 1, 2).$$

Here Q_{max} and Q_{min} are the maximal and minimal layer thicknesses at the end of the process. Thus,

$$\overline{Q}' = 0.5 (Q_1' + Q_2').$$

The difference in the mass of layers between substrates is defined by

$$\Delta Q = \frac{\max (Q_1, Q_2) - \min (Q_1, Q_2)}{Q_1 + Q_2} \cdot 100\%,$$

where Q_1 and Q_2 are dimensionless amounts of substance deposited on the substrates at the end of the process.

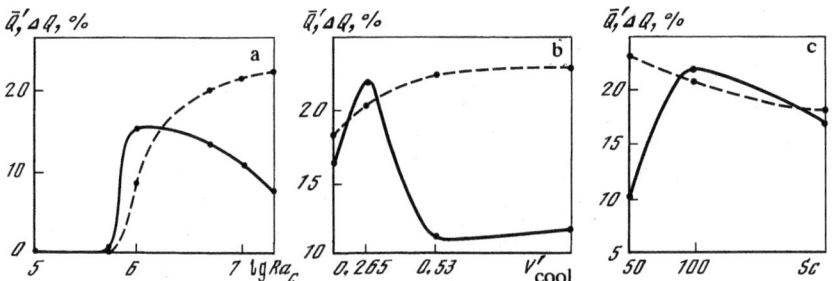

Fig. 2. Dependence of geometric nonuniformity of layers $\bar{Q}'$ (solid line) and difference in thickness between layers ΔQ (dashed line) on Ra_c (a), V'_{cool} (b), and Sc (c) for horizontal substrates. $V_{cool} = 1°C/min$ (a), $Ra_c = 10^7$, $V'_{cool} = 0.53$ corresponds to $V_{cool} = 1°C/min$ (b), $V_{cool} = 0.5°C/min$ (c). $D = const$, $\Delta T = 30°C$.

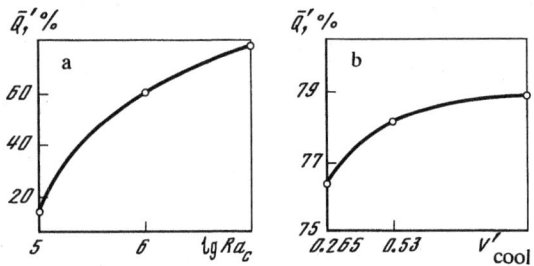

Fig. 3. Dependence of geometric nonuniformity on Ra_c (a) and V_{cool} (b) for vertical substrates. $V_{cool} = 1°C/min$ (a), $Ra_c = 10^7$ (b).

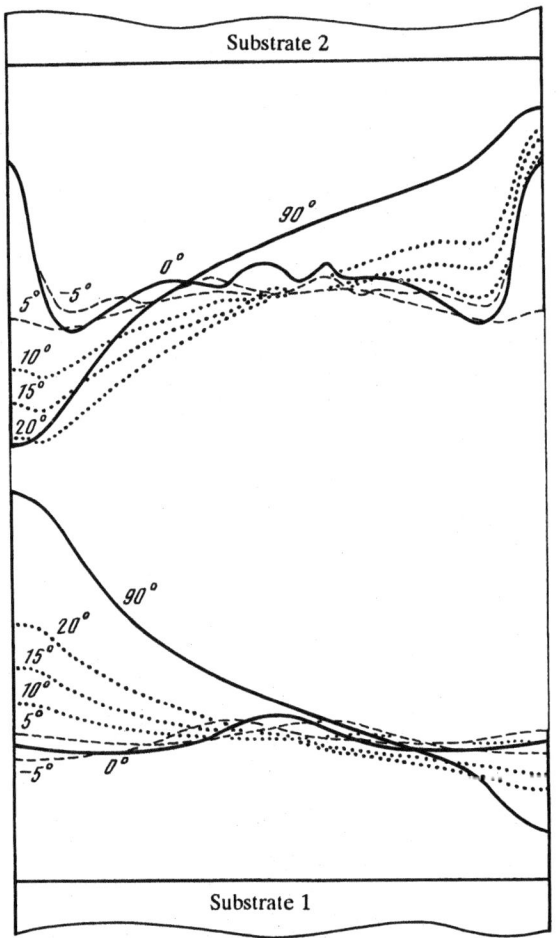

Fig. 4. Configuration of layers for various angles φ. $Ra_c = 10^7$, $V_{cool} = 1°C/min$.

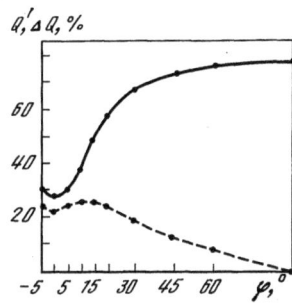

Fig. 5. Dependence of geometric parameters of layers Q' (solid line) and ΔQ (dashed line) on angle φ. $Ra_c = 10^7$, $V_{cool} = 1°C/min$.

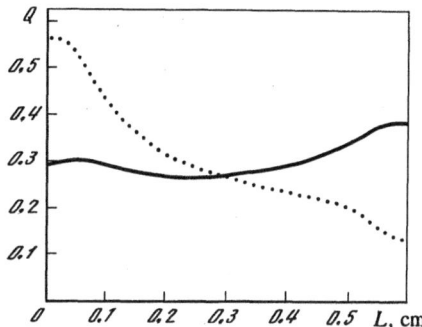

Fig. 6. Configuration of layers as a function of LPE region orientation. $Ra_c = 10^6$, $V_{cool} = 1°C/min$. The dotted line corresponds to $\varphi = 90°$; solid line, $\varphi = 90°$ up to a Fourier diffusion number $Fo_D = Dt/H^2 = 0.25$ and $\varphi = -90°$ up to $Fo_D = 0.5$.

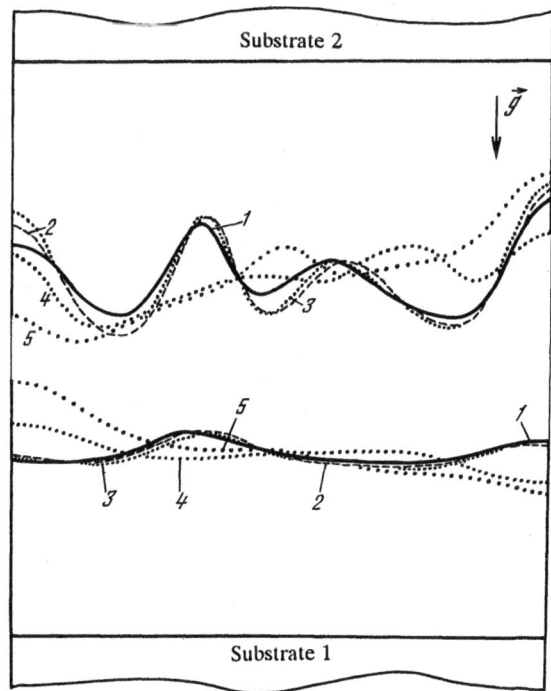

Fig. 7. Influence of perturbations on layer configuration for horizontal substrates. $Ra_c = 10^6$, $V_{cool} = 1°C/min$. $\varphi = 0$ (1), $\varphi = 5°$ up to $Fo_D = 0.15$ (2), $\varphi = 90°$ up to $Fo_D = 0.05$ (3), $\varphi = 90°$ up to $Fo_D = 0.15$ (4), $\varphi = 90°$ up to $Fo_D = 0.25$ (5).

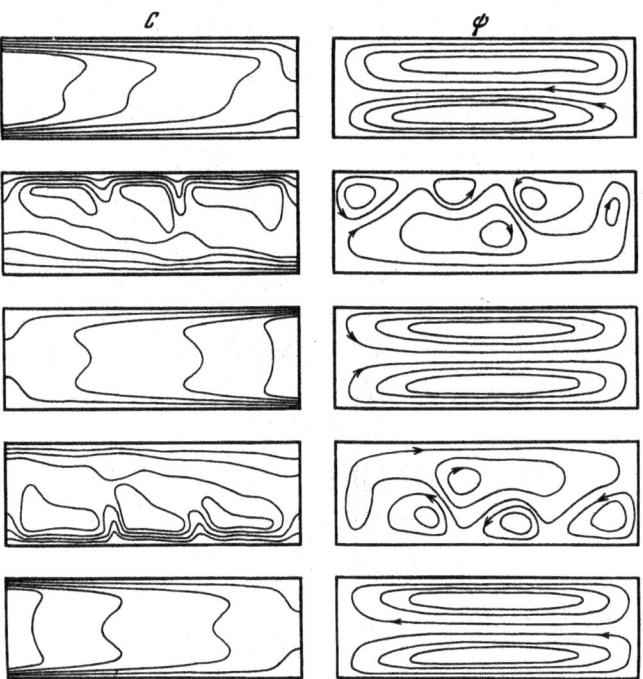

Fig. 8. Change of concentration distribution C and motion regimes ψ for a rotating LPE region. $\mathrm{Ra}_c = 10^7$, $V_{cool} = 1°\mathrm{C/min}$, $\Omega = 2$ rph, $\varphi_0 = 90°$.

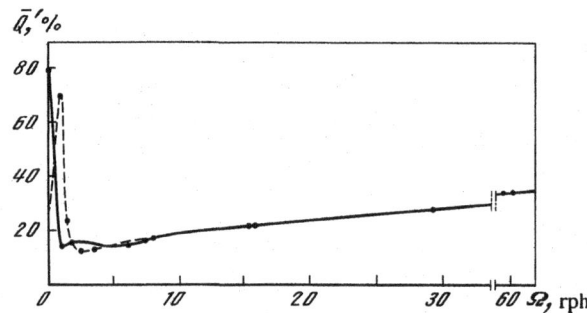

Fig. 9. Dependence of geometric nonuniformity of layers on container rotation rate. $\mathrm{Ra}_c = 10^7$, $V_{cool} = 1°\mathrm{C/min}$, $\varphi_0 = 90°$ (solid line), $\varphi_0 = 0$ (dashed line).

Slow rotation of the container during growth increases the geometric uniformity of the layers. The motion regimes change constantly, changing the flow of dissolved component responsible for the layer configuration on the growing surface (Fig. 8). The dependence of the nonuniformity on rotation rate for various starting substrate positions is plotted in Fig. 9.

The horizontal process was described in detail in [27]. The following Rayleigh—Bènard convection regimes are possible depending on the Ra_c number: regular steady-state regime ($\mathrm{Ra}_c \lesssim 5 \cdot 10^5$), regular non-steady-state [$\mathrm{Ra}_c \approx (1\text{-}5) \cdot 10^6$], and irregular non-steady-state ($\mathrm{Ra}_c \gtrsim 10^7$).

Layers with the most uneven thickness on the upper and lower substrates grow under the regular non-steady-state regime (Fig. 1), for which the structure of motion into the LPE region does not change with time. This is a result of the principle of maximum concentration and thermal layering in closed volumes presented in [28]. The dependence of geometric parameters of layers on Ra_c is plotted in Fig. 2a. The nonlinear nature of the function $\dot{Q}'(\mathrm{Ra_c})$ is related to the change of motion regimes in a region of this geometry. The geometric nonuniformity of layers on both substrates is decreased by carrying out LPE under weightlessness. This was confirmed in experiments performed in space [29, 30]. Increasing the cooling rate also increases irregular flow in the region (Fig. 2b). Changing the flux viscosity is equivalent to changing the cooling rate while keeping the diffusion coefficient constant (Fig. 2c).

Vertical substrates produce inclined layers due to plane-parallel motion in the region. The dependences of geometric nonuniformity of the epitaxial layers on $\mathrm{Ra_c}$ and on cooling rate are plotted in Fig. 3.

Changing the LPE region orientation changes the motion regimes. The configuration of the layers produced and the generalized dependence of the geometric parameters on angle φ are shown in Figs. 4 and 5.

Changing the LPE region orientation can be a controlling factor in the process. Rotating the substrates by 180° in the vertical reactor (Fig. 6) and combining the vertical and horizontal variants (Fig. 7) can decrease the nonuniformity.

Thus, changing the container orientation during LPE and carrying out the process in space can control the geometric parameters of the epitaxial layers.

REFERENCES

1. J. J. Hsieh, "Thickness and surface morphology of GaAs LPE layers grown by supercooling, step-cooling, equilibrium-cooling and two-phase solution techniques," *J. Cryst. Growth*, **27**, 49-61 (1974).
2. M. B. Small and J. Crossley, "The physical processes occurring during liquid phase epitaxial growth," *J. Cryst. Growth*, **27**, 35-48 (1974).
3. M. B. Small and J. F. Barnes, "The distribution of solvent in an unstirred melt under the conditions of crystal growth by liquid epitaxy and its effect on the rate of growth," *J. Cryst. Growth*, **5**, 9-12 (1969).
4. T. Bryskiewicz, "Investigation of the mechanism and kinetics of growth of LPE GaAs," *J. Cryst. Growth*, **43**, 101-114 (1978).
5. R. Ghez and M. B. Small, "Growth and dissolution of ternary alloys of III—V compounds by liquid phase epitaxy and the formation of heterostructures," *J. Cryst. Growth*, **52**, 699-709 (1981).
6. D. Dutartre, "LPE growth rate in $Al_xGa_{1-x}As$ system; theoretical and experimental analysis," *J. Cryst. Growth*, **64**, 268-274 (1983).
7. W. R. Wilcox, "Computer simulation of growth of thick layers from solutions of finite solute concentration without convection," *J. Cryst. Growth*, **56**, 690-698 (1982).
8. S. I. Chikichev and V. M. Ioffe, "Calculation of impurity profiles during nonisothermal liquid-phase epitaxy of GaAs," *Élektron. Tekh. Ser. 6. Mater.*, No. 1 (186), 34-37 (1984).
9. T. Bryskiewicz, "Peltier-induced growth kinetics of liquid phase epitaxial GaAs," *J. Cryst. Growth*, **43**, 567-571 (1978).
10. B. L. Meiler, L. Ya. Zolotarevskii, A. A. Paat, and I. A. Orenshtein, "Growth and morphology of thick GaAs layers grown by liquid-phase epitaxy," in: *Semiconductors and Heterojunctions* [in Russian], Valgus, Tallinn (1987), pp. 24-26.
11. A. Mottram and A. P. Peaker, "The growth of gallium phosphide layers of high surface quality by liquid phase epitaxy," *J. Cryst. Growth*, **27**, 193-204 (1974).
12. R. L. Moon, "The influence of growth solution thickness on the LPE layer thickness and constitutional supercooling requirement for diffusion-limited growth," *J. Cryst. Growth*, **27**, 62-69 (1974).
13. J. Crossley and M. B. Small, "Some observations of the surface morphologies of GaAs layers growth by liquid phase epitaxy," *J. Cryst. Growth*, **19**, 160-168 (1973).
14. H. Meinders, "An alternative method for liquid phase epitaxy," *J. Cryst. Growth*, **26**, 180-182 (1974).
15. V. V. Avdeeva, N. A. Verezub, N. V. Mal'kova, and T. G. Yugova, "Defects of meniscus lines in multilayered heterocomposites of the InP (substr.)—InP—GaInAsP—InP type," *Élektron. Tekh. Ser. 6. Mater.*, No. 1 (200), 34-37 (1985).
16. E. A. D. White and J. D. C. Wood, "Heat and mass transfer in LPE processes," *J. Cryst. Growth*, **17**, 315-321 (1972).
17. M. G. Mil'vidskii, V. P. Orlov, and V. G. Tsepilevich, "Features of mass transfer processes during liquid epitaxy," *Izv. Akad. Nauk SSSR, Neorg. Mater.*, **16**, No. 7, 1159-1163 (1980).
18. G. I. Zhovnir and I. E. Maronchuk, "Mass transfer processes during preparation of epitaxial structures of $A^{III}B^V$ structures from the liquid phase," *Avtometriya*, No. 6, 22-32 (1980).
19. L. A. Dmitrieva, O. S. Mazhorova, Yu. P. Popov, et al., "Numerical study of convective mass transfer during preparation of semiconducting structures by liquid-phase epitaxy," in: *Mathematical Modelling: Preparation of Single Crystals and Semiconducting Structures* [in Russian], Nauka, Moscow (1986), pp. 84-101.
20. G. Z. Gershuni and E. M. Zhukhovitskii, *Convective Stability of Uncompressed Liquid* [in Russian], Nauka, Moscow (1972).
21. E. P. Kostogorov, É. A. Shtessel', and A. G. Merzhanov, "Dynamic natural convection in cooling liquids," *Inzh.-Fiz. Zh.*, **37**, No. 1, 5-12 (1979).
22. K.-R. Kirchartz, "Time-dependent convection under reduced gravity," *Z. Flugwiss. Weltraumforsch.*, **6**, No. 5, 300-309 (1982).

23.   J. E. Hart, "Stability of the flow in a differentially heated inclined box," *J. Fluid Mech.*, **47**, 547-576 (1971).

24.   R. M. Clever and F. H. Busse, "Instabilities of longitudinal convection rolls in an inclined layer," *J. Fluid Mech.*, **81**, 107-127 (1977).

25.   M. Ya. Dashevskii, "On the relation of GaAs surface properties to growth processes of its doped crystals," in: *Surface Effects in Melts* [in Russian], Naukova Dumka, Kiev (1968), pp. 73-84.

26.   V. I. Polezhaev, A. V. Bune, N. A. Verezub, et al., *Mathematical Modeling of Convective Heat and Mass Transfer Based on the Navier—Stokes Equations* [in Russian], Nauka, Moscow (1987).

27.   N. A. Verezub and V. I. Polezhaev, "Mathematical modelling of convection and concentration fields during growth of epitaxial layers," in: *Mathematical Modelling: Preparation of Single Crystals and Semiconducting Structures* [in Russian], Nauka, Moscow (1986), pp. 101-112.

28.   V. I. Polezhaev and A. I. Fedyushkin, "Hydrodynamic effects of concentrational layering in closed volumes," *Izv. Akad. Nauk SSSR, Mekh. Zhidk. Gaza*, No. 3, 11-18 (1980).

29.   V. I. Polezhaev and N. A. Verezub, "Numerical study of liquid epitaxy under weightlessness," in: *Gagarin Scientific Studies of Astronautics and Aviation (1983, 1984)* [in Russian], Nauka, Moscow (1985), pp. 235-239.

30.   N. A. Verezub, É. S. Kopeliovich, V. I. Polezhaev, and V. V. Rakov, "Features of heat and mass transfer processes in melts of certain elemental $A^{III}B^{V}$ semiconductors under weightlessness," in: *Technical Experiments in Weightlessness* [in Russian], Ural Scientific Center, Academy of Sciences of the USSR, Sverdlovsk (1977), pp. 79-94.

Part IV

GROWTH OF CRYSTALS FROM THE MELT

AGGREGATION OF POINT DEFECTS IN SILICON CRYSTALS GROWING FROM THE MELT

V. V. Voronkov

Structural microdefects peculiar to dislocation-free silicon single crystals became problematical during the scale-up to mass production [1-3]. These growth microdefects are obviously due to intrinsic point defects (IPD) generated at the crystallization temperature T_0 that persist during cooling and are not annealed at dislocation pile-ups but form various aggregates (clusters). The IPD equilibrium concentration in Si is relatively low [4], of the order of 10^{14} cm^{-3}, i.e., much less than the concentration of certain residual impurities in even the most pure crystals. Thus, the principal residual impurity, oxygen, has a concentration 10^{15}-10^{16} cm^{-3} for Si grown without a crucible and 10^{18} cm^{-3} for that grown by the Czochralski method [5]. The carbon concentration [6] varies over a wide range and can reach $5 \cdot 10^{17}$ cm^{-3}. Rapidly diffusing metallic impurities (iron and others) can have concentrations 10^{16} cm^{-3} and greater [7].

The high concentration of impurities compared with IPD raises the question of why the IPD play the decisive role in formation of growth microdefects. The answer is that the supersaturation and not the absolute point defect concentration determines the ability to form small but macroscopic aggregates (particles of the impurity phase). The impurity solution near T_0 is unsaturated for all the impurities listed above and remains so during cooling within a range of hundreds of degrees (this range is comparatively small, ~100 K, only for oxygen in crystals grown in crucibles). With respect to the IPD solution, its supersaturation increases very steeply during cooling due to the high formation energy of these defects (about 4.5 eV according to [4], which is much greater than the dissolution energy for known impurities in Si). Namely for this reason the IPD remain in the minority and play a decisive role in aggregation although impurities can also participate. Thus, the pivotal question regarding the formation of growth microdefects is what type of IPD dominate in the crystal, vacancies or interstitial atoms.

RECOMBINATION OR DIFFUSION

The nature of growth microdefects observed in Si changes dramatically if the growth or cooling conditions are varied [1-3, 8-10]. These data suggest [4] that the dominant type of IPD is not fixed but depends on the ratio of growth rate V to axial temperature gradient G. If this ratio is less than a certain universal value $\xi_t \approx 3.3 \cdot 10^{-5}$ cm^2/(sec $\cdot$ K), then Si$_i$ atoms dominate; for $V/G > \xi_t$, vacancies. Recombination or diffusion in a narrow region near the crystallization front determines the choice. The scale of this region l is determined by the drop of equilibrium concentrations of both types of IPD [4]:

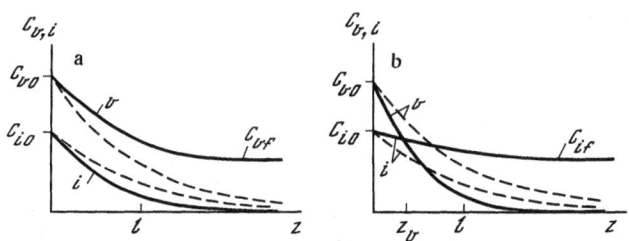

Fig. 1. Concentration profile of vacancies v and interstitial atoms i near the crystallization front. Dashed lines represent the equilibrium concentrations of these defects, for vacancy growth ($V/G > \xi_i$) (a), for interstitial growth ($V/G < \xi_i$) (b).

$$l = 2kT_0^2 / [(E_i + E_v)G], \tag{1}$$

where E_i and E_v are the formation energies of Si$_i$ and vacancies, respectively ($E_i \approx E_v \approx 4.5$ eV). The melt zone without a crucible has $l \approx 2$ mm; for the Czochralski method, $l \approx 1$ cm.

In the simplest model that does not account for thermal diffusion (which will be examined below), the persistence of IPD after recombination depends on their initial equilibrium concentrations C_{i0} and C_{v0} (at crystallization temperature T_0) and diffusion coefficients D_i and D_v (which are practically constant at high T due to the low migration energy). If vacancies dominate at T_0 (i.e., $C_{v0} > C_{i0}$) but are less mobile (so that $D_v C_{v0} < D_i C_{i0}$), the outcome of the recombination depends substantially on V/G.

Where V/G is large enough, diffusion is insignificant and recombinations outlive vacancies. At large V/G, their residual concentration C_{vf} is $C_{v0} - C_{i0}$. The characteristic profiles of the two IPD near the crystallization front are plotted in Fig. 1a. The dominant defects (vacancies) form a supersaturated solution (the actual concentration C_v is greater than the equilibrium value C_{ve}) whereas the Si$_i$ solution is unsaturated ($C_i < C_{ie}$) and C_i decreases exponentially with distance from the front.

At sufficiently small V/G, the loss of Si$_i$ to recombination can be compensated for by their diffusion away from the front. This effect is weaker for vacancies. As a result, Si$_i$ persist at a certain final concentration C_{if}. They form a supersaturated solution whereas the vacancy solution is unsaturated. The vacancy concentration decreases exponentially (Fig. 1b). The principal feature of the IPD profiles for this case (i.e., for interstitial growth, $V/G < \xi_i$) is that a band of width z_v in which vacancies dominate exists near the front although Si$_i$ dominate in the crystal bulk.

The choice of recombination or diffusion at the very start of cooling determines the type of dominant IPD, i.e., the subsequent pattern of defect formation (aggregation). Let us now examine the two principal aspects of this definitive step in defect formation, the characteristic IPD recombination time and their thermal diffusion.

RECOMBINATION TIME

Recombination–generation equilibrium between vacancies and Si$_i$ is established after a certain recombination time τ_r. A quantitative selection theory [4] was based on the assumption that τ_r near T_0 is small compared with the residence time of IPD in the recombination zone, $l/V \approx 1$ min. For this case, the product of the actual concentrations $C_i C_v$ can be considered to be equal to the product $C_{ie} C_{ve}$.

Recombination, which is limited only by diffusive approach of a pair of defects, is known [11] to be described by

$$\dot{C}_v = \dot{C}_i = -4\pi r (D_i + D_v)(C_i C_v - C_{ie} C_{ve}), \tag{2}$$

where $r \approx 3$ Å is the interaction radius. If a barrier ΔF_r to the reaction in the right part of Eq. (2) is present, a barrier factor $\exp[-\Delta F_r/(kT)]$ must be introduced. Considering that $D_i > D_v$, and recognizing that $C_i \gtrsim C_v$ (interstitial growth), we obtain

$$1/\tau_r \approx 4\pi r D_i C_i \exp(-\Delta F_r/(kT)). \tag{3}$$

The time τ_r at T_0 is very small (tenths of a microsecond) for barrierless recombination but can increase greatly due to ΔF_r.

The quantity τ_r at 1100°C (i.e., substantially below T_0) was estimated indirectly in [12], where the time dependence of the Sb diffusion coefficient in Si was measured during thermal oxidation producing an injection of Si_i. The Sb diffusion coefficient at first decreased but began to increase after 30 min. It was thought that Sb diffuses by a vacancy mechanism. The concentration of Si_i injected decreases with time whereas the corresponding vacancy concentration increases after $\tau_r \approx 30$ min. Such a long recombination time corresponds to a rather high $\Delta F_r = 1.2$ eV according to Eq. (3) (the ratio C_i/C_{ie} is close to three [12]).

The interesting value τ_r at T_0 is obtained by extrapolation from 1100°C using Eq. (3). The exponential decay of τ_r with increasing T is characterized by the total activation energy of self-diffusion (about 5 eV) and the energy component of the barrier ΔE_r (ΔF_r is composed of energy and entropy terms, $\Delta F_r = \Delta E_r - T\Delta S_r$). Even if it is assumed that the barrier is purely entropic [13], i.e., $\Delta E_r = 0$, the decrease of τ_r is very steep. At T_0 we obtain $\tau_r \approx 3$ sec (or 0.3 sec if $\Delta E_r = \Delta F_r = 1.2$ eV). This means that the requirement that τ_r is smaller than l/V is fulfilled.

THERMAL DIFFUSION OF Si_i

Thermal diffusion and ordinary diffusion (proportional to the concentration gradient) occur in a nonisothermal crystal. Defects drift at a rate proportional to the temperature gradient and equal to αG, where α is the thermal diffusion coefficient (if $\alpha > 0$, drift is along the gradient, i.e., from the cold part to the hot). The coefficients of diffusion D and thermal diffusion α can be expressed through kinetic coefficients with two thermodynamic forces represented by the gradients μ/T and $1/T$, where μ is the chemical potential of the defects [14]. The thermodynamics of irreversible processes impose certain limitations on these kinetic coefficients. However, this does not limit the sign and size of α [14]. A few experimental data for different substances conform to the empirical rule [15] that α/D is less than the modulus of $E_m/(kT^2)$, where E_m is the activation energy of diffusion. If this rule were also valid for Si, then the drift rate αG would be small compared with V so that thermal diffusion would not be important. However, the experimental and theoretical estimates presented below for α_i of Si_i are consistent with anomalously fast thermal diffusion in violation of this rule. Crystals of Si grown at a very slow rate [16], i.e., in the interstitial regime, have A-type microdefects only at the end of the ingot at $l_a \approx 6.3$ mm. The microdefect distribution is diagrammed in Fig. 2. This can be explained qualitatively by the consumption of Si_i at the sides becoming significant at small V so that C_i decreases exponentially along the z axis and microdefects do not arise in the crystal bulk. After the growth period (and extracting the crystal from the melt), microdefects appear in the zone where C_i is greater than a certain threshold value C_a. The steady-state concentration field is $C_i(r, z)$, where r is the radial coordinate and approximately satisfies the zero-th limiting condition at the side (at $r = R$) since C_{ie} is small at $z \gtrsim l_a$. If C_i is expanded into Bessel functions, then only the first term of the series is important at $z \gtrsim l_a$ [4]:

$$C_i/C_{if} \approx A_1 J_0(\lambda_1 r/R)\exp(-\lambda'_1 z/R), \tag{4}$$

where C_{if} is the residual Si_i concentration after recombination (Fig. 1b), $A_1 = 1.4$, $\lambda_1 = 2.4$, and

$$\lambda'_1 = -VR/2D_i + \sqrt{(VR/2D_i)^2 + \lambda_1^2}. \tag{5}$$

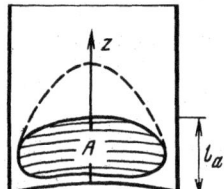

Fig. 2. Distribution of A-defects in a
crystal grown at a slow rate.

The position of the upper limit of the A-defect zone in Fig. 2 is determined by $C_i(r, z) = C_a$. The threshold concentration C_a could be determined [4] from the experimental dependence of the lower limit of the zone on growth rate (in Fig. 2 this limit is near the front but shifts greatly upward with increasing V). It was found that $C_a/C_{if} \approx 0.18$. The upper limit calculated using Eqs. (4) and (5) is shown by the dashed line in Fig. 2. The actual limit is located much below this. It is nearly planar near the ingot axis. Both of these differences can be explained by thermal diffusion of Si_i toward the crystallization front (i.e., $\alpha_i > 0$). In order to account for thermal diffusion, V in Eq. (5) must be replaced by $V - \alpha_i G$, i.e., by the actual transfer rate of defects along z. The observed length of the defect zone l_a corresponds to $\alpha_i/D_i \approx 0.009$ K^{-1}. Since $D_i \approx 5 \cdot 10^{-4}$ cm^2/sec [4], $\alpha_i \approx 4.3 \cdot 10^{-6}$ cm^2/(sec$\cdot$K).

On the other hand, α_i/D_i can be calculated using the Si_i atom as the smallest melt drop size [17]. Since the melt density ρ_m is greater than that of solid Si ρ_c, then a drop of n_m atoms is a defect containing Δn excess Si atoms and $\Delta n = n_m(1 - \rho_c/\rho_m)$. An interstitial Si atom is a defect with $\Delta n = 1$, i.e., it corresponds to a drop of size $n_m = 11$. If a spherical drop of radius r_0 is placed in a uniform temperature field with gradient G, then the temperature field disappears near the drop. This is completely analogous to the electrostatic problem of polarization of a sphere in an electric field [18]. The expression for the temperature distribution along the drop—melt interface can immediately be written. The variable part of this temperature is

$$\delta T = \frac{3\kappa_c}{2\kappa_c + \kappa_m} Gr_0\cos\theta, \qquad (6)$$

where κ_c and κ_m are thermal conductivity coefficients of the crystal and melt and θ is the angle between the radius vector and the gradient. Movement of the interface in the simplest case is limited by its kinetics. Heat evolved by the phase transition can be neglected. The radial rate of the interface in the direction V_r is $\beta\delta\mu$, where β is the kinetic coefficient and $\delta\mu$ is the difference of chemical potentials of the solid and liquid ($\delta\mu = H\delta T/T$, where H is the heat of fusion of Si per atom). According to Eq. (6), V_r is proportional to $\cos\theta$. This corresponds to movement of the drop as a whole along the temperature gradient at rate $\alpha_i G$, where the thermal diffusion coefficient is

$$\alpha_i = \beta r_0 \frac{H}{T} \frac{3\kappa_c}{2\kappa_c + \kappa_m}. \qquad (7)$$

The drop diffusion coefficient D_i can also be expressed in terms of β. For this, we will examine drop movement in an isothermal crystal under the influence of some external uniform field directed along z and providing the correction $-az$ to the liquid chemical potential (a is an arbitrary coefficient). The difference of chemical potentials at the interface $\delta\mu$ is $ar_0 \cos\theta$. The local interface rate $V_r = \beta\delta\mu$ is proportional to $\cos\theta$. The drop moves as a whole along z at $v = \beta r_0 a$. The ratio v/D_i, according to the Einstein equation, is equal to the derivative of $-F(z)/(kT)$ with respect to z, where $F(z)$ is the free energy of the system as a function of the drop position. Since $F(z) = -n_m az$,

$$D_i = \beta r_0 / n_m .$$

(8)

The ratio of coefficients of thermal diffusion and diffusion does not contain the kinetic coefficient β and is expressed only in terms of known constants of Si:

$$\frac{\alpha_i}{D_i} = \frac{H}{kT^2} \frac{3\kappa_c}{2\kappa_c + \kappa_m} n_m .$$

(9)

Setting $n_m = 11$, $\kappa_m/\kappa_c = 2$, and $H = 0.52$ eV, we find that $\alpha_i = 0.017$ K^{-1}. This is approximately twice as large as the experimental value given above. If evolution (absorption) of heat at the dynamic drop—crystal interface is considered in deriving Eqs. (7) and (8), the right parts of these equations must be separated into the identical quantity $1 + \beta r_0 H Q / [(2\kappa_c + \kappa_m)T]$, where Q is the heat of fusion per unit crystal volume. This does not change Eq. (9).

INFLUENCE OF THERMAL DIFFUSION ON TYPE OF DOMINANT IPD

Thermal drift of interstitial atoms at rate $\alpha_i G$ and vacancies at rate $\alpha_v C$ can substantially affect the selection of dominant IPD. The total flux of Si$_i$ from the crystallization front into the crystal along z is written as

$$\mathcal{J}_i = -D_i \partial C_i / \partial z - \alpha_i G C_i + V C_i .$$

(10)

An analogous expression is valid for the vacancy flux J_v. It was proposed [19] simply to replace the actual concentration gradient by the difference $C_i - C_{ie}(T)$ instead of adding the thermal diffusion flux to the diffusion flux [the first term in Eq. (10)] in order to describe defect transfer in a nonisothermal crystal. However, such an expression is erroneous since the flux would remain finite even in the absence of defects, i.e., at $C_i = 0$.

The steady-state profiles $C_i(z)$ and $C_v(z)$ are described by the conservation of matter equation $J_i - J_v = $ const and the recombination—generation equilibrium equation

$$C_i(z)C_v(z) = C_{ie}C_{ve} = C_{i0}C_{v0}\exp(-2z/l).$$

(11)

After changing to the dimensionless length z/l, V and G enter this system of two equations only in the ratio V/G, i.e., the type and concentration of dominant IPD depends only on V/G. In particular, the threshold value of this equation ξ_t at which the type of dominant (residual) IPD changes from interstitial to vacancy corresponds to the coalescence of fluxes J_i and J_v. For this, C_i and C_v should depend identically on z, i.e., according to Eq. (11) $C_i = C_{i0} \exp(-z/l)$ and $C_v = C_{v0} \exp(-z/l)$. Substituting these profiles into $J_i = J_v$, we find

$$\xi_t = \frac{C_{i0}(\epsilon D_i - \alpha_i) - C_{v0}(\epsilon D_v - \alpha_v)}{C_{v0} - C_{i0}} ,$$

(12)

where $\epsilon = (E_i + E_v)/(2kT_0^2) \approx 0.018$ K^{-1}. So that the type of residual IPD actually changes from vacancies, dominant at T_0, to interstitials as V/G decreases, the numerator in Eq. (12) should be positive. Without taking into account thermal diffusion, this is equivalent to the inequality $D_i C_{i0} > D_v C_{v0}$ (denoting the predominant contribution of Si$_i$ to self-diffusion). This condition now becomes optional. The change of IPD type can occur due to the inequality $\alpha_v C_{v0} > \alpha_i C_{i0}$. In particular, if C_{v0} is much greater than C_{i0} and the vacancy thermal diffusion coefficient α_v is sufficiently large, ξ_t is close to α_v. In this case Si$_i$ dominate over vacancies as V/G decreases since the vacancies stop penetrating into the crystal bulk as a result of thermal drift in the opposite direction, to the crystallization front ($\alpha_v G > V$).

Since consideration of thermal diffusion does not require the dominant defect at T_0 to have lower mobility, the model for choosing the dominant IPD becomes symmetric relative to the two types of IPD. It could be thought that Si$_i$ dominate at T_0 but relinquish the dominant role to vacancies at $V/G < \xi_t$ as a result of thermal

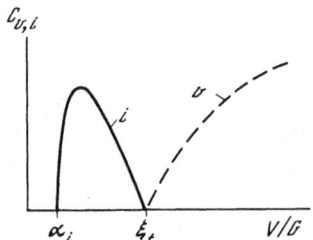

Fig. 3. Dependence on V/G of IPD concentration in a growing crystal due to recombination, diffusion, and thermal diffusion ($\alpha_v < \xi_t$). Solid line, Si_i; dashed line, vacancies.

drift to the front. This alternative assumption is rejected mainly because the A-type growth microdefects observed at $V/G < \xi_t$ are interstitial dislocation loops [20, 21]. Residual IPD at $V/G < \xi_t$ (Si_i in the model examined) diffuse thermally toward the crystallization front. The thermal diffusion coefficient α_i was estimated in the preceding section. This effect has an important practical consequence. At $V/G < \alpha_i$, IPD do not penetrate into the crystal bulk even if they are not consumed at the side of the ingot (i.e., for crystals of any diameter). The total dependence of IPD concentration in the crystal bulk [obtained on the basis of a numerical calculation of the profiles $C_i(z)$ and $C_v(z)$] is shown in Fig. 3 for $\alpha_v < \xi_t$. In principle, α_v can also exceed ξ_t. The vacancy region in Fig. 3 then will be located at $V/G > \alpha_v$ whereas from ξ_t to α_v (as for $V/G > \alpha_i$) IPD do not penetrate into the crystal bulk.

EXPERIMENTAL EVIDENCE FOR A CHANGE OF DOMINANT IPD TYPE

The model for selecting the dominant IPD leads to a general conclusion that does not need to be checked by identifying the aggregation processes involving residual IPD as the crystal is cooled further. The conclusion is that if V/G changes during growth and passes through a threshold value ξ_t, then the crystal separates into two zones. In one of these (where $V/G < \xi_t$) Si_i dominate; in the other (where $V/G > \xi_t$), vacancies. The two types of IPD almost completely mutually annihilate at the interface of these two zones. The simplest example is a crystal grown at a gradually increasing V (or with a gradually decreasing temperature gradient G at the crystallization front that can be achieved using a supplemental heater). In this case, different types of growth microdefects arise in the first half of the ingot (where Si_i dominate) and in the second half (where vacancies dominate). These two defect zones will be divided by a defect-free buffer zone. Just such a three-dimensional picture (shown schematically in Fig. 4a) was observed in crystals grown both without a crucible [2, 8] and by the Czochralski method [3]. So-called A- and B-type swirl defects were seen in the upper (interstitial) zone. Microdefects of the lower (vacancy) zone vary widely for both types of crystals and are denoted as D- and A'-defects, respectively.

The microdefect distributions can become very complicated at a constant growth rate if G is not controlled since G is not constant along the front and changes as the ingot length increases on going from the seed to the cylindrical part. The value G usually reaches an absolute minimum G_m at the crystal axis in the start of the cylindrical part [22]. Therefore, if the growth rate slightly exceeds $\xi_t G_m$, a local vacancy zone should arise near this minimum whereas the remainder of the ingot will have an interstitial zone (Fig. 4b). Further from the seed G no longer depends on the ingot length (for zone melting without a crucible). However, it can change substantially along the front. The gradient in the center of the front (G_c) is usually slightly less than at the side (G_s). If V is intermediate between $\xi_t G_c$ and $\xi_t G_s$, a vacancy zone forms near the ingot axis and an interstitial zone near the side (Fig. 4c). Microdefect distributions of the type in Fig. 4b and c are in fact observed [22] at the corresponding growth rates.

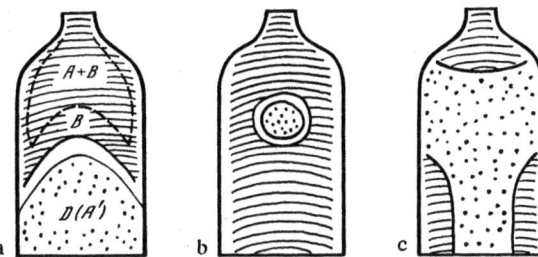

Fig. 4. Microdefect distribution along the longitudinal cross section due to changing V/G. Gradual increase of V along the ingot length (a), different constant V along the length (b, c). The value V in c is greater than in b.

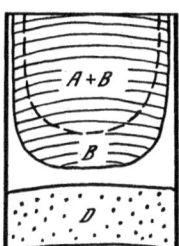

Fig. 5. Microdefects in a crystal grown in the interstitial regime at V/G close to the transitional value ξ_t.

The selection of dominant IPD near the crystallization front is also confirmed by the microdefect distribution in crystals quenched by removing from the melt. If the crystal was growing in the interstitial regime, then a vacancy zone of width z_v exists near the front, as noted above (cf. Fig. 1b). At small V/G, the zone is narrow and can disappear during quenching due to recombination of vacancies with Si_i from the adjacent interstitial zone. However, as V/G approaches the transitional value ξ_t and the final concentration C_{if} decreases, z_v increases greatly. In this case the vacancy zone is retained and the microdefect distribution should have the shape shown in Fig. 5. A quenched zone of D-defects of the shape in Fig. 5 was observed in [23] at V/G close to ξ_t but was absent at small V/G.

AGGREGATION OF POINT DEFECTS DURING INTERSTITIAL GROWTH

The impurity solution becomes supersaturated during cooling if the impurity concentration is high enough (for example, oxygen in crystals grown in crucibles). The actual impurity concentration C exceeds its solubility $C_e(T)$ corresponding to the Si—unstressed impurity equilibrium. As each impurity atom transfers from solution into the unstressed impurity phase, the total free energy of the system decreases by

$$f = kT \ln(C/C_e). \tag{13}$$

However, the inequality $C > C_e$ still does not mean that impurity can precipitate in the crystal. The unstressed impurity phase (silicide) has a certain impurity atom density q and Si atom density q_1, with q_1 much less than the density of Si in the matrix q_0. Therefore, an unstressed impurity phase can be formed from n interstitial impurity atoms only by removing "excess" Si through emission into the interstices or by absorbing the same amount of vacancies from the IPD solution. The number of such excess Si atoms per single aggregated impurity atom will be denoted γ, equal to $(q_0 - q_1)/q$ (for example, γ is close to 0.5 for SiO_2).

Thus, impurity precipitation (aggregation) is closely related to the IPD solution state. For the interstitial growth regime, a certain number n_i of Si_i should be emitted into the supersaturated solution as particles from n impurity atoms. The emission decreases the elastic energy of the particle but increases the free energy of the Si_i solution by $f_i = kT\ln(C_i/C_{ie})$ calculated for each emitted atom. The total free energy of the incoherent precipitate has a minimum at a certain emission coefficient n_i/n [24]:

$$n_i/n = \gamma(1 - \gamma f_i/2w),\tag{14}$$

where w is the elastic energy per single impurity atom for a particle formed without emission.

The free energy increase during particle formation consists of a surface term (due to the interface) and a bulk term $f_{ef}n$. Impurity can precipitate for $f_{ef} > 0$. The optimal emission coefficient from Eq. (14) is given by the effective driving force of precipitation f_{ef}, equal to [24]

$$f_{ef} = f - \gamma f_i + \gamma^2 f_i^2/(4w).\tag{15}$$

The last term in Eq. (15) is usually a small correction. The impurity precipitation condition $f_{ef} > 0$ gives the inequality $f > \gamma f_i$ or

$$C/C_e > (C_i/C_{ie})^\gamma.\tag{16}$$

The temperature dependence of C_e is characterized by the energy of dissolution E; C_{ie}, by the energy of formation $E_i \approx 4.5$ eV. Since E_i is known to be greater than recombination E/γ for known impurities (including oxygen), C/C_e at reduced T increases less steeply than the right part of inequality (16). Therefore, inequality (16) is not fulfilled and precipitation of impurity is thermodynamically unfavorable. The decrease of free energy due to removal of impurity from the supersaturated impurity solution is less than the increase of free energy due to emission of Si_i into the supersaturated solution.

Under these conditions, aggregation is only possible by forming clusters of Si_i. A quantitative theory of Si_i aggregation was examined in [4, 24]. Simple formulas are found for the cluster concentration N_i and number of Si_i in one cluster m_i (these values are interrelated, $m_i N_i = C_i$) assuming that clusters are spherical and their growth is limited by diffusion of Si_i. The concentration N_i is proportional to $|\dot{T}|^{3/2}$, where $|\dot{T}|$ is the cooling rate, equal to the product of growth rate and axial temperature gradient at the cluster nucleation temperature. The quantity $|\dot{T}|$ is about an order of magnitude greater for crystals grown without crucibles than for those grown in them. Thus, N_i in the first case ($\sim 10^6$ cm^{-3}) is much greater than in the second ($\sim 3 \cdot 10^4$ cm^{-3}). These estimates agree well with the observed A-type microdefect concentrations that probably arise through transformation of primary spherical clusters into dislocation loops.

In principle, interstitial impurity atoms (for example, carbon) as well as intrinsic interstitial atoms can participate in cluster formation. This can facilitate cluster formation, i.e., increase the nucleation temperature. However, if the impurity has no significant effect on C_i (due to reactions of Si_i and impurity atoms), the cluster growth rate and consequently m_i and N_i are practically constant.

The growth of interstitial microdefects (clusters and then loops) can greatly decrease C_i. This can satisfy the condition for impurity precipitation [Eq. (16)]. In this case, the crystal is populated by new microdefects, impurity particles. The growing precipitates emit atoms, the concentration C_i of which increases until a balance is attained between the flux absorbed by clusters and loops. Two qualitatively different variations of the process are possible:

(1)   The concentration C_i increases so much that the second generation of interstitial clusters begins to nucleate. Two types of microdefects, impurity particles (Si_i sources) and interstitial clusters (Si sinks), nucleate self-consistently;

(2) The concentration C_i remains rather low and new interstitial clusters are not generated.

AGGREGATION OF POINT DEFECTS DURING VACANCY GROWTH

Impurity particle nucleation is accompanied by absorption of vacancies from the supersaturated vacancy solution and therefore is thermodynamically favorable even if the impurity solution is unsaturated. The vacancy supersaturation is characterized by $f_v = kT\ln(C_v/C_{ve})$. The free energy increase of the IPD solution per single absorbed vacancy is $-f_v$, whereas in the previous case the analogous increase was f_i per single emitted Si_i. Therefore, the optimal ratio of absorbed vacancies n_v to the aggregated impurity atoms n and the corresponding driving force for precipitation f_{ef} are defined by Eqs. (14) and (15) if f_i is replaced by $-f_v$. The thermodynamic precipitation condition $f_{ef} > 0$ instead of Eq. (16) is now approximated by

$$C/C_e > (C_v/C_{ve})^{-\gamma}. \tag{17}$$

A problem specific to the vacancy regime through what mechanism will the vacancies be used up, arises, due to the conditions favorable for impurity precipitation, by a purely vacancy aggregation (as vacancy pores or loops) or by vacancy–impurity coaggregation (as impurity particles). In other words, which of these two competing aggregation processes begins earlier at higher temperature? The temperature of purely vacancy aggregation T_v is close to T_0, 300 K [24] (for a characteristic dimensionless vacancy concentration $C_v/C_{v0} \approx 0.3$). The ratio C_v/C_{ve} in this case is of the order of 0.003. The condition for precipitation [Eq. (17)] at T_v (and thus in a certain range of higher T) is satisfied for an oxygen impurity at $C \gtrsim 10^{16}$ cm^{-3} and for metallic interstitial impurities like iron with comparatively low solubility ($C_e \approx 10^{16}$ cm^{-3} [25]) and high concentration ($C \gtrsim 10^{14}$ cm^{-3}).

The temperature at which this type of aggregation occurs is determined by the work required to form the critical nucleus F_{cr}, which is decreasing with decreasing T and must reach a certain sufficiently small value (of the order of $60kT$ [24]). The macroscopic description of the spherical nucleus for a vacancy pore indicates that F_{cr} is proportional to $\sigma_0^3/(q_0^2 f_v^2)$, where σ_0 is the specific free energy of the crystal–vacuum boundary. For an impurity particle, σ_0 must be replaced by the specific free energy σ of the interface; q_0, by the density of impurity atoms q; and f_v, by f_{ef}. The vacancy–impurity aggregation begins earlier than the purely vacancy type if F_{cr} for the first process at T_v is less, i.e., for

$$1 + f/(\gamma f_v) > (1 - q_1/q_0)^{-1}(\sigma/\sigma_0)^{3/2}. \tag{18}$$

This criterion can be fulfilled even for an unsaturated impurity solution ($f < 0$) if the surface energy σ for the boundary of the two solids (Si and the impurity) is much less than that of Si, σ_0. In particular, $f/(\gamma f_v) \approx 1$ at T_v and criterion (18) is known to be fulfilled for an oxygen impurity in crystals grown in crucibles, i.e., vacancy–oxygen aggregation begins earlier than the purely vacancy type. In this case, the concentration of growth microdefects (SiO_2 particles) is estimated as $N \approx 10^7$ cm^{-3} [24]. This agrees with the observed concentration of A'-defects. The quantity N is expressed in terms of the product of the impurity diffusion coefficient D and its concentration C. For metallic impurities (for example, iron), the product DC is of the same order of magnitude as that for oxygen. Therefore, the A'-defects can be interpreted either as oxygen precipitates or precipitates of other impurities at lower concentration but greater mobility. It is possible that several types of A'-defects are formed. These correspond to various impurities. The predominant type depends on C_v.

The vacancy concentration drops quickly at a practically constant impurity concentration due to absorption of vacancies and impurity by the growing microdefects since $C_v \ll C$. Equilibrium is established between the impurity particles and the solid solution at the limit. This is described by $f_{ef} = 0$, or $f_i = -f_v \approx f/\gamma$. The state of the IPD solution is characterized either by interstitial supersaturation (at $f > 0$) or vacancy supersaturation (at $f < 0$). In the first case, n_i/n according to Eq. (14) is less than γ. This corresponds to complete relief of stresses, i.e., the impurity particles are compressed and create interstitial deformation in the surrounding matrix. The quantity f can be negative at the microdefect formation temperature T_g (for metallic impurities). However, it becomes positive at lower T. The final state of the microdefects in this case depends on the cooling rate and the impurity mobility. The distended particle state that existed at T_g is preserved with sufficiently rapid cooling. Equilibrium is maintained between the particles and the solid solution and the particles become compressed with sufficiently slow cooling. At even lower T, f_i can be so great that interstitial clusters begin to nucleate, i.e., as during interstitial growth, interstitial clusters (Si_i outlets) and impurity particles (Si_i sources) will form and grow in interrelated processes.

NATURE OF *D*-DEFECTS

The *D*-type microdefects observed in Si crystals grown without crucibles at $V/G > \xi_t$ (in the vacancy regime) have a very high concentration (at least 10^{13} cm^{-3}) and a very small size (diameter of the order of 50 Å corresponding to several hundreds of aggregated point defects) [26, 27]. The microdefects create interstitial deformation. However, as V increases near the crystal axis (where V/G is maximal) microdefects with vacancy deformation also appear [27]. This result is consistent with the general conclusion reached in the previous section about the possible types of deformation for vacancy–impurity aggregates. Direct-resolution electron microscopy demonstrated [28] that *D*-defects with interstitial deformation are also divided into two types, ordered and amorphous (the second type dominates at large V/G).

It was noted above that vacancy–oxygen aggregation in crystals grown without crucibles with oxygen concentrations $C \gtrsim 10^{16}$ cm^{-3} can compete with the purely vacancy type. Therefore, it seemed natural to consider at least part of the *D*-defects as small SiO_2 particles [24]. Their concentration is roughly estimated to be of the order of 10^{13} cm^{-3}. However, this possibility should be rejected since DC for oxygen is too small to explain the observed size of the *D*-defects. In fact, the highest possible frequency at which oxygen joins an aggregate of radius r is limited by diffusion to $\omega = 4\pi r DC$. The oxygen diffusion coefficient D decreases rapidly with decreasing T [29], as $\exp[-E_m/(kT)]$, where $E_m = 2.5$ eV. The effective diffusion time determined from the decrease of D by e times is $\tau_d = kT^2/(E_m|T|)$. Setting $|T| \approx VG \approx 3$ K/sec and $r \approx 3$ Å (for the aggregate–nucleus), we find that $\omega \approx 0.3$ sec^{-1} and $\tau_d \approx 20$ sec at the aggregation temperature (near T_v), i.e., the maximal number of oxygen atoms joined to one aggregate is $\omega \tau_d \approx 6$.

Thus, it can be presumed that *D*-defects are formed mainly by a rapidly diffusing interstitial impurity that is highly mobile up to such low T that the impurity solution becomes supersaturated. Precipitation of impurity onto the particles changes the type of deformation from vacancy to interstitial. If several types of similar impurities are present, the vacancy–impurity aggregation temperature T_g for each of these is an increasing function of C_v, i.e., of V/G. The curves $T_g(C_v)$ for different impurities can intersect. Then, as V/G increases the type of impurity dominating in the aggregate will change, i.e., one type of *D*-defect is gradually replaced by another.

REFERENCES

1.   A. J. R. DeKock, "Microdefects in dislocation-free silicon crystals," *Philips Res. Rep.*, Suppl. 1, 1-105 (1973).

2.   N. V. Veselovskaya, É. G. Sheikhet, K. N. Neimark, and É. S. Fal'kevich, "Cluster defects in Si single crystals," in: *Growth and Doping of Semiconducting Crystals and Films*, Part 2 [in Russian], Nauka, Novosibirsk (1977), 284-288.

3.   A. M. Éidenzon and N. I. Puzanov, "Effect of growth rate on the swirl-defects of large dislocation-free Si crystals grown by the Czochralski method," *Kristallografiya*, **30**, No. 5, 992-998 (1985).

4.   V. V. Voronow, "The mechanism of swirl defect formation in silicon," *J. Cryst. Growth*, **59**, No. 3, 625-643 (1982).

5.   J. R. Patel, "Oxygen in silicon," in: *Semiconductor Silicon*, Electrochem. Soc., Princeton (1977), pp. 521-545.

6.   B. O. Kolbesen and A. Muhlbauer, "Carbon in silicon: Properties and impact on devices," *Solid State Electron.*, **25**, No. 8, 759-775 (1982).

7.   A. Mayer, "The quality of starting silicon," *Solid State Technol.*, No. 4, 38-45 (1972).

8. N. V. Veselovskaya, Yu. V. Dankovskii, and D. I. Levinzon, "On spiral-banded contrast revealed by etching in dislocation-free Si," in: Abstracts of the Sixth International Conf. on Crystal Growth, Vol. 4 [in Russian], Izd. Akad. Nauk SSSR, Moscow (1980), pp. 289-291.

9. P. J. Roksnoer and M. M. B. Van den Boom, "Microdefects in a non-striated distribution in floating-zone silicon crystal," *J. Cryst. Growth*, **53**, No. 3, 563-573 (1981).

10. T. Abe, H. Harada, and J. Chikawa, "Swirl defects in float-zoned silicon crystals," *Physica B (Amsterdam)*, **116**, No. 1, 139-147 (1983).

11. T. R. Waite, "Theoretical treatment of the kinetics of diffusion-limited reactions," *Phys. Rev.*, **107**, No. 2, 463-470 (1957).

12. D. A. Antoniadis and I. Moskowitz, "Diffusion of substitutional impurities in silicon at short oxidation times: An insight into point defect kinetics," *J. Appl. Phys.*, **53**, No. 10, 6788-6796 (1982).

13. U. Gösele, W. Frank, and A. Seeger, "An entropy barrier against vacancy–interstitial recombination in silicon," *Solid State Commun.*, **45**, No. 1, 31-33 (1983).

14. S. R. de Groot and P. Mazur, *Nonequilibrium Thermodynamics*, Interscience Pubs., New York (1962).

15. P. Sh'yumon, *Diffusion in Solids* [Russian translation], Metallurgiya, Moscow (1966).

16. P. J. Roksnoer, W. J. Bartels, and C. W. T. Bulle, "Effect of low cooling rates on swirls and striations in dislocation-free silicon crystals," *J. Cryst. Growth*, **35**, No. 2, 245-248 (1976).

17. A. Seeger and K. P. Chik, "Diffusion mechanism and point defects in silicon and germanium," *Phys. Status Solidi*, **29**, No. 2, 455-542 (1968).

18. L. D. Landau and E. M. Lifshits, *Electrodynamics of Continuous Media* [in Russian], Gostekhteoretizdat, Moscow (1957).

19. T. Y. Tan and U. Gösele, "Point defects, diffusion processes, and swirl defect formation in silicon," *Appl. Phys. A*, **37**, No. 1, 1-17 (1985).

20. P. M. Petroff and A. J. DeKock, "Characterization of swirl defects in floating-zone silicon crystals," *J. Cryst. Growth*, **30**, No. 1, 117-124 (1975).

21. H. Föll and B. O. Kolbesen, "Formation and nature of swirl defects in silicon," *Appl. Phys.*, **8**, No. 4, 319-331 (1975).

22. V. V. Voronkov, G. I. Voronkova, N. V. Veselovskaya, et al., "Influence of growth rate and temperature gradient on the type of microdefects in dislocation-free silicon," *Kristallografiya*, **20**, No. 6, 1176-1181 (1984).

23. H. K. Kuiken and P. J. Roksnoer, "Analysis of the temperature distribution in FZ silicon crystals," *J. Cryst. Growth*, **47**, No. 1, 29-42 (1979).

24. V. V. Voronkov and M. G. Mil'vidskii, "Role of oxygen in formation of microdefects during growth of dislocation-free silicon single crystals," *Kristallografiya*, **33**, No. 2, 471-477 (1988).

25. V. M. Glazov and V. S. Zemskov, *Physicochemical Principles of Semiconductor Doping* [in Russian], Nauka, Moscow (1967).

26. A. A. Sitnikova, L. M. Sorokin, I. E. Talanin, et al., "Microscopic study of microdefects in silicon single crystals, grown at high speed," *Phys. Status Solidi A*, **81**, No. 2, 433-438 (1984).

27. A. A. Sitnikova, L. M. Sorokin, I. E. Talanin, et al., "Vacancy type microdefects in dislocation-free silicon single crystals," *Phys. Status Solidi A*, **90**, No. 1, K31-K33 (1985).

28. A. A. Sitnikova, L. M. Sorokin, and É. G. Sheikhet, "Study of *D*-type microdefects in silicon single crystals," *Fiz. Tverd. Tela*, **29**, No. 9, 2623-2628 (1987).

29. J. C. Mikkelsen, "Diffusivity of oxygen in silicon during steam oxidation," *Appl. Phys. Lett.*, **40**, No. 4, 336-337 (1982).

INTERACTION OF CRYSTALS GROWING IN THE MELT

WITH INCLUSIONS AND CONCENTRATION INHOMOGENEITIES

O. P. Fedorov

Conditions under which foreign inclusions are captured or repulsed by a growing crystal have been studied in great detail. Depending on the ratio of crystal, melt, and particle surface energies γ_{sl}, γ_{sp}, γ_{lp}, the particles can be steadily repulsed by the crystal [1, 2]. There exists a certain critical velocity V_c below which repulsion is possible and above which the particle can be captured. The absolute value of V_c depends substantially on the melt viscosity [2, 3], the ratio of heat conductivities of the particle and melt [4], the material and dimensions of the particle [2, 3], and the surface state of the crystal [5]. The thermocapillary force for certain temperature gradients in the melt can determine whether attraction occurs [6]. Gravity [7] as well as the particle size are also important. The Brownian motion of submicron particles ($R < 10^{-4}$ cm) must be considered [8].

Quantitative analysis of the crystal—particle interaction is based on the forces leading to repulsion. Uhlmann et al. [2] suggested that the surface free energy of the interface changes from $\gamma_{sl} + \gamma_{sp}$ (for a thick melt layer) to γ_{sp} (at contact) and the attraction $\gamma_{sp} < \gamma_{sl} + \gamma_{lp}$. An approach was developed in [1, 9, 10] for correctly accounting for the dependence of repulsive forces on the shape of the gap. The forces are determined using the disjoining pressure $\Pi(h)$, where h is the thickness of the melt film.

According to [1, 9], steady repulsion is possible as long as the repulsive force due to a positive disjoining pressure in the melt layer between the particle and crystal compensates the hydrodynamic attractive force. The Gibbs—Thomson effect can prevent collapse of the front and a decrease of $\Pi(h)$ during normal growth. Therefore, the resulting expression for V_c contains the surface free energy γ:

$$V_c = 0{,}1(B_3/(\eta R))(2\gamma/(B_3 R))^{1/3}, \tag{1}$$

where η is the dynamic melt viscosity, B_3 is the disjoining pressure constant, and R is the particle radius.

For layered growth, the anisotropy of the kinetic coefficient $\beta(\varphi)$ stabilizes the front. This increases the rate of protruding parts of the front and R_c acquires the form

$$R_c \sim (\Delta S \Delta T)^{1/3} B_3/(6\eta V(\Delta T)), \tag{2}$$

where ΔS is the entropy of fusion, $V(\Delta T)$ is the growth rate depending on the melt supercooling, and Ω is the volume per single particle.

Table 1. Critical Velocities of Crystals

Substance	$L/(kT_0)$	Growth mechanism	$\eta \cdot 10^2$, Pa · sec	$V_c \cdot 10^6$, m/sec		B_3^{***}, J
				$R=10\ \mu m$	$R=15\ \mu m$	Experiment
Cyclohexanol	0.7	Normal	8.0	<0.02	–	–
Camphene	1.15	Normal	0.25	0.5	–	$1.2 \cdot 10^{-21}$
Succinonitrile	1.35	Normal	0.27	1,2	0.5	$1.2 \cdot 10^{-21}$
Salol	7.0	Layered	1.0	–	1 (110)	$1.0 \cdot 10^{-20}$
					2 (111)	–
Benzophenone	7,2	Layered	0.75	–	5*	$1.0 \cdot 10^{-20}$
					20**	–

*Slow-growing face.
**Most rapidly growing face at a given ΔT.
***Estimated 10^{-21}-10^{-22}

Equations (1) and (2) were derived for isothermal conditions in the experiments examined. Such experiments make it possible to study the interaction of the crystallization front with particles, neglecting extraneous factors due to imposition of the temperature gradient. The different faces of freely growing crystals can also be observed.

Substances with normal and layered crystal growth mechanisms were used (Table 1). One of the goals of the work was to compare the calculated V_c [1, 9, 12] obtained for a normal growth mechanism with the experimental values and to reveal features of the various growth mechanisms exhibited during capture and repulsion of foreign particles and inclusions by the crystals. A second goal of the work was to investigate the influence of captured particles on the morphology of the growing crystal. This phenomenon has not been studied sufficiently, especially for dendritic growth, although indirect data do exist for the break up of the dendrite structure of metals on adding particles to the melt that do not cause nucleation of new crystals [11]. The third task of the present investigation was to study capture of gas and liquid inclusions, in particular, the growth and transformation of crystals near a soluble inclusion (concentration inhomogeneity) existing in the melt before crystallization begins or forming during growth.

Direct observation and cinematography were used to study interaction of the front with particles (inclusions) [12].

NORMAL GROWTH MECHANISM

1. Attraction and Repulsion of Particles

Stable repulsion and attraction processes are plotted in Fig. 1a and b. The region below the open circles represents repulsion of particles at a constant rate to a distance much greater than their diameter (Fig. 2a and b). (The maximal rate is denoted V_c.) The front collapses and particles are repulsed to a small distance and then attracted in the narrow region between the open and filled circles. The width of the unstable repulsion region for which results are nonreproducible is 10% of V_c. This region is apparently due to variable temperature in the system (the temperature change for succinonitrile of 0.01 K corresponds to growth rate variations of 3-5% V_c). All particles of a given size are captured (Fig. 2c and d) at front rates greater than those denoted in Fig. 1 by filled circles. For viscous cyclohexanol (Table 1), only unstable repulsion and attraction were observed at $V \gtrsim 0.02$ μm/sec. This agrees qualitatively with the role of viscosity predicted by Eq. (1). Lines of the calculated functions [1, 9] were plotted so that the value of B_3 in Eq. (1) gave the best fit with experiment. These B_3 for succinonitrile and camphene agree in order of magnitude with $B_3 = 10^{-21}$-10^{-22} J estimated from the heats of sublimation [1, 9]. Refinement of the experimental $V_c = K(R)^n$ by least squares supports the satisfactory agreement of experiment

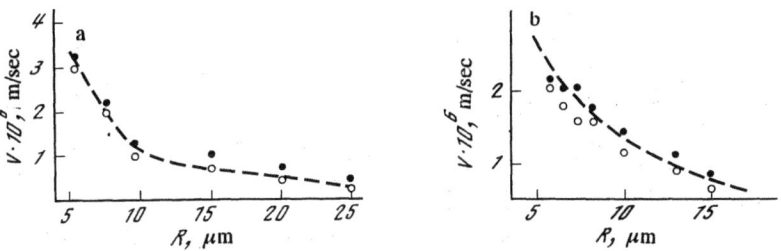

Fig. 1. Dependence of repulsion and attraction rates of nickel particles on particle radius by the crystallization front of succinonitrile (*a*) and camphene (*b*). Attraction (●), repulsion (○). The dashed line was calculated from Eq. (1) for $B_3 = 1.2 \cdot 10^{-21}$ J.

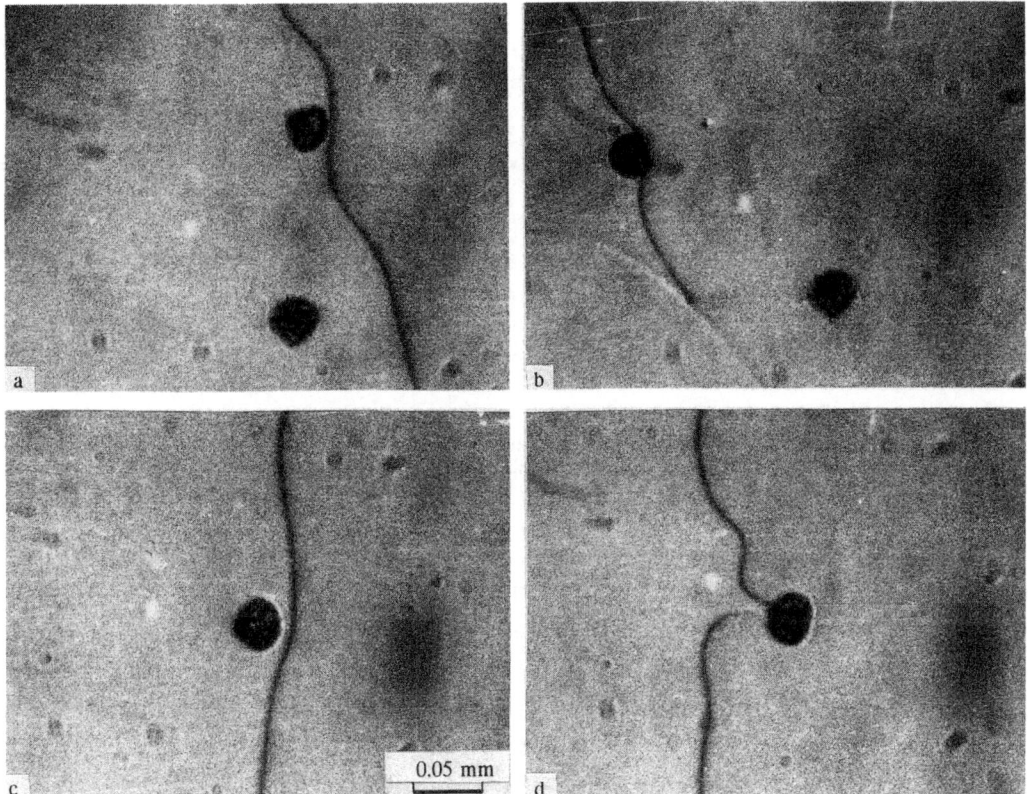

Fig. 2. Repulsion and capture of a lycopodium particle by the crystallization front of succinonitrile. $V = 9.4 \cdot 10^{-8}$ (*a, b*), $6 \cdot 10^{-7}$ m/sec (*c, d*).

with theory. The values *n* obtained this way for succinonitrile and camphene lie in the range $n = 1.4$-1.8, whereas theory gives 1.33.

A small amount of soluble impurity that reduces the melting point by 0.2-0.3 K makes it impossible to observe stable repulsion. This agrees qualitatively with the conclusions of [10]. However, the influence of a soluble impurity on V_c cannot be quantitatively studied.

2. Morphological Consequences of Particle Capture

Morphological changes occur when particles are captured by the crystal. The extent of the changes is determined by the melt supercooling and the relative sizes of the particles and the front structure. A distortion (Fig. 2c and *d*) slowly healing with time remains at the planar front after particle capture at $V > V_c$. Such a scenario is characteristic of a pure substance for which the melting point depression is less than 0.05 K. (The

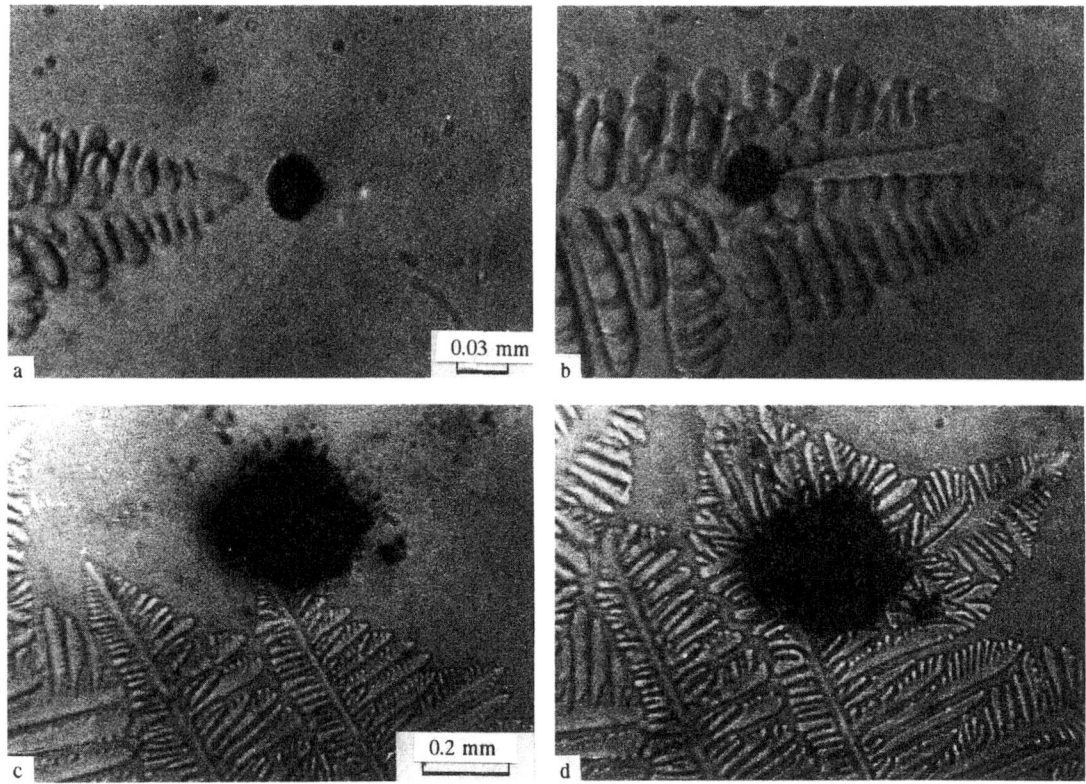

Fig. 3. Stages of succinonitrile dendrite splitting during capture of lycopodium particles (*a, b*) and during growth into a region saturated by Fe_2O_3 particles (*c, d*).

"planar" front in this case means the surface of a large round crystal that has a radius much greater than that of the particle and that grows in a small uncontrolled positive temperature gradient.)

As the supercooling increases, cells and dendrites form with a radius of curvature (R_d) that decreases as ΔT increases. At $R_d \ll R$, the particle blocks part of the front (one or several dendrites) and causes adjacent dendrites to branch again. At $R_d \lesssim R$, the dendrite splits, i.e., another stem is formed with an orientation close to the starting one (Fig. 3*a* and *b*). If the distortion is healed as the planar front stably advances, then the development of the distortion becomes reversible in the unstable growth region and new structures, cells, or dendrites are formed.

A large number of particles in the melt causes multiple splittings. Dendrites are generated in the branched crystals. Their configuration is controlled not by the temperature field but by the geometry of the particle placement. Growth of a dendrite into a region saturated by particles is shown in Fig. 3*c*. The practically isotropic growth in this region results in the emergence of many branches growing in different directions (Fig. 3*d*).

3. Influence of a Soluble Impurity. Concentration Inhomogeneities

Besides the decrease of V_c mentioned above, a soluble impurity greatly affects the morphology of particle capture. The impurity distribution in the cyclohexanol—water system containing lycopodium particles was studied by slow fusion. Impurities repulsed by the front congregate near the particles for a planar or cellular (dendritic) front where $R_d \gg R$. The gaps between dendrites are enriched in impurity where dispersed dendrites grow at $R_d < R$. Impurities do not appreciably congregate near the particles. Thus, foreign particles aid formation of concentration inhomogeneities. This determines the morphology of the front during capture.

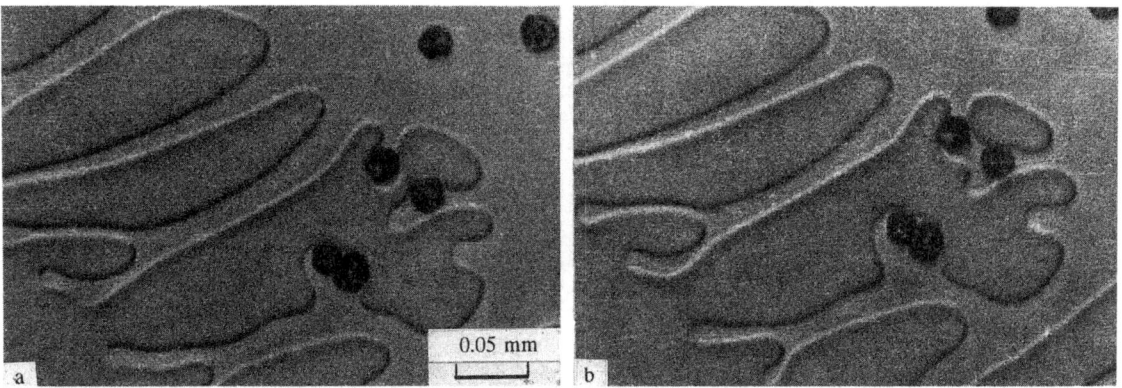

Fig. 4. Interaction of dendrites with lycopodium particles in the cyclohexanol—water system. The time between *a* and *b* is 30 sec.

Under the same conditions as for pure succinonitrile, the distortion at the planar front does not smooth out. It does form a persistent channel from the particle to the front, i.e., the planar front loses stability and cells are formed. The cell constants of the structure in this case are due only to the position of the particles. The same planar front is found in the absence of particles, as is observed in the pure compound containing particles.

Concentration inhomogeneities near the particles also determine their interaction with cells and dendrites. Their splitting is seen more clearly in the two-component melt since the dendrites (cells) are separated by a channel enriched in impurity.

Growth of a dendrite in a region saturated by particles where local concentration gradients are large is shown in Fig. 4. Under such conditions the dendrite branch not only splits into two fragments but also dissolves one while the other continues to grow as a separate crystal. A similar effect was observed for a dendrite growing near a wall [13]. It is important in principle that the splitting of a dendrite branch from the stem, i.e., crystal multiplication, occurs during crystal growth. Separation of dendrite branches from the stem controls the formation of a zone of crystals with the same axis in a metallic ingot. Branch splitting has been studied many times. The experimental data are often contradictory. The direct experimental data obtained in the present work for the succinonitrile—salol and cyclohexanol—water systems demonstrate the decisive role of concentration inhomogeneities in branch splitting.

The growth of crystals and their subsequent transformation under equilibrium conditions between the liquidus and solidus temperatures $T_L < T < T_S$ with a uniform component distribution cause the branches to smooth out and coalesce without splitting from the stem [13]. The substance redistributes between the parts with different curvature. The branches separate only in the preparation with a thickness comparable to that of the growing crystals.

A concentration gradient forms near the barrier during growth in this experiment and in the case shown in Fig. 4. The component can also be unevenly distributed by adding a drop of difficultly soluble camphene to the succinonitrile melt. After the crystals stop growing, the solid partially fuses near the liquid inclusion (under isothermal conditions) (Fig. 5a and b). Not only the shafts of the branches but also the branches themselves dissolve if the growth is dendritic (Fig. 5c). As a result, a finely crystalline structure is formed from a single dendritic crystal. The initial sharp concentration gradient evens out and the secondary dissolution effect decreases and then disappears after several growth and fusion cycles. The volume dissolving after crystallization stops depends on the crystal growth rate (supercooling). The crystals do not dissolve at all below a certain threshold rate. The growth—dissolution effect near the concentration gradient is apparently explained by the diffusionless nature of growth in a certain region of the phase diagram [14, 15]. The role of the concentration inhomogeneity

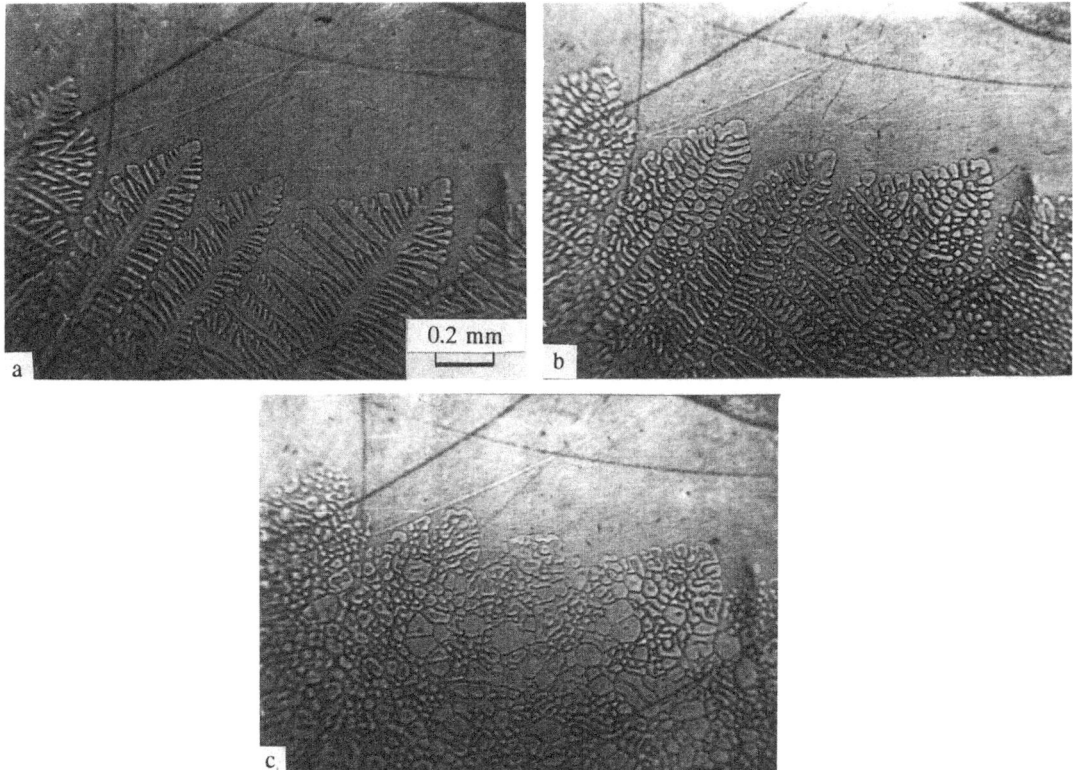

Fig. 5. Transformation of the dendritic structure near a region of increased camphene concentration. $t = 0$ (a), 10 (b), 60 sec (c).

suggests that the crystal from the diffusionless growth region minimizes diffusional growth and contacts the melt of equilibrium composition. The crystallized volume exceeds its equilibrium value. The solid dissolves under isothermal conditions. An analogous effect has been observed previously [16] in a system with another crystal morphology.

It should be noted that temperature fluctuations during growth under nonisothermal conditions can cause a dendritic crystal to fuse and form fragments. The data presented are consistent with the important contribution of concentration inhomogeneities in breaking up the dendritic structure.

4. Gaseous Inclusions

Interaction of growing crystals with air bubbles in the melt has been studied for facetted crystals and spherolites (salol and benzophenone) as well as for rounded cells and dendrites (succinonitrile and cyclohexanol). Capture of gas bubbles by crystals with planar faces has been studied in detail [17, 18]. It is basically analogous to capture of solids (cf. below). An experiment under isothermal conditions near the equilibrium temperature shows that the gas bubbles always stick to both the facetted and the round crystallization fronts and are always captured by the growing front. The bulge of the rounded front near a gas bubble at $T \approx T_0$ is shown in Fig. 6. For facetted crystals, this phenomenon is discussed in [19]. However, it must be pointed out that advance of the front in a melt saturated with gas (in particular, salol) is accompanied by visible repulsion of a bubble due to absorption of dissolved gas, i.e., a simple volume increase. As demonstrated in [17], this effect accompanies capture, i.e., there is no melt layer between the bubble and the face.

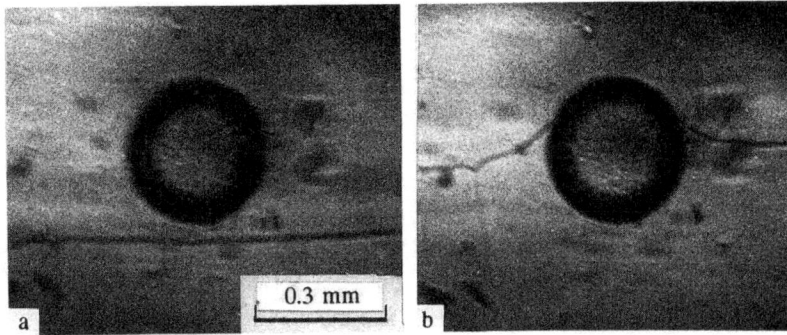

Fig. 6. Stages of interaction (*a*, *b*) of the succinonitrile crystallization front with an air bubble under near equilibrium conditions. Time between *a* and *b*, 60 sec.

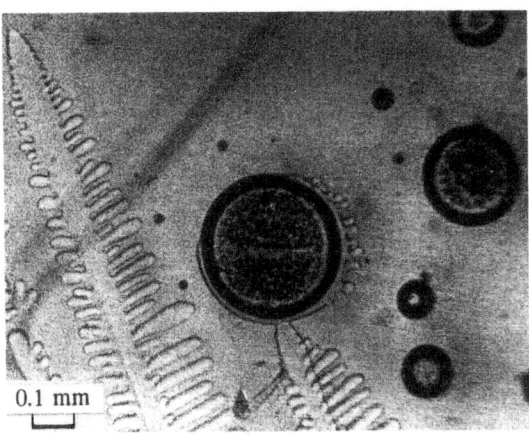

Fig. 7. Succinonitrile dendrites passing an air bubble.

Increasing the melt supercooling, which leads to formation of cells and dendrites for the normal growth mechanism, causes the crystals to grow more rapidly along the melt—gas interface (Fig. 7). As ΔT increases and R_d decreases, the difference between the crystal rates along the boundary with the gas V_g and in the pure melt V increases. The growing dendrite passes the air bubble both in the thin (of the order of the crystal thickness) and in the bulk sample. The rate increase did not depend on the bubble radius and within uncertainty limits the measurements did not change for a planar melt—gas interface. Adding a soluble impurity at small concentrations C did not qualitatively change the capture scenario. Beginning at a certain C, the growing crystals avoid the bubble region enriched in impurity.

Facetted crystals do not exhibit accelerated growth at the melt—gas boundary. They capture bubbles and form macrosteps. However, at large ΔT, where the facetted growth changes to spherolitic, the front clearly passes bubbles (Fig. 8*a* and *b*). The region where this effect occurs corresponds to the descending portion of $V(\Delta T)$ (past the maximal crystal growth rate). Figure 8 shows a thin sample where the gas bubble is compressed by the cuvette walls. If the bubble has its maximum volume and the melt is highly supercooled so that the spherolite front is almost stationary, crystals grow only at the boundary with the bubble. The bubble advances absorbing air dissolved in the melt. When it stops against the walls, a picture analogous to that shown in Fig. 8*b* is seen.

In the present work, the role of thermal effects in accelerated crystal growth was analyzed since it seems natural to explain the passage of the bubble using the thermal field near it [20].

However, an analysis of the experimental data suggests that thermal effects are not involved in this phenomenon. The first and most important argument is that the passage is observed both at small ΔT (dendrites)

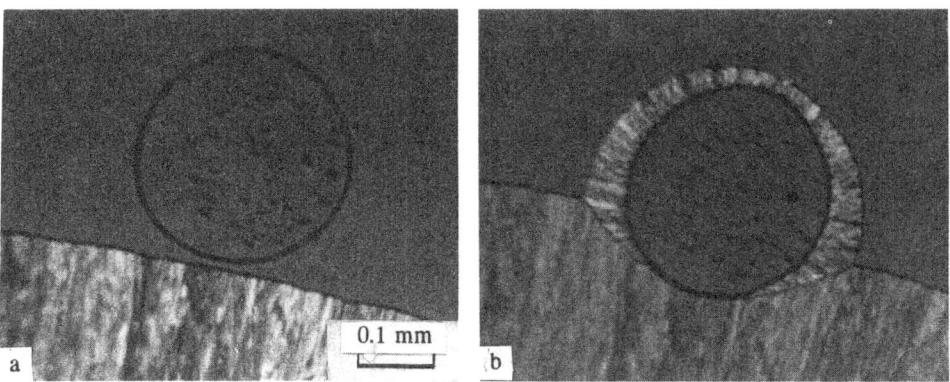

Fig. 8. Stages of accelerated growth (*a*, *b*) of benzophenone spherolites along the melt–gas boundary.

and at large ones after the maximum in the function $V(\Delta T)$. If heat were to accumulate (or dissipate) near the bubble during growth, this would lead to the opposite effects. In one case V would increase; in the other, decrease. Accelerated growth is observed in both experimental cases. Moreover, use of substrates with different thermal conductivities and prolonged isothermal storage did not change the ratio of rates V and V_g. Therefore, the described effect is evidently explained by the ratio of surface energies at the melt–crystal–gas interface.

Since the bubble adheres to the front (Fig. 6), it follows that

$$W - \gamma_{VS} - \gamma_{VL} - \gamma_{LS} < 0, \tag{3}$$

where W is the spreading coefficient. Another driving force exists at the melt–gas interface that is related to favorable energetics for forming a solid at this boundary compared with its formation in the melt. In other words, the inequality (3) denotes incomplete wetting of the crystal by its own melt. This was found experimentally for a number of substances [19]. The existence of an additional driving force can be correlated with a certain additional supercooling at the front

$$\Delta T_W = 2WT_0\kappa/(\rho_S\Delta H), \tag{4}$$

where κ is the contact surface curvature, ΔH is the heat of fusion, and ρ_S is the solid density. By measuring the crystal growth rate and the interface curvature and knowing the kinetic coefficients [21], ΔT_W can be estimated from accelerated growth data. For succinonitrile, $W \approx -3$ mJ/m^2. This agrees in order of magnitude with the W obtained in [19] for organic crystals.

LAYERED CRYSTAL GROWTH

1. Repulsion and Attraction of Solids

The dependence of V_c on R for various salol faces shows that the critical velocities are very anisotropic (Fig. 9). However, V_c is the same for the (111) and (110) faces for particle sizes $R < 150$ μm and $V < 5 \cdot 10^{-7}$ m/sec. The growth rates are slightly different for this range of parameters. Their value can change abruptly. The faces themselves are morphologically identical.

Adding a soluble impurity (10 mass % succinonitrile) does not affect the anisotropy of V_c and does not noticeably change the shape of the function $V_c(R)$ (Fig. 10).

A common feature occurs in the experiments with salol and benzophenone. The rapidly growing faces have even higher critical velocities and the anisotropy of V_c starts at a certain growth rate. This fact is unexpected considering the capture mechanism examined above [Eq. (2)]. As already shown in [12], the result obtained is ex-

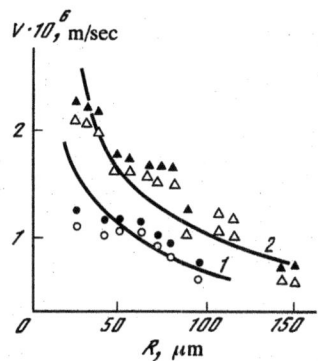

Fig. 9. Repulsion and attraction of nickel particles by salol crystals to the (110) (triangles) and (111) (circles) faces. Attraction (△, ●), repulsion (△, ○). Calculated using Eq. (2) for $B_3 = 10^{-20}$ J (1), for $\cos \varphi = 0.55$ (2).

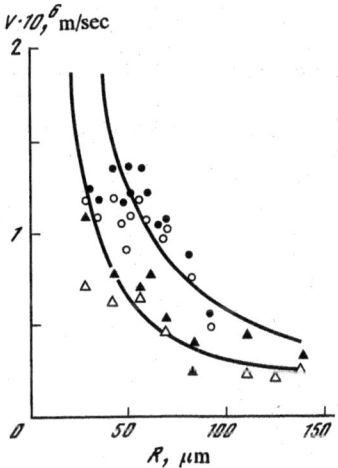

Fig. 10. Attraction (△, ●) and repulsion (△, ○) of nickel particles by the (111) face of salol in pure solution (○, ●) and that containing 10% succinonitrile (△, △).

explained by the fact that repulsion of the particle at $V < V_c$ is caused not by the face itself but by the facetted macrosteps formed under the particle (Figs. 11 and 12). The values R and V given above for salol starting from those at which anisotropy disappears correspond to conditions at which macrosteps are not seen using an optical microscope.

The observed morphology of macrosteps on the (111) and (110) faces of salol crystals is diagrammed in Fig. 12. In both cases the macrosteps are bounded by slow growing faces (111). However, these lie at different angles to the principal face.

The shape of the boundary varies. For the (110) face it is a symmetric valley; for (111), a step. These two morphological features of the boundary under the particle cause the anisotropy of V_c.

The macrosteps formed on the (111) face produce a velocity component in the face plane that does not change the normal repulsion rate. For the (110) face the situation is substantially different, $V_{c_{110}} = V_{c_{111}}/\cos \varphi$ (cf. Fig. 12). For the actual example, $\cos \varphi = 0.55$ and the critical velocities calculated this way satisfactorily describe the experiment. In the general case V_c depends on the amount of macrofaces that are involved in the repulsion. According to [5]

$$V_{c_1} : V_{c_2} : V_{c_3} : \dots : V_{c_n} = 1 : \sqrt{2} : \sqrt{3} : \dots : \sqrt{n},$$

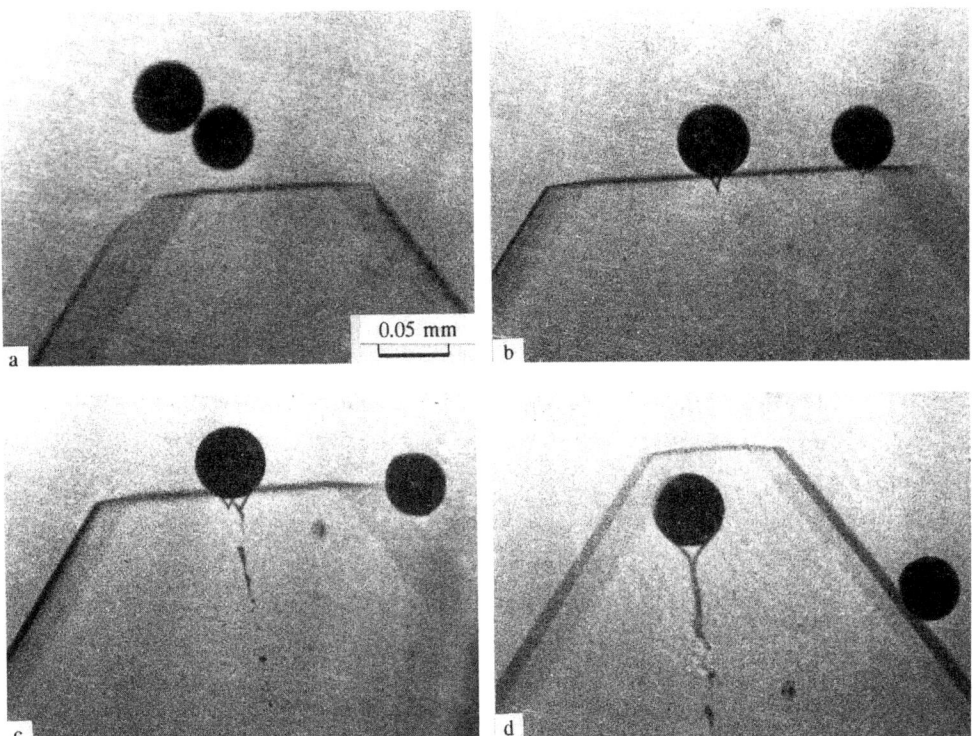

Fig. 11. Repulsion and capture of nickel particles by the (110) face of a salol crystal during gradual melt cooling. $V = 1.5 \cdot 10^{-6}$ (*a*), $1.7 \cdot 10^{-6}$ (*b*), $1.9 \cdot 10^{-6}$ (*c*), $2.1 \cdot 10^{-6}$ m/sec (*d*).

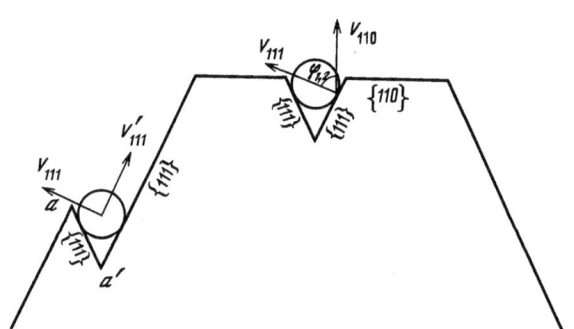

Fig. 12. Diagram of macrostep formation on various salol crystal faces during particle repulsion. The quantity v'_{111} is the component of the particle velocity due to advance of the aa' portion of the {111} face.

where $n = 1, 2, 3, \ldots$ is the number of faces involved in the repulsion.

The value n for a boundary gap is determined by the number of faces (edges) below the particle and the symmetry of their positions relative to the axis perpendicular to the face. The maximal rate normal to the face is seen for a strict symmetry (cf. Fig. 12). This ideal case is possible where the step sources are distributed evenly over the face and the formation of defects during interaction of the face with the particles does not increase the relative rate of any part of the face.

A situation close to that described occurs for the (110) face. Let us note that the gap for this face in a certain range of R and $\Delta T(V)$ and the tangential component of the repulsion rate due to it are asymmetric (cf. Fig. 11). Moreover, the maximal salol crystal growth rates in the range of ΔT studied follow an exponential dependence characteristic of a nucleation growth mechanism [22]. On the other hand, interaction of the particle with the (111)

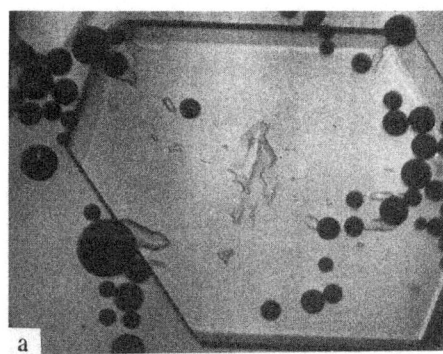 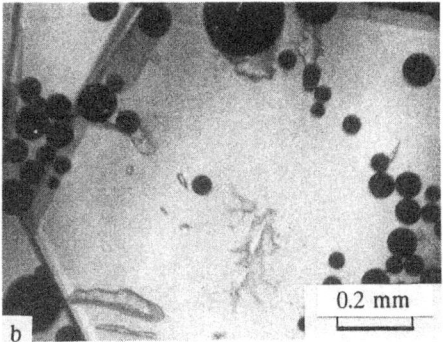

Fig. 13. Formation of inclusions during repulsion of nickel particles by a salol crystal from a melt containing 10 mass % succinonitrile. The interval between *a* and *b* is 100 sec.

face causes a jump in the growth rate of one of the parts of the face. The corresponding function $V(\Delta T)$ is parabolic. This suggests a dislocation growth mechanism. Therefore, steps form in the latter case and not a symmetric gap. Thus, the anisotropy of V_c due to the different morphology of the macrosteps forming the gap under the particle are due to the different growth mechanisms (dislocation or nucleation) of the different salol faces. Anisotropy of V_c is seen at growth rates greater than a certain threshold. It is a kinetic effect, i.e., it cannot be correctly described using equilibrium parameters (for example, surface free energies of the different faces).

The function $R_c(V)$ predicted by Eq. (2) correctly describes the experimental data (Fig. 9). However, in contrast with the normal growth mechanism, B_3 fit by the present experiment differs from that calculated independently (cf. Table 1) by 1-2 orders of magnitude.

Like for the normal growth mechanism, the region near the particle is enriched in impurity. Inclusions are formed in the crystal even during particle repulsion, $V < V_c$ (Fig. 13a and -b). As seen from Figs. 12 and 13, the shape of the inclusions is determined by the morphology of the boundary gap. Inclusions captured by the (110) face are extended perpendicular to it whereas they always make an angle to the (111) face and become parallel to it when captured.

2. Shape of Crystals Capturing Solid Particles

Depending on R, ΔT, and the presence of an impurity, different morphological changes can occur when crystals capture particles. Three characteristic ΔT ranges regarding the shape can be identified for salol crystals, which have been studied in most detail.

1. Lower limit ΔT (0-0.5 K, $V \lesssim 5 \cdot 10^{-7}$ m/sec). Unique macrosteps form as a result of particle capture ($R > R_c$) by the (110) and (111) faces. The profiles of these define the portions of the same faces. Interaction with the particle accelerates the macrostep growth relative to the starting face and crystal splitting (Fig. 14a and b). Moreover, the relative face rates can fluctuate causing the step initially parallel to the starting (111) face to transform into an independent crystal with a well developed (110) face (Fig. 14a and b). In this range of ΔT, the split face does not rejoin;

2. Small $\Delta T = 0.5$-2 K. The face growth rate fluctuates little and the growth rate and V_c are clearly anisotropic. Macrosteps are formed as a result of particle capture. They are different for different faces, as shown in Fig. 12. The step on and parallel to the (111) face quickly rejoins after the capture. For a soluble impurity, inclusions remain, the axes of which are parallel to the face (Fig. 15a and b). Such inclusions form depending on the particle size. An inclusion practically perpendicular to the face forms for a small particle (but with $R > R_c$). Capture of particles by the (111) face causes an inclusion perpendicular to the face to form;

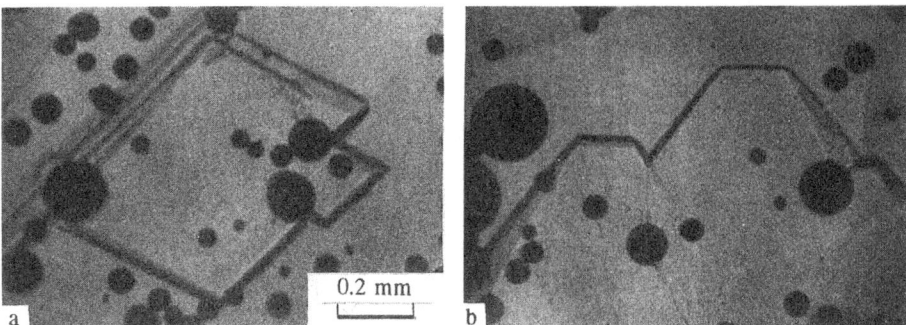

Fig. 14. Splitting of salol crystals capturing nickel particles at $v < 5 \cdot 10^{-7}$ m/sec. (100) face (*a*), (110) face (*b*).

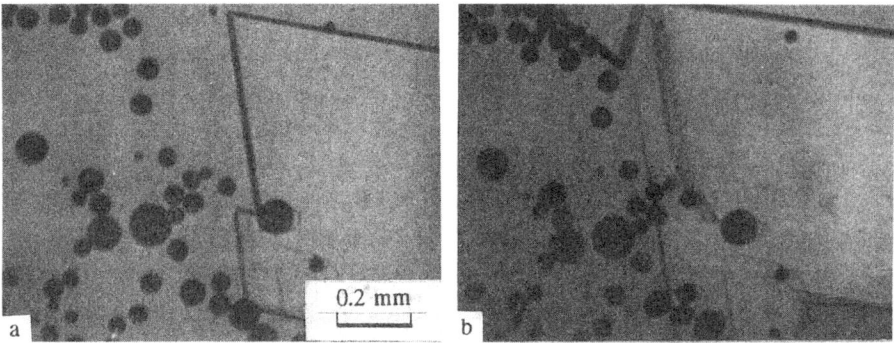

Fig. 15. Stages of capture (*a, b*) of nickel particles by salol crystals growing at $v = 1.3 \cdot 10^{-6}$ m/sec from a melt containing 10% succinonitrile.

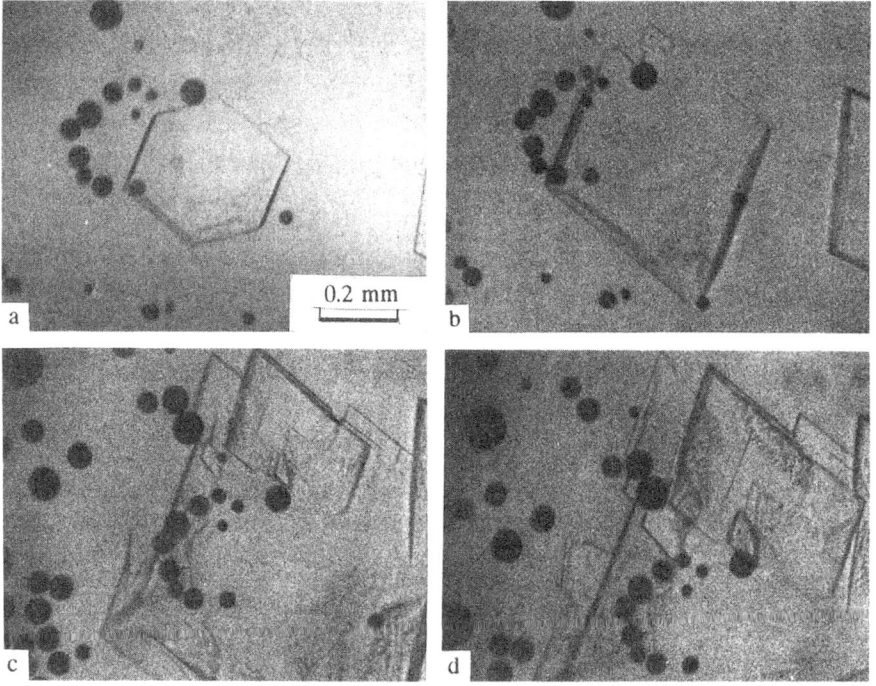

Fig. 16. Salol crystal growth in a melt containing 10% succinonitrile and nickel particles. $v = 1.2 \cdot 10^{-5}$ m/sec. (110) face splitting (*a, b*), formation of numerous macrosteps (*c, d*).

3. large $\Delta T = 2$-9 K. The crystal shape is determined by the (111) faces since the (110) faces are truncated at the start of growth. As the crystal size increases, the initial polyhedral shape loses stability. Numerous macrosteps developing in separate crystals are formed on the (111) faces and the (100) base. These form a jagged polycrystalline front. Particles capable of forming macrosteps increase the transition temperature to polycrystalline growth (Fig. 16). Large particles or their pile-ups can change the direction of macrostep growth (Fig. 16c and d). New crystals do not nucleate in advance of the front in this case. However, they do form from macrosteps.

Direct observation of the interaction of solid particles with freely growing planar-front crystals revealed the influence of macrosteps on morphological and kinetic features of particle capture. They are responsible for the anisotropy of V_c and the different shapes of the inclusions formed during repulsion and capture of particles by the different faces. The nature of the macrosteps formed and the tendency of facetted crystals to split when a particle is captured are directly related to the relative contribution of nucleation and dislocation mechanisms to the growth rate of the particular face. The role of the size of the captured particle is important to macrostep formation in the processes examined. An analysis showed that the macrostep is bounded by faces with a direction more and more removed from that of the starting singular face up to formation of an overhanging step depending on the driving force and, consequently, the inhomogeneity of the supersaturation (during growth from solution). The variety of macrostep forms described above is due to the fact that the collapse of the face and the formation of macrosteps are due to both the driving force and the size of the particle captured.

CONCLUSIONS

1. Experimental data for the critical velocities are satisfactorily described by the Chernov—Temkin theory. A substantial difference in the shape of the function $V_c(R)$ was not seen for normal and layered growth. The previous theory of the decrease of V_c with increasing impurity concentration and melt viscosity is confirmed qualitatively.

2. The particle capture processes and the concentration inhomogeneities during both growth mechanisms substantially affect the crystal shape. For dendrites and cells (normal growth mechanism), splitting, blocked growth, and separation of branches from the stem are characteristic during growth of crystals with concentration inhomogeneity. Introduction of such inhomogeneities to the melt facilitates formation of a finely crystalline structure in a certain range of growth parameters. The growing crystal characteristically splits during layered growth when particles are captured at lower limit and small ΔT.

3. The anisotropy of V_c during layered growth and the formation of inclusions during particle repulsion and capture are determined by the nature of the macrosteps formed. This in turn is related to the face growth mechanism (dislocation and nucleation).

4. Air bubbles under isothermal conditions are captured both by facetted and rounded crystals. Dendrites and spherolites exhibit accelerated growth along the melt—gas interface. This is due to incomplete wetting of the crystal by the melt itself and is not due to thermal effects.

We express sincere thanks to G. P. Chemerinskii for help in preparing and performing the experiments and to D. E. Ovsienko for useful discussions of the results.

REFERENCES

1. A. A. Chernov and D. E. Temkin, "Capture of inclusions in crystal growth," *Curr. Top. Mater. Sci.*, 2 (1976 Cryst. Growth Mater.), 3-77 (1977).

2. D. R. Uhlmann, B. Chalmers, and K. A. Jackson, "Interaction between particles and a solid—liquid interface," *J. Appl. Phys.*, 35, No. 10, 2986-2993 (1964).

3. S. N. Omenyi, A. W. Neuman, W. W. Martin, et al., "Attraction and repulsion of solid particles by solidification front," *J. Appl. Phys.*, 52, No. 2, 789-802 (1981).

4.   A. M. Zubko, V. G. Lobanov, and V. V. Nikonova, "Interaction of foreign particles with a crystallization front," *Kristallografiya*, **18**, No. 2, 385-389 (1973).

5.   G. F. Bolling and J. Cisse, "A theory for the interaction of particles with a solidifying front," *J. Cryst. Growth*, **10**, 56-66 (1971).

6.   A. S. Dzyuba and Yu Y. Zu, "Interaction of gas bubbles with the crystallization front of a melt," *Kristallografiya*, **30**, No. 6, 1177-1180 (1985).

7.   S. G. Grigoryan, A. S. Oganesyan, and A. G. Sarkisyan, "On the mechanism of macroparticle capture during crystal growth from a melt," *Kristallografiya*, **28**, No. 4, 782-785 (1983).

8.   G. G. Devyatykh, G. M. Vorotyntsev, and V. M. Malyshev, "On the transition of submicron particles from a melt into a crystal," *Dokl. Akad. Nauk SSSR*, **278**, 396-399 (1984).

9.   A. A. Chernov, D. E. Temkin, and A. M. Mel'nikova, "Theory of entrapment of solid inclusions during crystal growth from a melt," *Kristallografiya*, **21**, No. 4, 652-660 (1976).

10.   D. E. Temkin, A. A. Chernov, and A. M. Mel'nikova, "Capture of foreign particles by a crystal growing from a melt with impurities," *Kristallografiya*, **22**, No. 1, 27-35 (1977).

11.   Yu. Z. Babaskin, D. E. Ovsienko, A. A. Rostovskaya, and G. A. Alfintsev, "Modification of steel by refractory particles and its effect on the structure and properties of castings," *Litein. Pr'vo*, No. 2, 17-19 (1975).

12.   D. E. Ovsienko, D. E. Temkin, O. P. Fedorov, and G. P. Chemerinskii, "Interaction of solid particles with crystals growing from a melt," *Kristallografiya*, **32**, No. 5, 1246-1252 (1987).

13.   O. P. Fedorov, D. E. Ovsienko, and M. B. Krivoshei, "On the mechanism of transformation of dendrite branches of a two-component melt under isothermal conditions," *Metallofizika*, **9**, No. 2, 68-75 (1987).

14.   O. P. Fedorov and D. E. Ovsienko, "Mechanism of dendritic side-branches coarsening under isothermal conditions," *Cryst. Res. Technol.*, **23**, No. 4, 489-497 (1988).

15.   D. S. Kamenetskaya, "Study of crystallization of supercooled two-component liquids," Doctoral Dissertation in Physical-Mathematical Sciences, Moscow (1966).

16.   Ya. V. Grechnyi, "On nonselective nucleation and growth of crystals of a solid solution," in: *Crystallization and Phase Transitions* [in Russian], Izd. Akad. Nauk BSSR, Minsk (1962), pp. 156-168.

17.   Ya. E. Geguzin and A. S. Dzyuba, "Separation of gas, formation and capture of gas bubbles at the crystallization front from a melt," *Kristallografiya*, **22**, No. 2, 348-353 (1977).

18.   A. S. Dzyuba, "Features of formation of gas inclusions during crystal growth from a melt," *Kristallografiya*, **27**, No. 3, 551-556 (1982).

19.   Yu. V. Naidich, V. M. Perevertailo, and N. F. Grigorenko, *Capillary Effects in Growth Processes and Crystal Fusion* [in Russian], Naukova Dumka, Kiev (1983).

20.   D. E. Ovsienko, G. A. Alfintsev, and G. P. Chemerinskii, "Capture of air bubbles by benzophenone crystals," *Cryst. Res. Technol.*, **17**, K9-K11 (1982).

21.   G. A. Alfintsev and O. P. Fedorov, "Kinetic growth of crystals with low entropy of fusion," *Metallofizika*, **3**, No. 4, 115-118 (1981).

22.   D. E. Ovsienko and G. A. Alfintsev, "On the mechanism and shapes of crystal growth from a melt," in: *Metals, Electrons, and the Lattice* [in Russian], Naukova Dumka, Kiev (1977), pp. 144-167.

23.   A. A. Chernov and S. I. Budurov, "Types of growth of macroscopic steps in crystallization-development of faces at the end surfaces of steps," *Kristallografiya*, **9**, No. 3, 388-395 (1964).

ROLE OF GROWTH DISLOCATIONS IN FORMING INHOMOGENEOUS PROPERTIES IN GALLIUM ARSENIDE SINGLE CRYSTALS

A. V. Markov, M. G. Mil'vidskii, and V. G. Osvenskii

Fabrication of very large ultrafast integrated circuits has sharpened the need for highly homogeneous semiconductors. The great promise of microelectronics is tied to semi-insulating GaAs. However, it has been found that single crystals grown from the melt have inhomogeneous bulk properties. This inhomogeneity correlates on the macro- and microscale with the dislocation distribution. The problem of preparing large dislocation-free GaAs single crystals has not yet been solved even under laboratory conditions. Mass production of single crystals with a relatively high dislocation density (10^4-10^5 cm^{-2}) will require methods for decreasing the inhomogeneity of properties. For this, the reasons and mechanisms of formation of the inhomogeneities due to growth dislocations will have to be understood. In the present work, experimental data are used to analyze critically previously proposed mechanisms of formation of the inhomogeneity. A model is presented for formation of micro- and macroinhomogeneous crystal properties. The model is based on the assumption that intrinsic point defects (IPD) recombine quickly near dislocations.

INHOMOGENEOUS SINGLE CRYSTAL PROPERTIES DUE TO DISLOCATIONS
AND VARIOUS MODELS OF THEIR FORMATION

The first experiments, published in 1982, demonstrated that the surface intensity distribution of various photoluminescence bands [1] and leakage currents and resistivity [2-4] on wafers cut from single crystals of undoped GaAs or those lightly doped with Cr correlate with the dislocation density distribution N_d. Apparently the work of Nanishi et al. [5] had a decisive influence on future studies of such a correlation. They demonstrated that N_d is correlated with the parameters of field effect transistors prepared by direct ion implantation of a donor in wafers of semi-insulating GaAs. Later works, approaching 100 reports, not only confirmed the conclusions made in [5] but also demonstrated that the distribution of the majority of measured physical properties of semi-insulating GaAs crystals correlate with N_d. Examples of such a correlation for the usual parameters characterizing the functional readiness of the semi-insulating material (specific resistivity, carrier concentration and mobility at room temperature) and for the photoelectrochemical current are plotted in Fig. 1.

All models explaining the correlation of the physical properties of semi-insulating GaAs crystals with the distribution of growth dislocations (i.e., formed during growth) can be divided into four groups based on the following physical criteria:

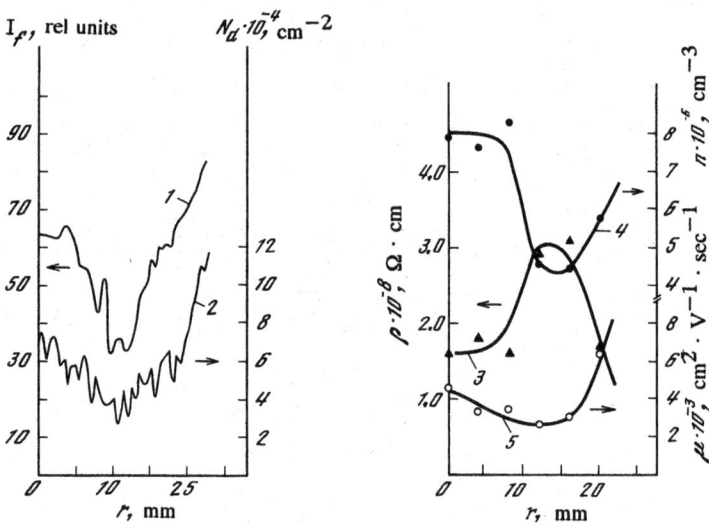

Fig. 1. Radial distributions of photoelectrochemical current (1), dislocation density (2), specific resistivity (3), electron concentration (4), and electron mobility (5) in the ⟨110⟩ direction in an undoped GaAs film.

1) the presence of electrically active centers at the center of the dislocation [4];

2) generation of IPD during creep [6] or slippage [7] of dislocations;

3) inhomogeneous thermal stresses in the ingot bulk [6];

4) redistribution of impurity and/or IPD due to atmosphere formation at dislocations [3, 8].

Obviously the model correctly describing the observed effect should be chosen on the basis of all of the experimental material. In this respect, we think that the dependence of the effect of growth dislocations on the crystal properties starting at the dislocation formation temperature during growth [9] and the detection of extended regions with variable properties around the individual dislocations or dislocation pile-ups [10, 11] are most important.

During growth of undoped GaAs single crystals by the Czochralski method, dislocations can be formed over a wide temperature range from the crystallization temperature to ~700°C. A method was proposed in [12] enabling growth dislocations to be categorized according to their formation temperatures and the morphology of dislocations revealed by selective etching using Abrahams—Buiocchi (AB) etchant [13]. In most cases "high-temperature" dislocations formed above 1100°C dominate in the crystals grown. However, crystals can be grown under the corresponding thermal conditions in which relatively "low-temperature" dislocations with the same or a different distribution dominate. Thus, a crystal with dense central dislocation pile-ups with a formation temperature of the order of 800°C and high-temperature dislocations was studied in [9]. The photoelectrochemical current in this crystal correlated with the distribution only of the high-temperature dislocations and not with the total N_d. This suggested that low-temperature growth dislocations (generally formed below 1000°C) are not important in forming macroscopically inhomogeneous properties in semi-insulating crystals.

Extensive regions with variable properties around dislocations in semi-insulating GaAs crystals were observed by microcathodoluminescence [10] and selective chemical etching [11] (Fig. 2a). It was found that such regions are formed during growth only around high-temperature dislocations [11]. The sizes of these regions measured by selective etching, luminescence methods, or IR absorption are usually 20-50 μm in crystals with relatively high growth dislocation densities (10^4-10^5 cm^{-2}). The sizes increase to hundreds of micrometers in near-

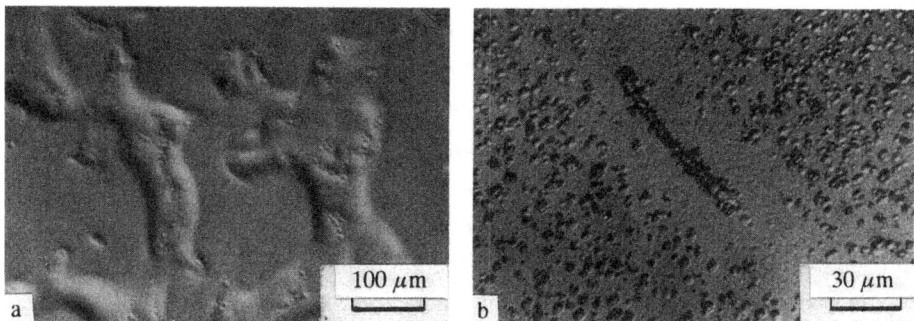

Fig. 2. Regions with variable properties around high-temperature growth dislocations visualized with AB etchant in undoped GaAs crystals (*a*) and those heavily doped with Te (*b*).

ly dislocation-free ($<10^2$ cm^{-2}) crystals doped with In [14, 15]. The sizes of the regions around dislocation pile-ups (walls of dislocation cells, small-angle boundaries, etc.) are several times greater than those observed around separate dislocations.

Regions free of microdefects with the same extent as the regions found in semi-insulating crystals [16] (Fig. 2*b*) are formed around high-temperature growth dislocations in GaAs crystals heavily doped with various electrically active impurities. These regions are revealed well by other methods, for example, microcathodoluminescence [17] or induced current [18]. Correlation of the formation temperatures and sizes of the regions in crystals with high and low impurity content suggests that the inhomogeneities found have a common nature and origin. It should be noted that the extended regions around the dislocations are formed regardless of the type of dislocation. They are practically symmetric relative to the dislocation lines.

The most important feature of these regions is their complicated nature. It can be seen in Fig. 2*b* that a smaller region in which the microdefect density is greater than in the matrix far from the dislocation exists near the dislocation along with a region free of microdefects. This is confirmed by transmission electron microscopy [18]. The cathodo- and photoluminescence intensities were recorded in concentric regions around the dislocations in semi-insulating single crystals [10, 14, 19]. It was found in [20] that the concentration of EL2 centers in the wall of the dislocation cell that determine the semi-insulating properties of undoped GaAs is greater whereas that in the intermediate region adjoining the wall is less than in the cell center, which is free of dislocations. The examples given do not nearly cover the existing experimental data and indicate that the regions with variable properties around the high-temperature growth dislocations consist of two concentric areas (at least several micrometers) in which some crystal properties change in the opposite fashion relative to those at a greater distance from the dislocation.

The experimental data noted above provide a basis for analyzing the models proposed for explaining the correlation between the various crystal properties and N_d.

Obviously the electrically active centers in the dislocation nucleus cannot be the ultimate reason for the correlation since the bulk concentration of these centers at the growth dislocation densities at which the correlation is clearly seen ($N_d = 10^4$-10^5 cm^{-2}) cannot exceed 10^{12}-10^{13} cm^{-3}. This is 2-3 orders of magnitude less than the concentrations of centers controlling the properties of semi-insulating crystals. The presence of charged centers in the dislocation nucleus presumes that charged regions with electrical properties different than those of the matrix are formed around the dislocation lines. However, the sizes of these charged regions are determined by the concentration of free charge carriers. Therefore, they should differ in order of magnitude in semi-insulating crystals and those heavily doped with electrically active impurities. As noted above, the sizes of the regions with variable properties in undoped crystals or heavily doped ones are practically the same. Thus, a common explanation for the macro- and microinhomogeneous effects due to dislocations in crystals with various impurity content cannot be given on the basis of the electrically active centers in the dislocation nucleus.

Models based on generation of electrically active centers (more accurately IPD with which they are associated) during dislocation advance are most prevalent in the literature. In our opinion, this is due to the following experimental observations. Firstly, regions in semi-insulating crystals with higher N_d and those around individual dislocations are enriched with EL2 deep centers [6, 15, 20]. Secondly, it was demonstrated in [7] that plastic deformation of GaAs single crystals at relative low temperatures generates EL2 centers. Thus, models were proposed to explain the enrichment with EL2 centers by generation of antistructural As_{Ga} defects (with which EL2 centers are as a rule associated) during creep [6] or slippage [7] of dislocations.

A correlation of the crystal properties with the distribution of only the high-temperature dislocations indicates that advance of dislocations is not the reason for the effects observed at moderate deformation. As demonstrated in [7], the concentration of EL2 centers generated is substantial (10^{15}-10^{16} cm^{-3}) only at several percent deformation. At 400-500°C, such deformations are usually due to dislocation densities of at least 10^8 cm^{-2}. This is much greater than the growth dislocation densities observed by us. The symmetry of the regions with variable properties (including regions with increased EL2 concentrations) relative to the dislocation lines indicates that these regions form after the dislocations stop. Thus, although the advance of dislocations can also change the point defects of the crystal (cf. for example, [21]), the effects observed are not related to dislocation migration.

A model connecting the observed correlation with an inhomogeneous distribution of thermal stresses in the crystal bulk does not explain the fact that macro- and microcorrelations exist only for high-temperature growth dislocations. A total of 1% of stresses of the order of 10^7 N/m^2 are needed to change the equilibrium concentration of IPD at temperatures near the melting point. This makes it impossible to generate a significantly inhomogeneous distribution of defects if the stresses drop to less than 10^6 N/m^2, which are characteristic for actual conditions under which GaAs single crystals are grown by the Czochralski method.

The observed dependence of inhomogeneity on dislocation formation temperature and microregion symmetry relative to the dislocation lines correlates with the diffusive redistribution of point defects around a dislocation during cooling after crystallization. However, it is easy to demonstrate that formation of an array of point defects around a dislocation cannot generate concentric regions of such a large diameter due to the small sizes and the correspondingly small capacity of the atmosphere [16]. Thus, the "array" mechanism as examined earlier is not consistent with all of the experimental data. Firstly, it does not explain the sizes and the complicated structure of the regions with variable properties around the dislocations.

The observed effects were explained in [16] by invoking the assumption that the growth dislocations act as outlets during cooling for recombination of excess IPD after crystallization. The point defects do not accumulate at the dislocation. As a result, the size of the depleted region is not limited by the volume. We constructed a model on the basis of this hypothesis for the local redistribution of the crystal around a dislocation. This enabled the existing experimental data to be explained both for undoped crystals and those heavily doped with various impurities.

FORMATION OF COMPOSITION MICROINHOMOGENEITIES

The simplest redistribution mechanism can be illustrated for the case where the crystal composition is defined by the concentrations of only two IPD of one sublattice, i.e., the deviation from a stoichiometric composition is written as $\delta = c_i - c_v$, where c_i and c_v are the concentrations of interstitial atoms and vacancies in the anionic (or cationic) sublattice. Accelerated recombination of excess interstitial atoms and vacancies at a dislocation generates concentration gradients and diffusion of these defects to the dislocation. The concentration distribution profiles are determined by the diffusive mobility of the defects. The ratio of defect concentrations characteris-

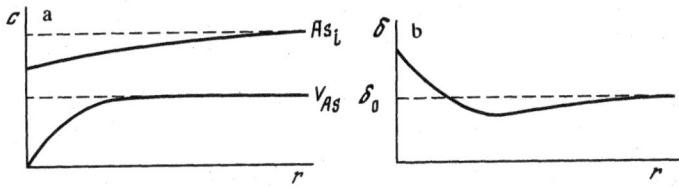

Fig. 3. Intrinsic point defect concentration distribution (*a*) and deviation of composition from stoichiometric (*b*) near a dislocation as a result of the outflow and recombination of point defects.

tic of the matrix at a large distance from the dislocation is disrupted if these mobilities are unequal near the dislocation. The composition redistribution for the case where the diffusive mobility of interstitial atoms is greater than the vacancy mobility ($D_i > D_v$) is diagrammed in Fig. 3. The region adjoining the dislocation is enriched with anions (or cations). The region surrounding it is depleted of this component compared with the starting composition δ_0. We emphasize that the composition redistribution is determined only by the ratio of defect mobilities and does not depend on the value and sign of δ_0. Thus, diffusion (i.e., a temperature-dependent process) causes two concentration regions to form around the dislocation. The composition of these (and, therefore, the properties depending on composition) changes in the opposite direction relative to the matrix composition. This agrees with experimental observations. Now the conditions under which the proposed model is valid must be determined. These conditions must be correlated with the actual situation in the crystal.

The concentration of the dominant IPD in GaAs crystals at temperatures near the melting point is $\sim 10^{19}$ cm^{-3} [22]. Thus, the average distance between a vacancy and an interstitial atom $r_{i\text{-}v}$ is less than 30 Å. If interaction of defects is neglected, then the recombination time of excess defects can be determined from $\tau \approx r_{i\text{-}v}^2/\bar{D}$, where $\bar{D}$ is the diffusion mobility of one of the recombining defects (if it is significantly greater than the mobility of the other). Evidently recombination in the bulk at such $r_{i\text{-}v}$ and $\bar{D}$ will occur practically instantaneously. Thus, a necessary condition for the proposed redistribution mechanism is the existence of a barrier to recombination of vacancies and interstitial atoms. The barrier height at rapid diffusion can be written $\Delta U = kT\ln(\nu\tau)$, where ν is the frequency of atomic jumps ($\approx 10^{13}$ sec^{-1}), T is the temperature, and k is Boltzmann's constant. An estimate shows that a potential barrier of about 5 eV is necessary for prolonged "freezing" of excess IPD (recombination time $\tau \approx 10^3$ sec).

The solution of the diffusion equation for a linear outflow can be approximated at $r < \sqrt{Dt}$ by

$$c = c_0 \, \frac{\lg(r/a)}{\lg(\sqrt{Dt}/a)} \, , \qquad (1)$$

where r is the distance from the dislocation, t is the time, c_0 is the matrix defect concentration, and a is the effective radius taken equal to the dislocation nucleus radius ($\sim 10^{-7}$ cm). We note that c depends very little on a, t, and D. Neglecting the interaction between dislocations and IPD, Eq. (1) yields an expression for the defect recombination rate at dislocations per unit volume Q (cm$^{-3} \cdot$sec^{-1})

$$Q = \frac{2\pi D' c_0' N_d}{\lg(\sqrt{Dt}/a)} \, , \qquad (2)$$

where c_0' and D' are the starting concentration and diffusion coefficient of the slowly diffusing defect, which determines the recombination rate by its rate of approach to the dislocation. The recombination time of the excess defects at dislocations is

$$\tau \simeq c_0'/Q \simeq (D' N_d)^{-1} . \qquad (3)$$

It is clear that recombination of defects at characteristic $N_d \approx 10^5$ cm^{-2} takes a long time ($\tau > 10^2$ sec) and is possible only for moderate diffusion coefficients ($D' < 10^{-7}$ cm^2/sec) or if a sufficiently high potential barrier to recombination exists.

Thus, the hypothesis of accelerated recombination of excess IPD at dislocations is consistent with the existence of a very significant ($\gtrsim 5$ eV) potential barrier for bulk defect recombination. The height decreases near the dislocation nucleus. A decrease of barrier height with decreasing temperature (at $T \lesssim 1100°C$) is also necessary. This accelerates bulk recombination and causes the inhomogeneous composition due to redistribution of excess defects to control the ratio of equilibrium defect concentrations at the given temperature. The composition tends to even out. If it does so during crystal growth, the defect diffusion coefficients in this temperature range will be small. The nature of the proposed potential barrier to recombination is difficult to determine at present. It may be due to the fact that the energetically favorable states of the IPD at high temperatures are split configurations. The existence of defects in such states hinders their recombination due to the so-called "entropic" barrier [24].

Incorporation of IPD into the nucleus, causing either splitting or creep of the dislocation, represents an alternate recombination at the dislocation. However, the fact that the concentration changes of IPD in regions around the dislocations should be large ($\sim 10^{18}$ cm^{-3}) relative to the matrix argues against this mechanism. In fact, the experimentally observed dislocation splittings in GaAs crystals of several tens of angstroms [25] are entirely insufficient to relax such large excess concentrations on the scale of the observed regions. Relaxation by creep also should have caused clearly noticeable effects such as a shift of the edge dislocations to a distance of hundreds of micrometers with the appearance of distinct asymmetry in the placement of regions with variable properties relative to dislocation lines and helicoidal twisting of screw dislocations. As was noted, regions with variable properties are symmetric relative to dislocation lines. Helicoidal twisting of screw dislocations is observed only in heavily doped GaAs crystals [26]. The twist parameters (pitch and radius) are such that the necessary excess IPD concentration cannot for certain be relaxed.

Starting with the sizes of the regions around the growth dislocations, the diffusion coefficient for the most mobile of the redistributing defects can be estimated. The average diffusion coefficient for the range from $\sim 1100°C$ to the melting point should be at least 10^{-4}-10^{-5} cm^2/sec. The low accuracy of the estimate is due to the weak dependence of c on D in Eq. (1). Microdefects form during growth in crystals heavily doped with Group VI impurities. A free zone near the surface of the ingot usually forms in their distribution. It extends for several millimeters. Considering the ingot surface as a planar boundary of the outflow, we find that the free zones observed can be formed at defect diffusion coefficients greater than 10^{-5} cm^2/sec. Finally, the diffusion coefficient estimated from [27] for interstitial atoms in Si at the melting point gave $\sim 5 \cdot 10^{-4}$ cm^2/sec. Thus, the composition redistribution model requires very high but apparently completely feasible diffusion coefficients for the IPD or at least one of them.

The concentration distribution profile determined from Eq. (1) at $r > 1$ μm (i.e., in the experimentally observed region) is very smooth. The concentration change is less than several tens of percent. However, since the starting defect concentrations are very high ($c_0 \approx 10^{19}$ cm^{-3}), the variations of δ are large enough to be comparable with the width of the GaAs homogeneous region [22].

The question of whether the recombination of such IPD causes the composition redistribution is important in principle. The regions with variable properties in very different crystals consist of two concentric regions. Evidently such a δ distribution profile in the general case can only arise with an outflow of defects that form one recombination pair i_{Ga}–v_{Ga} or i_{As}–v_{As}, i.e., disorder in one of the sublattices should dominate in the crystal. Further conclusions about the redistribution of δ can only be made by comparison with experimental data.

It was demonstrated in [20] that the EL2 concentration in semi-insulating crystals in the region adjoining dislocations is greater whereas that in the region surrounding it is lower than in the matrix at a distance from the dislocations. The concentration of EL2 centers is also known to increase with increasing As content in the crystal [28]. The distribution of microdefects around a dislocation in a crystal heavily doped with Te is shown in Fig. 2b. A region of increased density (10^{11}-10^{12} cm^{-3}) near the dislocation relative to the density in the matrix (10^9-10^{10} cm^{-3}) and a broad region free of microdefects are characteristic for microdefects of this type. Such microdefects disappear if GaAs(Te) single crystals are grown from nonstoichiometric melts enriched with Ga. The experimental facts presented agree well with the crystal composition redistribution model if it is assumed that the internal region adjoining the dislocation is enriched with As and the region surrounding it, with Ga, relative to the matrix. Other known results also agree with this type of composition redistribution. These results include, for example, the cathodoluminescence spectral distribution [29] and the fact that As inclusions form at dislocations in undoped and heavily doped crystals [18, 30], including in those grown from melts enriched with Ga [31].

The model can be made to agree with experimental data for one defect pair (i_{Ga}–v_{Ga}) or another (i_{As}–v_{As}) using the relative diffusion coefficients $D_v > D_i$ and $D_i > D_v$, respectively. In our opinion, the defect pairs in the anionic sublattice are preferred since this agrees with the conclusion about the decisive role of these defect pairs in forming solid solutions in GaAs that was made earlier by analyzing precise measurements of the lattice constant and the density [32].

Thus, not one of the arguments on which the local composition redistribution model is based contradict the existing experimental data. A model with single positions explains the different local effects observed in both undoped and heavily doped crystals.

Inhomogeneity in one property or another of the crystal around a dislocation is controlled by the dependence of this property on composition. A smooth monotonic dependence produces two concentric regions of variable property corresponding to regions of variable composition. A dependence with an extreme can produce three regions of variable property. On the other hand, only one region of variable property can be seen if the property does not depend on δ at compositions close to those of the matrix and such a dependence occurs near the dislocation. It should also be considered that the composition variations can pass through certain critical points, for example, through a stoichiometric composition ($\delta = 0$). The neighboring regions of the crystal will have on cooling a qualitatively different ensemble of IPD. This will produce a sharp inhomogeneity in certain properties. The microdefect distribution shown in Fig. 2b is an example of such an inhomogeneity.

FORMATION OF COMPOSITION MACROINHOMOGENEITIES

The quantity δ does not change if IPD recombine at a dislocation. Thus, the average composition in a broad region including the dislocation remains identical to that of the matrix. However, if the property does not depend linearly on composition, then the integral value of the property in the region surrounding the dislocation will not be equal to that in the matrix. If experimental measurements of such a property are averaged over a volume substantially greater than the size of the separate regions of variable composition then the measured quantity will depend on the dislocation density since this value depends on the fraction of the averaged volume occupied by regions of variable composition. At small dislocation densities, the dependence on N_d is linear. At large N_d, where the region of variable composition around the individual dislocations overlap greatly, the dependence becomes complicated. Thus, a nonlinear dependence of the property on crystal composition can cause a macroscopic correlation between the property and the dislocation density. Moreover, the experimental data indicate that another more general mechanism for the generation of such a correlation may exist.

A correlation between N_d and the quasi-forbidden x-ray intensity was found in [33] for GaAs crystals. The intensity is determined by the ratio of concentrations of Ga and As. Measurements made in [34] demonstrated that the so-called differential concentration of IPD ΔN, defined as the difference of the total concentrations of intersti-

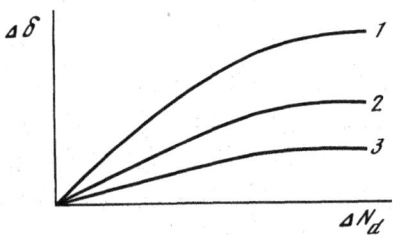

Fig. 4. Dependence of composition difference of neighboring crystal regions on drop of dislocation density for defect redistribution. $N_d^2 = 0$ (1), N_d^2 (2) $< N_d^2$ (3) (2, 3).

tial and vacancy defects, also correlates with N_d. The changes of ΔN observed in [34] may in principle be caused by creep due to absorption of a pair of vacancies from both sublattices. However, creep effects should be clearly observed in the crystal dislocation structure whereas they are in fact absent. This enables the observed changes of ΔN to be related to changes of crystal composition. The conclusion can be made that macroscopic composition inhomogeneities correlated with the distribution density of high-temperature growth dislocations exist in GaAs crystals. The formation of such a macroinhomogeneity can be explained by accelerated recombination of IPD at dislocations. Two possibilities should be considered. These are redistribution of defects between regions with different N_d and pumping of defects away from the crystallization front.

Redistribution of Defects between Regions with Different N_d. Recombination of excess defects at dislocations causes the defect concentrations between crystal regions with different dislocation densities to differ. If the mobility of just one of the recombining defects is large enough the defects redistribute between regions with different outflow density (N_d) causing the crystal composition to change. Assuming, as for the microscopic case, $\delta = c_{i_{As}} - c_{v_{As}}$ and $D_i > D_v$, we find that regions with greater dislocation densities should be enriched with As relative to regions with low N_d while the average crystal composition remains unchanged. The macroinhomogeneity observed in actual crystals can form if defects shift to distances of one quarter of the ingot diameter. This is possible for an average diffusion coefficient $D_i > 1 \cdot 10^{-4}$ cm^2/sec.

Using such a mechanism to examine the distribution of IPD between crystal regions with dislocation densities N_d^1 and N_d^2 $(N_d^1 = N_d^2 + \Delta N_d)$, it is easy to see that the composition difference $\Delta \delta$ depends not only on ΔN_d but also on the absolute values of the dislocation densities. This function is plotted qualitatively in Fig. 4. It is clear that the composition inhomogeneity under otherwise equal conditions (in particular, $D_i t = $ const) increases with increasing ΔN_d. However, the inhomogeneity at a given ΔN_d decreases with increasing average dislocation density $[(N_d^1 + N_d^2)/2]$.

Pumping of Defects Away from the Crystallization Front. The overwhelming majority of high-temperature dislocations are formed through the action of thermoelastic stresses. As a result, the cross-sectional distribution of N_d characteristic of the crystals grown takes shape at a certain distance l_z from the crystallization front that exceeds at least several dislocation cells (i.e., ~1 mm). A steady-state concentration of point defects is maintained at the crystallization front. Therefore, a defect concentration gradient arises between the prefrontal region and the region containing dislocations. The gradient is greater the more rapidly the dislocations recombine, i.e., the greater N_d. Thus, the region with greater N_d is enriched with defects of greater mobility. Such a mechanism is analogous to that proposed in [27] to explain the nature of microdefects formed in Si single crystals.

Still assuming that $\delta = c_{i_{As}} - c_{v_{As}}$ and $D_i > D_v$, we find that, like in the previous case, the As content will be greater in regions with higher N_d. However, pumping of defects changes the average crystal composition. The crystal is more enriched with As as the average dislocation density increases. The path of the function $\delta(N_d)$ for pumping of defects away from the crystallization front is plotted qualitatively in Fig. 5. The shape of the func-

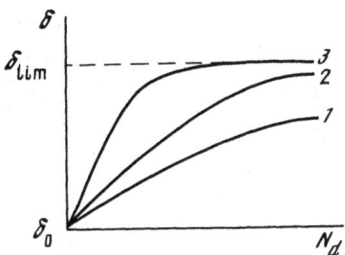

Fig. 5. Dependence of crystal composition on dislocation density for pumping of defects away from the crystallization front. The quantities δ_0 and δ_{lim} are the starting and limiting possible compositions. The product $D_i t$ increases from 1 to 3.

tion and the limiting composition δ_{lim} do not depend on the sign of δ_0. As N_d increases (and thus the recombination rate), the concentration of recombining defects approaches equilibrium whereas the concentration gradient between the crystallization front and the dislocation region and the defect flux caused by it become constant and do not depend on N_d. As a result, as for the case of defect redistribution, the composition inhomogeneity at large dislocation densities decreases.

Obviously the effectiveness of the proposed mechanisms is controlled by the distance to which the IPD should be displaced. For redistribution, this distance is about one quarter of the ingot diameter (10-15 mm in typical cases). Since the axial temperature gradients near the crystallization front often exceed 100°C/cm whereas the high-temperature dislocations form above 1100°C, it can be confirmed that $l_z < 10\text{-}15$ mm. However, it should be kept in mind that the typical Czochralski growth rates of 0.1-0.7 mm/min are comparable with the proposed (at $D_i \approx 10^{-4}$ cm²/sec) defect diffusion mobility. Therefore, the rapid withdrawal of the crystallization front even at small l_z will prevent the pumping mechanism.

The contribution of the two possible mechanisms for inhomogeneity formation can be estimated experimentally based on the facts that redistribution implies that the average composition is retained whereas pumping changes it. Thus, the cross-sectional distributions of EL2 centers in an ingot of semi-insulating undoped GaAs were studied. The average dislocation density increased significantly at relatively small ingot lengths (~2 cm) due to a sharp change of the thermal conditions during growth [35]. Figure 6 shows that an increase of the average dislocation density decreases the inhomogeneity of the EL2 distribution (the relative mean-square deviations for sections 1 and 2 are 17.0 and 10.5%, respectively) regardless of the increase of absolute decreases of the cross-sectional N_d. Moreover, the average EL2 concentration is practically constant: $1.69 \cdot 10^{16}$ cm^{-3} in the first section and $1.65 \cdot 10^{16}$ cm^{-3} in the second. The result obtained is consistent with predominance of defect redistribution over pumping in forming an inhomogeneity correlating with the N_d distribution. The pumping effect can apparently be found only at vary large changes of N_d. For example, it can cause the concentration of EL2 centers to increase from $5 \cdot 10^{15}$ to $1 \cdot 10^{16}$ cm^{-3} on changing N_d from 0 to 10^6 cm^{-2}, as found in [36].

Regardless of the formation mechanism of the composition macroinhomogeneity, the crystal region with increased dislocation density has a greater As content than that with low N_d. Experimental determinations of the composition inhomogeneity and ΔN made in [33, 34] agree with such a conclusion. Like for the microinhomogeneities, the macrodistribution of one property or another is controlled by its dependence on crystal composition.

The concentration of EL2 centers increases as the As content in the crystal increases [28]. However, these centers are distributed along the diameter in phase with the N_d distribution (cf. Fig. 6). Analogous dependences are observed for other deep centers in semi-insulating GaAs [37]. As the As content increases, the concentration of electron traps with ionization energy $E_c = -0.35$ eV increases whereas that of the hole traps with ionization

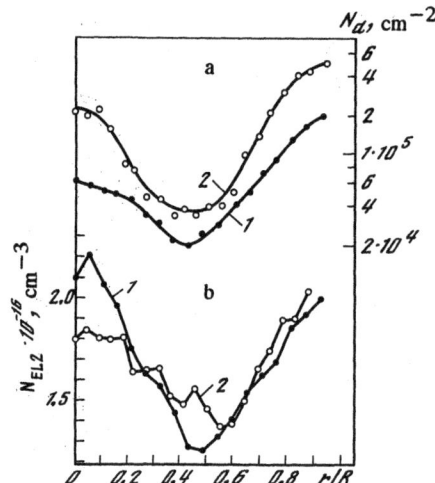

Fig. 6. Radial dislocation density distribution (*a*) and EL2 center concentrations (*b*) in two cross sections (1, 2) of the same crystal.

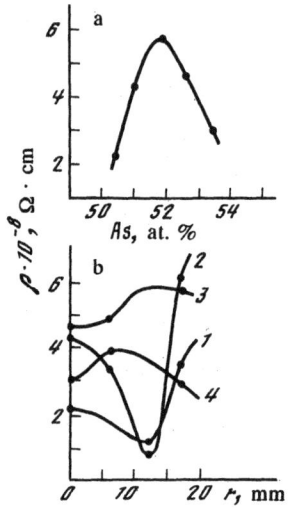

Fig. 7. Dependence of specific resistivity of undoped GaAs crystals on melt composition (*a*) and radial distribution of ρ (*b*) in samples grown from melts containing 50.4 (1), 51.3 (2), 52.6 (3), and 53.6 at. % As (4).

energy $E_v = +0.25$ eV decreases. An increase of N_d leads to exactly the same effect. Changing N_d causes large changes in the concentrations of certain deep centers up to the disappearance of some and the appearance of other traps. This may be related to a transition of the composition through a critical value (for example, $\delta = 0$).

The specific resistivity of undoped GaAs crystals is known to depend on melt composition. There is a maximum near the compensation point (Fig. 7a). In agreement with the proposed model for distribution of ρ along the diameter, N_d should change from W- to M-shaped as the As content increases in crystals with a W-shaped distribution. This was observed experimentally (Fig. 7b).

For implantation of Si in semi-insulating GaAs substrate, the impurity activation increases with increasing As content in the starting material [38]. This is also observed if N_d increases, as a result of which the threshold potentials of FET decrease in regions with a high dislocation density [5, 8, 39].

Fig. 8. Dependence of composition inhomogeneity on average dislocation density.

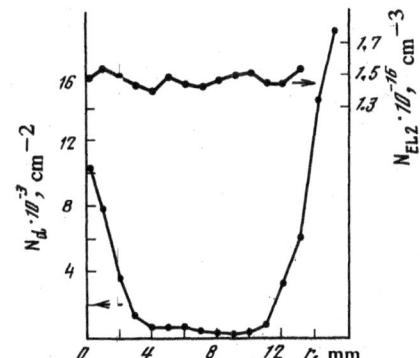

Fig. 9. Radial distributions of dislocation density and EL2 center concentration in a GaAs crystal with high In content (of the order of $1 \cdot 10^{20}$ at/cm^3). Average dislocation density $\bar{N}_d < 5 \cdot 10^3$ cm^{-2}.

Besides the facts listed above, there are many more known for the macrodistribution of various properties of GaAs crystals. The relation between the macrodistribution, which correlates with N_d, and the dependence on composition satisfies in all cases the proposed model. Several practically important conclusions can be made on the basis of the model.

1. The inhomogeneity of properties in single crystals grown can be reduced by using a range of melt compositions for which the weakest dependence of property on composition is observed.

2. Under otherwise other equal conditions, the dependence of composition (property) inhomogeneity on average dislocation density $\bar{N}_d$ has an extreme (Fig. 8). The macroscopic inhomogeneity should be small at low and high $\bar{N}_d$ and should reach a maximum at intermediate dislocation densities. The correlation of properties with N_d for grown crystals practically disappears at $\bar{N}_d \lesssim 5 \cdot 10^3$ cm^{-2} (Fig. 9) and is weak at $\bar{N}_d \gtrsim 1 \cdot 10^5$ cm^{-2}. Crystals with average dislocation densities of the order of $(1-5) \cdot 10^4$ cm^{-2} (for a typical W-shaped N_d distribution) have the largest inhomogeneities.

3. The inhomogeneity should decrease as the ingot diameter increases due to the increasing distance between regions exchanging IPD. Redistribution of IPD near the free surface of the ingot is important for crystals with small diameter (up to 20 mm). The outflow of defects to the surface may be due not only to their recombination but also to evaporation of As at high temperatures.

Besides these possibilities for decreasing the inhomogeneity of crystals grown, which do not follow directly from the conclusions made, heat treatment of single crystals as ingots or plates also is attractive. Parameters such

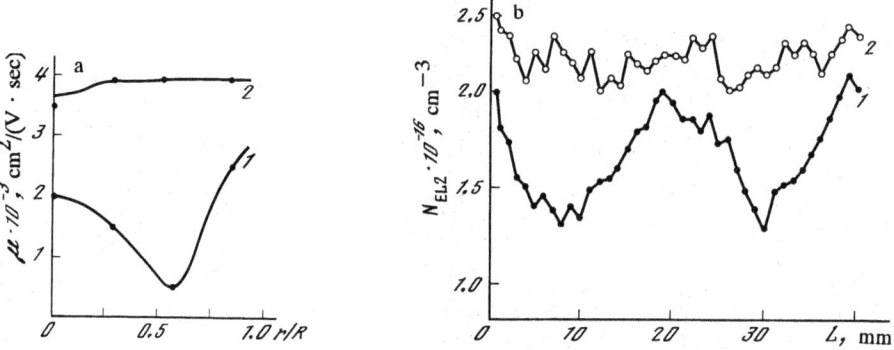

Fig. 10. Effect of annealing plates of undoped GaAs at 850°C on Hall mobility along the normal radius (*a*) and the EL2 center concentration along the diameter L (*b*). Starting distribution (1), distribution after annealing (2).

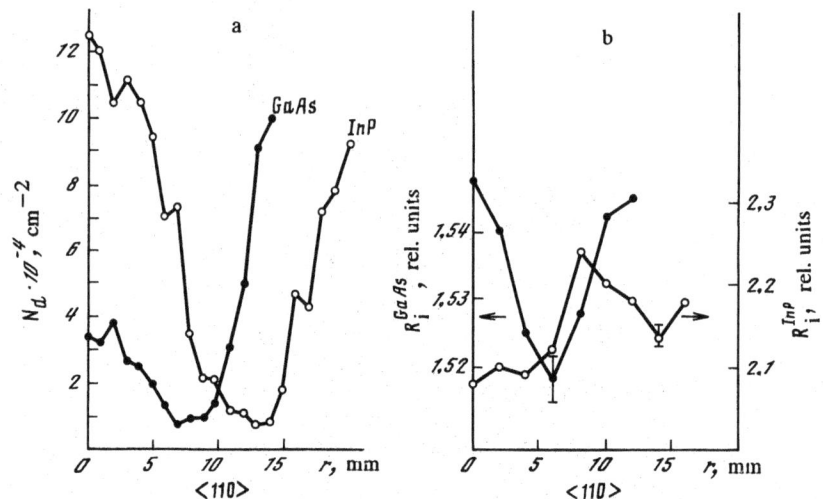

Fig. 11. Correlation of radial distributions of dislocation density (*a*) and integrated scattered x-ray intensity from (600) (*b*) in GaAs and InP crystals.

as the EL2 center concentration [40, 41], the specific resistivity and carrier mobility [41-43], and the photoluminescence intensity [43] were substantially evened out by annealing at 800-1000°C. Examples of the homogenization of properties of semi-insulating crystals by annealing are given in Fig. 10. The processes occurring during heat treatment by far have not been completely studied. However, existing data suggest that the observed homogenization of certain properties results from generation and recombination of IPD. This changes the nature of the dependence of certain properties on composition. The composition at these annealing temperatures and times, which reach several tens of hours, is apparently not evened out.

In conclusion, the described formation mechanism of micro- and macroinhomogeneities due to dislocations should occur to a certain extent in single crystals of other semiconductors grown from the melt. It causes formation of inhomogeneous composition in crystals of compounds — inhomogeneous density in elemental semiconductors. This conclusion is consistent with the existence of broad regions free of microdefects around dislocations in GaP and Si crystals. Radial distributions of N_d and the integrated intensity of quasi-forbidden x-ray reflections for undoped GaAs and InP crystals are presented in Fig. 11.* In both cases the intensity distribution (and thus the composition) correlates with the N_d distribution. The nature of the correlation in InP is inversely

---

*Work of A. G. Efimov.

related to that seen in GaAs crystals. Such an inversion is due to the fact that the scattering power of the cations and anions changes on going from GaAs to InP. This is consistent with defects in the anionic sublattice controlling the formation of the inhomogeneity in these single crystals.

REFERENCES

1.   M. Tajima, "Characterization of nonuniformity in semi-insulating LEC GaAs by photoluminescence spectroscopy," *Jpn. J. Appl. Phys. Lett.*, **21**, No. 4, 227-229 (1982).

2.   R. T. Blunt, S. Clark, and D. J. Stirland, "Dislocation density and sheet resistance variations across semi-insulating GaAs wafers," *IEEE Trans. Electron Devices*, **29**, No. 7, 1039-1044 (1982).

3.   V. Matsumoto and H. Watanabe, "Inhomogeneity in semi-insulating GaAs revealed by scanning leakage current measurements," *Jpn. J. Appl. Phys. Lett.*, **21**, No. 8, L515-L517 (1982).

4.   S. Miyazawa, T. Mizutani, and H. Yamazaki, "Leakage current I_L variation correlated with dislocation density in undoped, semi-insulating LEC GaAs," *Jpn. J. Appl. Phys. Lett.*, **21**, No. 9, L542-L544 (1982).

5.   Y. Nanishi, S. Ishida, T. Honda, et al., "Inhomogeneous GaAs FET threshold voltages related to dislocation distribution," *Jpn. J. Appl. Phys. Lett.*, **21**, No. 6, L335-L337 (1982).

6.   D. E. Holmes and R. T. Chen, "Contour maps of EL2 deep level in liquid-encapsulated Czochralski GaAs," *J. Appl. Phys.*, **55**, No. 10, 3588-3599 (1984).

7.   T. Figielski and T. Wosinski, "Properties and nature of the main electron trap in GaAs," *Czech. J. Phys. B*, **34**, No. 5, 403-408 (1984).

8.   S. Miyazawa, Y. Ishii, S. Ishida, and Y. Nanishi, "Direct observation of dislocation effects on threshold voltage of a GaAs field-effect transistor," *Appl. Phys. Lett.*, **43**, No. 9, 853-855 (1983).

9.   A. V. Markov, M. G. Mil'vidskii, and V. B. Osvenskii, "On the role of dislocations in forming properties of single crystals of semi-insulating GaAs," *Fiz. Tekh. Poluprovodn.*, **20**, No. 4, 634-640 (1986).

10.   A. K. Chin, A. R. Von Neida, and R. Caruso, "Spatially resolved cathodoluminescence study of semi-insulating GaAs substrates," *J. Electrochem. Soc.*, **129**, No. 10, 2386-2388 (1982).

11.   A. V. Markov, S. P. Grishina, M. G. Mil'vidskii, and S. S. Shifrin, "Complexation and thermal stability of electrophysical properties of single crystals of semi-insulating GaAs," *Fiz. Tekh. Poluprovodn.*, **18**, No. 3, 465-470 (1984).

12.   S. S. Shifrin, A. V. Markov, M. G. Mil'vidskii, and V. B. Osvenskii, "Possibilities of studying the relation of dislocation structure of single crystals of semiconductors grown from the melt," *Izv. Akad. Nauk SSSR, Ser. Fiz.*, **47**, No. 2, 295-301 (1983).

13.   M. S. Abrahams and C. J. Buiocchi, "Etching of dislocations on the low-index faces of GaAs," *J. Appl. Phys.*, **36**, No. 9, 2855-2863 (1965).

14.   A. T. Hunter, "Spatially resolved luminescence near dislocations in In-alloyed Czochralski-grown GaAs," *Appl. Phys. Lett.*, **47**, No. 12, 715-718 (1985).

15.   I. Fillard, P. Gall, M. Asgarinia, et al., "EL2° distribution in the vicinity of dislocations in GaAs—In materials," *Jpn. J. Appl. Phys. Lett.*, **27**, No. 5, 899-902 (1988).

16.   A. V. Markov, M. G. Mil'vidskii, and S. S. Shifrin, "Characteristics of the formation of microdefects near dislocations in GaAs crystals doped with various impurities," *Kristallografiya*, **29**, No. 2, 343-349 (1984).

17.   F. A. Gimel'farb, A. V. Govorkov, S. P. Grishina, et al., "Microcathodoluminescence study of the decoration of dislocations during growth of doped GaAs single crystals," *Kristallografiya*, **19**, No. 5, 1115-1117 (1974).

18.   M. G. Mil'vidskii, A. A. Kalinin, A. V. Markov, and A. N. Shershakov, "Role of intrinsic point defects in the formation of microdefects in doped GaAs crystals," *Fiz. Kristalliz.*, Kalinin, 3-11 (1986).

19.   C. A. Warwick and G. T. Brown, "Spatial distribution of 0.68 eV emission from undoped semi-insulating gallium arsenide revealed by high resolution luminescence imaging," *Appl. Phys. Lett.*, **46**, No. 6, 574-576 (1985).

20.   P. Dobrilla and D. C. Miller, "Correlation of the etching morphology with the main midgap donor distribution in undoped, semi-insulating GaAs," *J. Electrochem. Soc.*, **134**, No. 12, 3197-3199 (1987).

21.   I. E. Bondarenko, V. G. Eremenko, B. Ya. Farber, et al., "On the real structure of monocrystalline silicon near dislocation slip planes," *Phys. Status Solidi A*, **68**, No. 1, 53-60 (1981).

22.   V. T. Bublik, M. G. Mil'vidskii, and V. B. Osvenskii, "Nature and behavior of point defects in doped single crystals of $A^{III}B^V$ compounds," *Izv. Vyssh. Uchebn. Zaved. Fiz.*, No. 1, 7-22 (1980).

23.   M. G. Mil'vidskii, O. V. Pelevin, and B. A. Sakharov, *Physicochemical Principles of Obtaining Dissociating Semiconductor Compounds. (As Illustrated by Gallium Arsenide)* [in Russian], Metallurgiya, Moscow (1974).

24.   U. Gösele, W. Frank, and A. Seeger, "An entropy barrier against vacancy-interstitial recombination in silicon," *Solid State Commun.*, **45**, No. 1, 31-33 (1983).

25.   A. M. Gomez and P. B. Hirsch, "The dissociation of dislocations in GaAs," *Philos. Mag. A*, **38**, No. 6, 733-737 (1978).

26.   S. S. Shifrin and A. V. Markov, "Helicoidal dislocations in GaAs single crystals," *Kristallografiya*, **25**, No. 5, 1089-1093 (1980).

27.   V. V. Voronkov, "The mechanism of swirl defect formation in silicon," *J. Cryst. Growth*, **59**, No. 3, 625-643 (1982).

28.   J. Lagowski, H. C. Gatos, J. M. Parsey, et al., "Origin of the 0.82 eV electron trap in GaAs and its annihilation by shallow donors," *Appl. Phys. Lett.*, **40**, No. 4, 342-344 (1982).

29.   A. V. Govorkov and L. I. Kolesnik, "Microcathodoluminescence study of the effect of defect structure on emissive recombination in GaAs," *Fiz. Tekh. Poluprovodn.*, 12, No. 3, 448-452 (1978).

30.   A. G. Cullis, P. D. Augustus, and D. J. Stirland, "Arsenic precipitation at dislocations in GaAs substrate material," *J. Appl. Phys.*, 51, No. 5, 2556-2560 (1980).

31.   I. A. Koval'chuk, A. V. Markov, and M. G. Mil'vidskii, "Effect of melt composition on electrophysical properties and structure of undoped GaAs single crystals," *Izv. Akad. Nauk SSSR, Neorg. Mater.*, 24, No. 2, 324-326 (1988).

32.   V. T. Bublik, V. V. Karataev, R. S. Kulagin, et al., "Nature of point defects in GaAs single crystals as a function of melt composition during growth," *Kristallografiya*, 18, No. 2, 353-356 (1973).

33.   I. Fujimoto, "Characterization of stoichiometry in GaAs by x-ray intensity measurements of quasi-forbidden reflections," *Jpn. J. Appl. Phys. Lett.*, 23, No. 5, 287-289 (1984).

34.   A. V. Markov and A. N. Morozov, "Reasons for macroscopic inhomogeneity of GaAs single crystals," *Fiz. Tekh. Poluprovodn.*, 20, No. 1, 154-157 (1986).

35.   A. V. Kartavykh and A. V. Markov, "Relation of concentration of EL2 deep centers and dislocation density in semi-insulating GaAs," *Fiz. Tekh. Poluprovodn.*, 22, No. 9, 1702-1704 (1988).

36.   D. A. O. Hope, M. S. Scolnick, B. Cockayne, et al., "A comparison of the deep donor (EL2)° and strain distributions in dislocated and dislocation-free semi-insulating undoped GaAs," *J. Cryst. Growth*, 71, No. 3, 795-798 (1985).

37.   A. V. Markov, É. M. Omel'yanovskii, A. Ya. Polyakov, et al., "Effect of dislocations on distribution of deep centers in semi-insulating GaAs," *Fiz. Tekh. Poluprovodn.*, 22, No. 1, 44-48 (1988).

38.   T. Sato, M. Nakajima, T. Fukuda, and K. Ishida, "Stoichiometry dependence of electrical activation efficiency in Si-implanted layers of undoped, semi-insulating GaAs," *Appl. Phys. Lett.*, 49, No. 23, 1599-1601 (1986).

39.   T. Egawa, Y. Sano, H. Nakamura, and K. Kaminishi, "Influence of annealing method on microscopic one-to-one correlation between threshold voltage of GaAs MESFET and dislocation," *Jpn. J. Appl. Phys. Lett.*, 25, No. 12, 973-975 (1986).

40.   D. E. Holmes, H. Kuwamoto, C. G. Kirkpatrick, and R. T. Chen, "Effect of thermal history on properties of LEC GaAs," in: *Semi-insulating III—V Materials*, Shiva, Nantwich (1984), pp. 118-125.

41.   T. Inada, T. Fujii, and T. Fukuda, "Native defect related inhomogeneity in characteristics of GaAs field-effect transistors fabricated on annealed dislocation-free substrates," *J. Appl. Phys.*, 61, No. 12, 5483-5485 (1987).

42.   D. Rumsby, I. R. Grant, M. R. Brozel, et al., "Electrical behavior of annealed LEC GaAs," in: *Semi-insulating III—V Materials*, Shiva, Nantwich (1984), pp. 165-170.

43.   T. Obokata, T. Matsumura, K. Terashima, et al., "Improved uniformity of resistivity distribution in LEC semi-insulating GaAs produced by annealing," *Jpn. J. Appl. Phys. Lett.*, 23, No. 8, 602-605 (1984).

MULTICOMPONENT FLUORIDE SINGLE CRYSTALS
(CURRENT STATUS OF THEIR SYNTHESIS AND PROSPECTS)

B. P. Sobolev

World production of single crystals using directed melt crystallization had up to the mid-1960's included less than ten one-component fluorides. This situation has persisted until now and has practically no indication of changing. The fluorides are mainly those of Li, Mg, Ca, Ba, and several others. The technical potential of these materials has also been exhausted. The chemical compositions have been determined. Thus, the tightly controlled properties limit the range of practical application of one-component fluoride crystals. The fluorides can be activated by addition of small amounts of impurities. As a rule, this does not substantially change the useful properties of the crystalline matrices themselves, including the thermal and chemical stability, mechanical, electrical, dielectric, and other properties. This method of modifying the individual properties (mainly laser) by adding small amounts of impurities will not be reviewed in the present article.

Many years of study of the chemistry of fluoride systems suggested to us that the growing demands on crystalline fluoride materials can be satisfied only by substantially changing their chemical composition while keeping the structure the same [e.g., 1, 2].

The transition from one- to multicomponent fluoride crystals, as has become completely evident, will ensure progress in this area of inorganic materials science. It will provide a large number of new materials with properties radically different from those known for the one-component prototypes and variable over wide limits.

Strictly speaking, addition of any amount of extraneous impurity, a second component, transforms the system from one- to two-component. For the present review, we were constrained to narrow the range of crystals considered multicomponent. This is due to the fact that not all two-component (and more complicated) materials can be prepared simply. We intended to examine changes of the chemical composition that produced substantial changes in the crystal properties. Such changes are obvious for a new double compound formed with a structural type different from the components. With respect to the composition changes where the crystal structural type is retained (i.e., as a result of solid solution formation), only a sufficiently high content of the second component can have a noticeable effect on the properties of compounds with largely ionic chemical bonding, such as fluorides of most metals. The relatively weak dependence of properties on composition means that only sufficiently concentrated solid solutions will be examined. Therefore, crystals with impurity concentrations up to 0.1-0.5 at. % were excluded (in principle also multicomponent).

A number of difficulties along with the complicated composition arises during the search for and the growth of crystals. Data on the chemical reactions of fluorides in double (or more complicated) systems are needed to choose the crystal compositions. Data on the phase diagrams of the systems are also required to choose growth methods, crystallization conditions, and permissible thermal regimes in the applications.

The necessity for a systematic approach to development of physicochemical and crystal chemical syntheses of new multicomponent fluorides is obvious. However, few examples can be found in the literature due to the labor intensiveness of studying phase equilibria. Physicochemical studies usually precede synthesis of the materials.

The goal of the present review is to communicate briefly the separate stages of the multiyear domestic program designed to develop scientific bases for preparing new multicomponent fluorides by crystallization from melts of complicated chemical composition. The program was formulated in the mid-1970's. It was founded by the scientific organizations of the Institute of Crystallography of the Academy of Sciences of the USSR (ICAS), the Chemistry Faculty of Moscow State University (MSU), and the Institute of Chemistry of the Academy of Sciences of the Tadzhik SSR. Studies on the growth of crystals and their physical properties along with physicochemical and crystal chemical investigations were conducted. Many academics, university faculties, and branch organizations as well as an Institute of the Czechoslovak Academy of Sciences participated. As a result, a rather rare situation in materials science has developed in which fundamental chemical and physical studies have outdistanced applied works and are determining the direction and level of the latter. The total number of domestic publications on only physicochemical and crystal chemical studies of multicomponent fluorides is several hundred. Hundreds of Soviet and foreign works on the physics of these crystals can also be added. It was impossible to cite all articles due to limitations on the review. We were limited to listing scientific groups contributing most to the areas of fluoride materials examined below and to citing the principal (mainly seminal) publications.

The following topics are examined successively in the review:

— the potential of a chemical class of fluorides as a source of one-component materials;
— two-component MF_m–RF_n systems as a source of the simplest (by composition) multicomponent materials;
— phase-formation processes in MF_m–RF_3 systems, the nature of phase melting, and their structural classification;
— high nonstoichiometry in fluorides and its role in creating new materials with controlled characteristics;
— preparation of two-component fluorides from melts;
— the current status and prospects of preparing new multicomponent fluorides by crystallization from melts and the principal directions of practical application, most obvious from the publications that involve a study of the properties.

POTENTIAL OF A CHEMICAL CLASS OF FLUORIDES

AS A SOURCE OF ONE-COMPONENT MATERIALS

The search for new fluoride crystal matrices that could eventually provide bases for multicomponent materials is based on several completely obvious main requirements:

— sufficiently high chemical, thermal, mechanical, radiation, and other stability to external influences during use;
— possibility of production from melts in single-crystalline form;
— significant isomorphous capacity relative to the component-impurity for effective control of the crystal properties through its chemical composition.

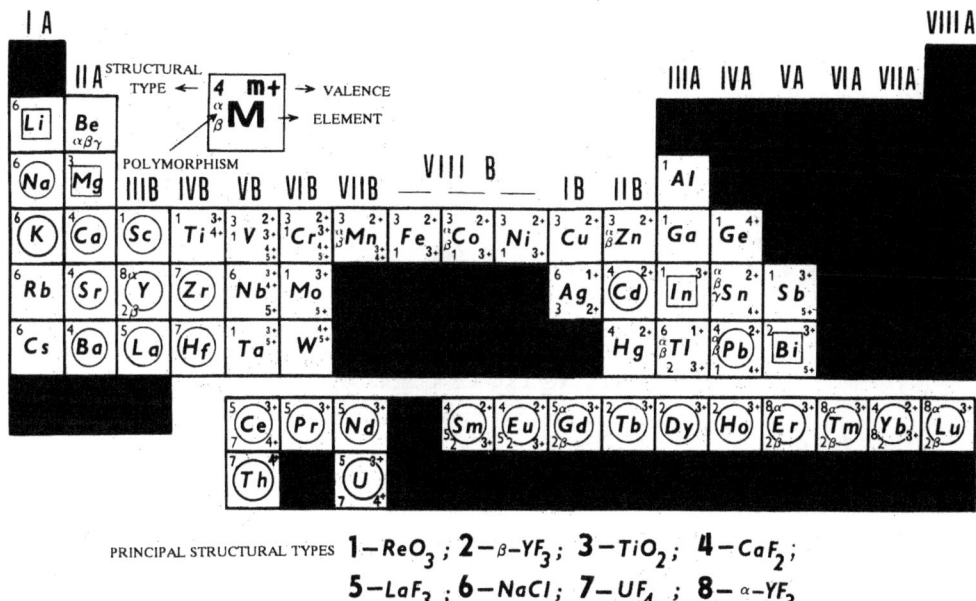

PRINCIPAL STRUCTURAL TYPES $1-ReO_3$; $2-\beta-YF_3$; $3-TiO_2$; $4-CaF_2$; $5-LaF_3$; $6-NaCl$; $7-UF_4$; $8-\alpha-YF_3$

Fig. 1. Principal data for fluorides of 55 metals.

Let us examine 55 fluorides formed by 55 metals (Fig. 1) that agree with these conditions. For these, compounds with F have been reliably identified and are not gaseous or liquid under standard conditions. The valence states (principal ones are larger) of the metals in their compounds with F are located to the right of the element symbols in the periodic table. The structural types of the corresponding fluorides (without considering distortions) are denoted by the numbers to the left, which are explained under the table. Polymorphism is signified by a Greek letter to the left of the symbol.

The variable valence leads to formation of 83 individual fluorides by the 55 metals. Polymorphism expands the number of crystallographically different phases to 101. Obviously, the choice itself of the most promising compounds is very complicated without quantitative evaluation of the criteria formulated above for choosing simple fluorides. According to our estimates, the conditions are satisfied best by fluorides of the 27 metals circled in Fig. 1. The fluorides of the four elements in squares are close but less satisfactory. The fluorides of 24 metals (excluding K, Th, and U from the list) were used in our program of physicochemical studies of the phase diagrams.

What is the chemical and structural relationship of the 27 simple fluorides chosen? First of all, it should be noted that most of them, 16, belong to the same family, rare earths (RE). The RE trifluorides crystallize in four structural types: LaF_3 (tysonite), β-YF_3 (cementite), α-YF_3 (α-UO_3), and ReO_3.* The 7 RE trifluorides have a reconstructive polymorphic transition that hinders crystal preparation.

The next most numerous group of compounds has the CaF_2 (fluorite) structure and contains the majority of alkaline earth (AE) fluorides and the difluorides of Cd and Pb. A total of 5 fluorides of Group IIA, IIB, and IVA metals are included. Only PbF_2 exhibits polymorphism. However, this first-order phase transition is very slow and does not affect the synthesis of single crystals.

The alkali metal (Li, Na, K) fluorides crystallize with the NaCl structure. This type has a low isomorphic capacity relative to all of the cations listed above. However, their chemical interactions in binary combinations leads to formation of practically interesting phases of complicated constant and variable chemical composition.

---

*This type includes one scandium fluoride that was not included in the subsequent examination due to its chemistry, distinctly different from that of the remaining RE trifluorides.

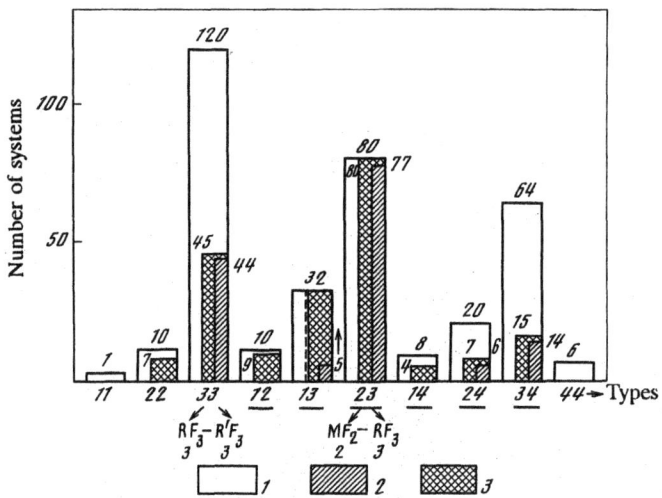

Fig. 2. Histograms of the types and status of phase diagrams of binary systems MF_m–RF_n ($m \leq n \leq$ 4), (M = Na, K; Ca, Sr, Ba, Cd, Pb; R = Sc, Y, La, Ce, Pr, Nd, Sm, Eu, Gd, Tb, Dy, Ho, Er, Tm, Yb, Lu; Zr, Hf, Th, U), 27/24. Total number of systems 351 (including 214 with $m \neq n$) (1); 200 studied (140 with $m \neq n$) (2); including 146 studied by us (102 with $m \neq n$) (3).

Finally, a small group of fluorides of +4 cations (Zr, Hf, Th, U) are highly volatile. This creates technical complications during growth of crystals from melts that are difficult to overcome.

This short analysis of the prospects of expanding the variety of single-crystalline fluorides using simple fluorides indicates that several unexplored possibilities exist. This primarily concerns RE and individual AE trifluorides, which have not yet been prepared industrially as single crystals. However, these few additions, about 10 one-component materials, will not solve all practical problems regardless of their promise.

The fact that the majority of the 27 selected simple fluorides (22 compounds) contain divalent and trivalent metals is interesting.

TWO-COMPONENT FLUORIDES AS SOURCES OF THE SIMPLEST

(BY COMPOSITION) MULTICOMPONENT FLUORIDES

Let us examine how changing from the 27 simple compounds to the two-component systems MF_m–RF_n formed by the components ($m \leq n \leq 4$) selected in the previous section affects the type of fluoride crystals. The total number of such systems is 351. Even if it is assumed that each system gives only one two-component material, the number of materials increases by more than an order of magnitude. Many of the systems contain 2-3 phases that can be prepared as single crystals.

Only a few phase diagrams of binary fluorides existed when we began to study them (1964). By 1987, 200 diagrams of solid MF_m–RF_n systems (Fig. 2) and about 20 with volatile components were known. About 75% of all the diagrams of solids were obtained for the first time in our program. Some of the systems studied earlier were revised with substantial corrections. Errors in the earlier works were due to uncontrolled pyrohydrolysis of fluorides under the physicochemical experimental conditions.

The phase diagrams are distributed by type in Fig. 2. Each type combines systems with a constant combination of cation valences denoted by the numbers under the columns. The column height corresponds to the total number of system of the given type (denoted by the number on top). The number of phase diagrams studied is shown by cross hatching, including those of our program by hatching. Combinations of cations with different valence are emphasized. Their fluoride systems have the potential of forming nonstoichiometric phases.

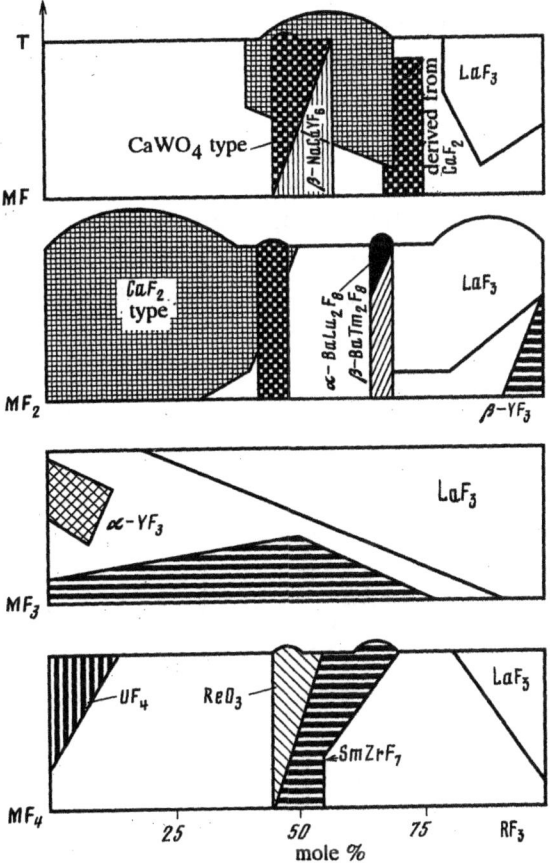

Fig. 3. Phase formation in MF_m–RF_3 systems and the phase fusibility
(schematic).

Systematic physicochemical studies of fluoride systems of those most promising as sources of crystalline materials have characterized the phase formation and thermal behavior for more than 60% of the MF_m and RF_n combinations. Phase diagrams are known for only 16% of the analogous combinations of fluorides of the 55 metals studied earlier.

The remarks made in the previous section about the chemical nature of the fluorides are evident in Fig. 2. The number of systems is clearly dominated by types in which the components are fluorides of di- or trivalent metals. Such systems constitute more than 90% of the total.

Another important feature of the systems examined, also predetermined by the choice of the 27 components, is the predominance of combinations in which only one of the components is a RE trifluoride. As a result, the high-temperature chemistry of the binary fluoride systems most promising for materials science is more than 80% that of the RE trifluorides (RF_3). We used this approximation to analyze the principal types of phases formed in MF_m–RF_3 systems.

PHASE FORMATION IN MF_m–RF_3 SYSTEMS

We will rely on the very approximate scheme shown in Fig. 3 since it would be impossible to present all of the phase diagrams in the present review. Homogeneous regions of typical phases of various structure are shown by cross hatching (schematically). The maximal component miscibilities for each type of system are usually chosen. The dome over the homogeneous regions arbitrarily denotes congruent melting of the separate solid solution compositions. As we will see below, the presence of such compositions is an exclusive feature of the thermal behavior of phases formed through heterovalent isomorphous substitutions. This feature is understood to be an extraordinarily valuable condition for preparing highly homogeneous two-component single crystals.

Of the four types of systems, three are formed by fluorides of elements with different valences. Phases crystallizing in the CaF_2 (and its ordered forms) and LaF_3 (two types $MF–RF_3$ and $MF_2–RF_3$, cf. Fig. 3) structural types have the broadest homogeneous regions. Phases with the fluorite structure for the most part melt congruently. The tysonite phases melt congruently only for a part of the $MF_2–RF_3$ systems. Some phases with the ReO_3 and $SmZrF_7$ (or their derivatives) structures in the $MF_4–RF_3$ systems also melt congruently. However, the majority of them undergo polymorphic transformations prohibiting single crystals from being prepared.

Another two structural types with a small number of representatives in systems with cations of different valence have practical significance. These are $LiRF_4$ (R = Sm–Lu, Y) with the $CaWO_4$ scheelite structure. The homogeneous region shown in Fig. 3 was taken from the literature. Solid solutions based on $LiRF_4$ were not observed during our recently completed revisions of the phase diagrams of $LiF–RF_3$ systems. The other family is BaR_2F_8, some of which melt congruently. They crystallize in two structural types, β-$BaTm_2F_8$ and α-$BaLu_2F_8$.

The majority of $MF_3–RF_3$ systems with isovalent cations form pairs with RE trifluorides. Phases of two structural types, LaF_3 and β-YF_3, form broad homogeneous regions. A third structural type, α-YF_3, has phases that decompose on cooling. Phases with the β-YF_3 structure melt either with decomposition or not at all.

The scheme examined is a rough picture of the principal features of phase formation and the fusibility of phases in $MF_m–RF_3$ systems. Nevertheless, it enables the potential of more than 200 systems to be evaluated as sources of two-component fluorides.

These materials are classified by structure into five basic families:

— phases of variable composition with the fluorite structure in $MF–RF_3$ (M = Na, K) and $MF_2–RF_3$ (M = Ca, Sr, Ba, Cd, Pb) systems, the majority of which melt congruently;

— phases of variable composition with the tysonite structure in $NaF–RF_3$, $MF_2–RF_3$ (M = see above), $MF_3–RF_3$ (M = In, Bi, RE), and $MF_4–RF_3$ (M = Zr, Hf) systems, many of which melt with decomposition;

— binary $LiRF_4$ with the scheelite structure for R = Tm–Lu melt congruently, the others melt with decomposition;

— binary BaR_2F_8 (two related structural types) melt congruently with R = Y, Ho–Lu, the others melt with decomposition;

— phases of variable composition with the orthorhombic β-YF_3 structure as a rule melt with decomposition or decompose in the solid to mainly $RF_3–R'F_3$ systems (R, R' = RE).

NEW MATERIALS WITH CONTROLLED PROPERTIES CREATED BY
HIGH NONSTOICHIOMETRY IN FLUORIDES

Fluorite and tysonite phases have a number of unique parameters among the five principal structural types formed in $MF_m–RF_3$ systems. Their distinguishing features are large homogeneous regions (composition and temperature), many compositions that melt congruently, and a large number and variety of cations that participate in their formation, all 27 examined (cf. Figs. 1 and 2) [2].

These two families are related by a common trait. The starting components of the CaF_2- and LaF_3-types are highly nonstoichiometric. Heterovalent isomorphous substitution produces the phases $M_{1-x}R_xF_{2+x}$,

$M_{0.5-x}R_{0.5+x}F_{2+2x}$, and $R_{1-y}M_yF_{3-y}$, with fluorite (the first two types) and tysonite defect structures. The atomic structure of the nonstoichiometric phases requires a separate treatment since much experimental data, some of which is contradictory, has been collected on it.

It is important that the concentration of structural defects in both CaF_2 and LaF_3 structures can be extraordinarily high. The degree of defectiveness can be controlled by the chemical composition. The defects begin to interact even with a few percent of a heterovalent ion. Practically all solid solutions reliably identified in MF_m–RF_n ($m \neq n$) systems by thermal and x-ray analyses are considered highly nonstoichiometric (due to the relatively low accuracies of these methods).

Representations of the high nonstoichiometry in these fluorides did not exist before the present program for studying phase diagrams of MF_m–RF_n systems was completed since ab initio prediction of the formation of such phases is impossible and the necessary experiment had not been performed. Our studies of the phase diagrams demonstrated that high nonstoichiometry is very common in binary fluorides. Nonstoichiometric phases were observed experimentally in 75% of the MF_m–RF_n systems where they were possible.

Since the phase diagrams of the majority of binary MF_m–RF_n systems promising for materials science have been studied, we come to the statistically meaningful conclusion that the high nonstoichiometry is most characteristic of phase formation in high-melting fluorides and typically results from their chemical interactions at high temperatures.

About half of the highly nonstoichiometric phases in MF_m–RF_n ($m < n \leq 4$) systems of those known in 1987 crystallize with the fluorite structure; another quarter, with the LaF_3-type.

Phases with these structures are most widely represented in MF_2–RF_3 systems formed by fluorite-type Ca, Sr, Ba, Cd, and Pb difluorides and RE trifluorides. The maximal homogeneous regions of heterovalent solid solutions $M_{1-x}R_xR_{2+x}$ and $R_{1-y}M_yF_{3-y}$ based on the first three fluorides are shown in Fig. 4 [3-5]. The maxima indicated by circles in the homogeneous regions on the fusibility curves of the nonstoichiometric phases are especially important for the growth of crystals. The open circles represent thermal analysis data; filled circles, distribution coefficients of RE impurity during growth of $M_{1-x}R_xF_{2+x}$ single crystals.

It is well known that high nonstoichiometry in particular is produced at high temperatures. However, fluoride systems in general and especially those examined in Fig. 4 have an exceptionally high tendency to retain a disordered (high-temperature) crystal structure in a metastable state down to liquid He temperatures. This is an exceedingly important condition that causes highly nonstoichiometric fluorides to decompose into crystalline materials with a partially disordered atomic structure. These materials are suitable for very different uses, including thermal cycling and cryogenic applications.

The uppermost rectangle in Fig. 4 shows the average homogeneous region for phases with the CaF_2 structure, ordered phases (ORD., including distorted tetragonal T; trigonal Rh, Rh', and R; and cubic C), tysonite LaF_3, and α-YF_3. The numbers 43, 4, 18, and 7 denote the average homogeneous regions of these structural types for the three series of MF_2–RF_3 given below. Thus, an average of 72% of the compositions exhibit nonstoichiometry.

Let us examine a separate example of how high nonstoichiometries affect crystal properties in order to estimate the value of this effect for inorganic materials science. The melting point of nonstoichiometric $Ba_{0.69}La_{0.31}F_{2.31}$ is significantly elevated (by 130°C) compared with pure BaF_2. The hardness increases by a factor of two. The ionic conductivity of $Ba_{1-x}R_xF_{2+x}$ crystals increases by 10 orders of magnitude. There are very many literature examples of a radical change of properties as a result of a disruption of the stoichiometry of multicomponent fluoride crystals. However, it seems to us that this particular example is sufficient to find the representative features of this family of materials:

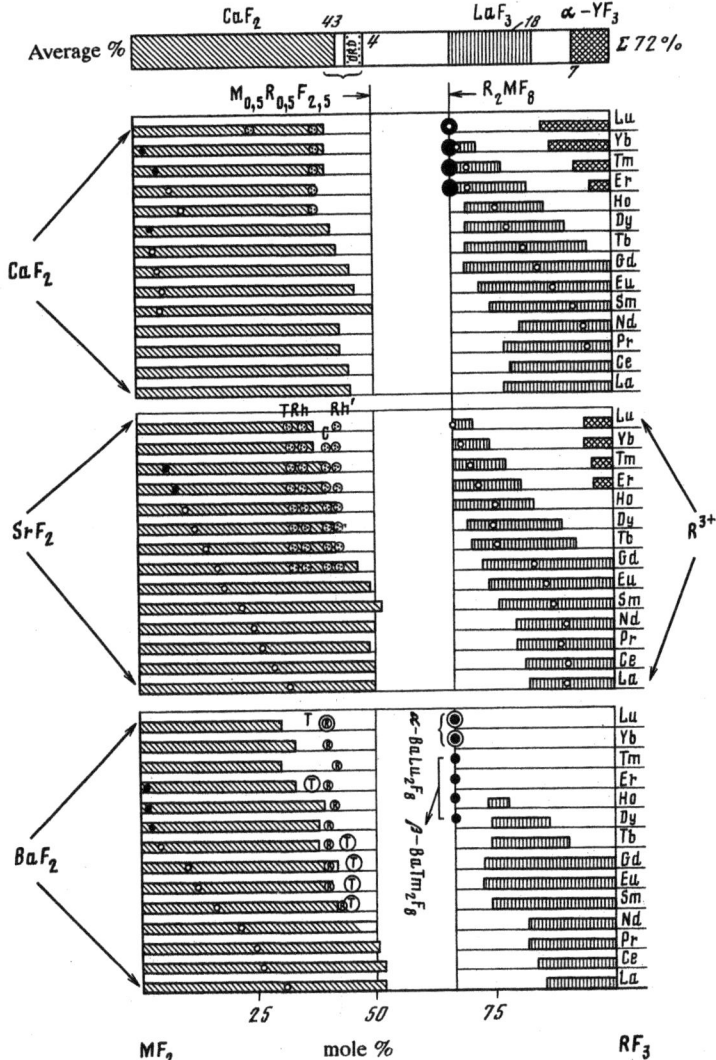

Fig. 4. Nonstoichiometry in MF$_2$–RF$_3$ systems (M = Ca, Sr, Ba; R = RE; $T > 800°$C).

— multicomponent nonstoichiometric fluorides differ substantially in physicochemical and physical properties from the one-component prototypes so that they can be viewed as a new broad class of inorganic materials;

— highly nonstoichiometric fluorides enable one of the most important and difficult problems of inorganic materials science to be solved, the creation of materials with useful properties controlled over wide limits and even with qualitatively new combinations of properties.

PREPARATION OF TWO-COMPONENT FLUORIDES FROM MELTS

Thus far, we have examined only the positive effects of going from one-component to multicomponent crystals.

However, significant difficulties are encountered on the path to practical application of fluorides. These are primarily due to the directional hardening of melts, which is used most widely to prepare fluorides. Most crystals are prepared by the Bridgman–Stockbarger method, with the Czochralski method in second place. In general, incongruent crystallization of a two-component melt produces a differentiated composition as it hardens.

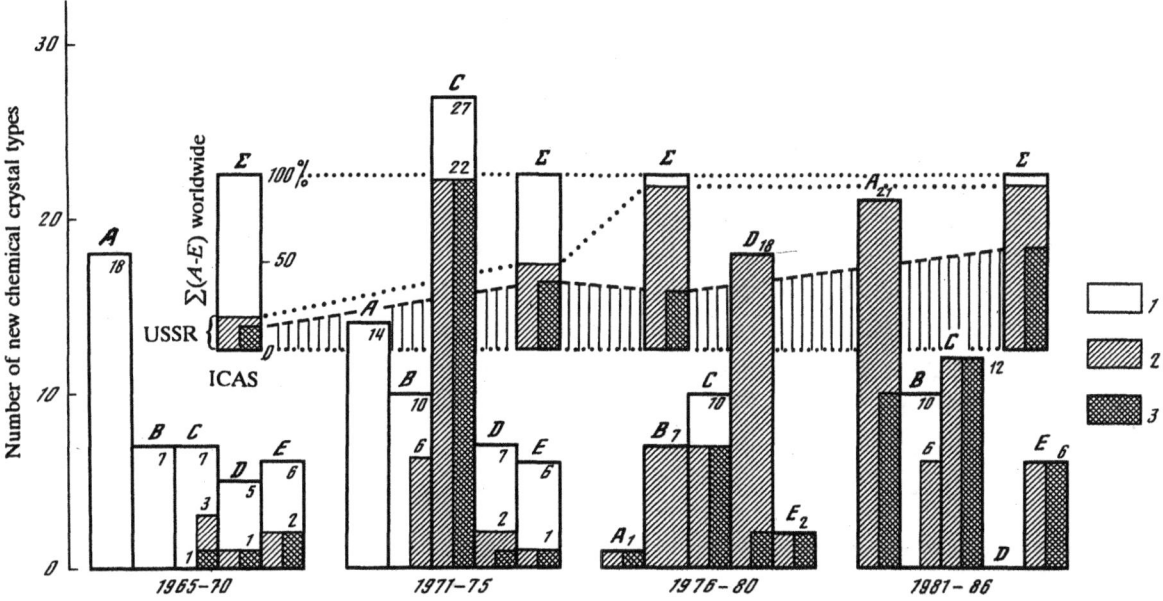

Fig. 5. Chronology of synthesis of single crystals formed in the systems MF_m–RF_3. Structural types: LaF_3, β-YF_3–$(R_{1-x}R'_x)F_3$ (A); $CaWO_4$–$Li(R,R')F_4$ (B); CaF_2–$M_{1-x}R_xF_{2+x}$ (C); LaF_3–$R_{1-y}M_yF_{3-y}$ (D); β-$BaTm_2F_8$–$Ba(R,R')_2F_8$ (E).

As a result, the axial and radial composition fluctuates in single crystals. This makes them unsuitable at least for optics, laser media, and other similar applications.

Recent reviews [6, 7] in the *Growth of Crystals* series contain analyses of the generation of inhomogeneities in highly nonstoichiometric fluorite crystals. We refer the interested reader to these articles. Preparation methods of high-quality crystals of complicated chemical composition due to variations of growth rate and temperature gradient are also given there. However, the values of these parameters achievable under actual growing conditions do not guarantee preparation of all heterovalent solid solutions as homogeneous single crystals. This is one of the most difficult obstacles to overcome in creating multicomponent materials with smoothly regulated properties. The difficulties related to melt methods can be surmounted under certain conditions such as hydrothermal, flux crystallization, and synthesis of optical ceramics by hot pressing.

Another important distinction of preparing one-component fluoride crystals is the technology requiring active fluorinating atmospheres [e.g., 8-10]. Industrial methods for preparing crystals of one-component fluorides uses a vacuum as the crystallization medium. Melts are purified from oxygen by adding lead or cadmium fluorides [e.g., 11, 12]. These measures are in general sufficient for growing fluorides of AE, which are relatively difficult to hydrolyze. However, stronger fluorinating agents are needed to suppress pyrohydrolysis of RE trifluorides or heavy metal tetrafluorides, which are much more reactive toward water vapor above 700°C. These can be anhydrous HF, F_2, CF_4, XeF_2, SF_6, BF_3, etc. A stream of fluorinating gas is often passed through the chamber, even bubbled through the melt, for better purification of the fluoride melts from oxygen. The oxygen content in fluorides prepared in active gas atmospheres can be decreased to tens of ppm (detection limit of the analysis). The groups at Bell Telephone [8], the Physics Institute of the Academy of Sciences of the USSR (PIAS) [9], and later the Hughes Research Laboratory [10] and others have played an important role in developing technology for high purification of fluorides from oxygen using active gas atmospheres. Purification from oxygen is one of the central problems affecting the quality of fluorides. It has a long history full of dramatic searches and misconceptions that is beyond the scope of the present article.

STATUS AND PROSPECTS FOR PREPARING NEW MULTICOMPONENT

FLUORIDES BY MELT CRYSTALLIZATION

The increase in the number of publications on growth of crystals of multicomponent fluorides from 8 during 1960-1965 to 70 in 1981-1986 attests to the growing interest in these materials. The chronology of these studies is presented in Fig. 5. Each of the five basic structural types of solid solutions and binary compounds were analyzed separately. Since crystals containing MF_4 were not grown in the period examined, we limited ourselves to MF_m–RF_3 systems with $m \leq 3$. Crystals of the families $LiRF_4$, $Na_{0.5-x}R_{0.5+x}F_{2+2x}$, and BaR_2F_8, which are more complicated than the two-component composition, were also included. Addition of the RE ions complicates the systems but is necessary to impart the required laser properties to the crystals.

The column height of the histograms in Fig. 5 represents the total number of crystals synthesized worldwide during the time period given on the horizontal axis and possessing qualitatively different chemical (elemental) composition. Hatched areas represent the total number of crystals prepared in the USSR; cross hatched areas, the contribution from the collaborative program between the ICAS and MSU.

Single Crystals of Concentrated Solid Solutions $(R_{1-x}R_x')F_3$ with LaF_3 and β-YF_3 Structures (Column A). Single crystals initially prepared were based on LaF_3 with various added RE [13-16] and others. Somewhat later, studies on crystallization of pure RE trifluorides [17-20] and others were published. These played an important role in defining the polymorphism and morphotropy of RF_3. These properties are very directly related to the effect of an oxygen impurity on the phase transition temperatures, the RF_3 crystalline form, and the optical quality. It would not be exaggerating to say that the pyrohydrolysis of the fluorides, studied in most detail for RF_3, has controlled the optical quality of existing fluorides. These studies are especially important for multicomponent fluorides containing RE trifluorides.

One of the leaders in supplying $(R_{1-x}R_x')F_3$ crystals is Varian (Palo Alto, USA). These materials have been studied extensively in the USSR for the last decade [e.g., 20-23].

$LiRF_4$ Crystals (YLF, according to the name of the first representative, yttrium lithium fluoride) (column B). Crystals of these compounds were grown only abroad during 1965-1975. The work began at Massachusetts Institute of Technology in connection with highly effective low-threshold lasers. They were the first commercial laser crystals of multicomponent fluorides. The compounds $LiRF_4$, discovered [32] during a study of phase diagrams of LiF–RF_3 systems, crystallize with the $CaWO_4$ scheelite structure and form in systems with R = Eu-Lu, Y.

The Czochralski method is the principal means of preparing $LiRF_4$ crystals [e.g., 24-27]. However, YLF crystallizes rather well using the Bridgman–Stockbarger method [28-31]. The Hughes Research Laboratory (USA) group then began to work with $LiRF_4$ crystallization processes [29]. Domestic efforts to prepare individual $LiRF_4$ compounds in order to characterize them physicochemically started during this same period. The majority of single crystals of pure $LiRF_4$ were prepared in the USSR by groups at the Vavilov State Institute of Optics [25]. Single crystals of $LiTbF_4$ were first synthesized at the Kazakh State University [31]. Many studies on the growth of crystals mention the change of nature of $LiRF_4$ melting from incongruent for the lighter RE to clearly congruent for fluorides of the later RE. Studies of the phase diagrams of LiF–RF_3 systems [32, 33] did not answer this question due to the low precision and partial pyrohydrolysis of RF_3. We recently revised the phase diagrams of these systems and demonstrated that these compounds melt congruently starting with $LiTmF_4$. The prototype $LiYF_4$ is apparently strictly intermediate (the 1:1 composition and the peritectic coincide), similar in behavior to $LiErF_4$ [34]. For this reason, the nature of melting of $LiRF_4$ crystals is strongly influenced by various factors (mainly impurities and the degree of supercooling), shifting it toward congruent in some cases and toward incongruent in others.

Nonstoichiometric Crystals with the Fluorite Structure (Column C). These were first synthesized in MF_2–RF_3 systems (M = Ca, Sr, Ba, Cd) [35]. Since 1965, when scientists at the PIAS and the ICAS produced stimulated radiation from an yttrium fluorite crystal doped with Nd [36], domestic efforts on the synthesis of multicomponent crystals of this type have been strenuous and have outdistanced foreign achievements (cf. Fig. 5). A detailed review of studies on the growth and preparation of new single crystals of $M_{1-x}R_xF_{2+x}$ appeared recently [6]. Therefore, we will not address these questions.

The highly nonstoichiometric phases $Na_{0.5-x}R_{0.5+x}F_{2+2x}$ formed in NaF–RF_3 systems with R = (Pr-Lu, Y) also crystallize with the fluorite structure. The growth and characterization of these phases are little studied. Solid solutions with R = Y, Ho-Lu were prepared as single crystals [37, 38].

Phase diagrams of NaF–RF_3 systems first studied in [39] need to be refined. This was partially done in [38, 40, 41], where mistakes of the previous studies were revealed.

Nonstoichiometric Crystals $R_{1-y}M_yF_{3-y}$ with the Tysonite Structure (Column D). Single crystals of $Y_{1-y}Ca_yF_{3-y}$ were first prepared by the Czochralski method [42] and shortly later by the Stockbarger method [43]. Tysonite phases of variable composition in Ca, Sr, Ba, and RE fluoride systems are most widely represented [3-5] (cf. Fig. 4). Heterovalent substitution of R^{3+} by M^{2+} in the LaF_3 structure thermally stabilizes this structure. As a result, many $R_{1-y}M_yF_{3-y}$ phases melt congruently in CaF_2 and SrF_2 systems. In the first series, single crystals of $R_{1-y}Ca_yF_{3-y}$ with R = (Y, Gd-Ho) were systematically prepared in [44]. Tysonite nonstoichiometric phases $R_{1-y}Sr_yF_{3-y}$ with all RE except Sc were prepared in [45-47]. They melt congruently, making possible preparation of highly homogeneous single crystals with compositions of the maxima on fusibility curves. Single crystals of tysonite phases in BaF_2–RF_3 systems were prepared only for a few RE [e.g., 48]. More than 80% of the new single crystals of $R_{1-y}M_yF_{3-y}$ were first prepared in the USSR.

A serious shortcoming of materials based on tysonite phases $R_{1-y}M_yF_{3-y}$ is the tendency to order or decompose at low temperatures.

BaR_2F_8 Compounds (Column D). These were first prepared as single crystals [42, 49] by the Bridgman and Czochralski methods. In the Bell Telephone studies [e.g., 50-52] crystals of this type were synthesized as IR–visible converters and as laser-active materials. A study of BaF_2–RF_3 phase diagrams [5] revealed two BaR_2F_8 polymorphic modifications with monoclinic β-$BaTm_2F_8$ and orthorhombic α-$BaLu_2F_8$ structures. We examined questions of the synthesis of this type of crystals in [53]. Growth of BaR_2F_8 by the Czochralski method was investigated more systematically in [54, 55]; by the Stockbarger method, in [56]. Melts from which BaR_2F_8 crystallizes greatly tend to supercool by up to 100-150°C. Crystalline matrices of monoclinic BaR_2F_8 with R = (Dy-Yb, Y) have a very high isomorphous capacity relative to heavier RE. However, RE impurities of the Ce subgroup can also be added to BaY_2F_8.

It is clear from Fig. 5 (column E) that domestic work on preparation of new single crystals with the β-$BaTm_2F_8$ structure with multiple activation by RE ions has been at the forefront.

Thus, we have reviewed the dynamic change of studies on growth of multicomponent fluorides during 1965-1986. Crystals of five families representing materials of the most practical interest were selected. The upper part of Fig. 5 shows the number of new types of multicomponent fluoride crystals adjusted to 100% for five-year periods with domestic contributions (hatched) and those of the ICAS and MGU (cross hatched) highlighted. The fraction of domestic efforts in this region has clearly increased substantially in the last decade.

We will now examine the prospects for synthesizing new single crystals based on phases formed in MF_m–RF_3 systems, where m = 1-3 and R = RE (Fig. 6). The majority of the crystals examined are basically two-component. However, practical demands necessitate changing to three-component (and more) materials for a number of crystalline matrices, as we have seen. In order to remedy this situation, we were forced to abandon the idea of examining only the simplest of the multicomponent crystals, the two-component ones. Such exceptions were made in Fig. 6 for crystals of YLF, fluorites $Na_{0.5-x}(R, R')_{0.5+x}F_{2+2x}$ and $Ba(R, R')_2F_8$. Data for these

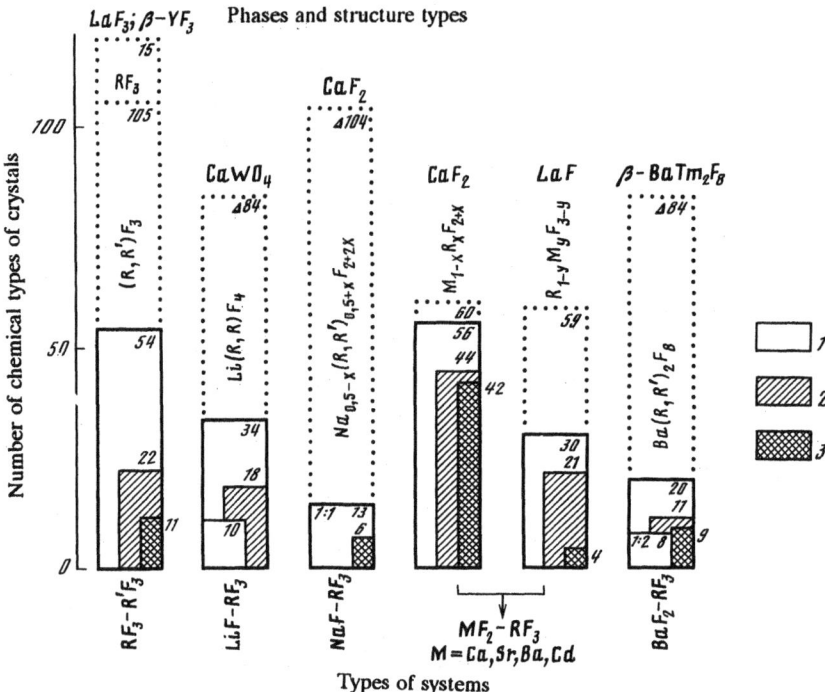

Fig. 6. Status and prospects for synthesizing crystals in MF_m–RF_3 ($m \leq 3$) systems. Total (1), in the USSR (2), at the ICAS and MGU (3).

three families are given for growth of three-component crystals. The total number (10, 13, 8) of two-component phases is shown in the lower part of the corresponding columns to the left. The three-component nature of the families listed are shown by triangles in the upper part of the dotted columns, the height of which corresponds to the number of possible chemical types of crystals.

More than half of the possible combinations were prepared as single crystals for $(R_{1-x}R'_x)F_3$ phases of variable composition with LaF_3 and β-YF_3 structures.

For three-component $Li(R, R')F_4$, 40% of all possible combinations were synthesized. Since YLF are used for lasers, the selection of RE ions added to the crystals is limited.

Nonstoichiometric $Na_{0.5-x}(R, R')_{0.5+x}F_{2+2x}$ fluorites, numbering 13 representatives (R = Pr–Lu, Y) in binary systems, have been prepared for half of the binary combinations and for a few ternary ones. Analogous phases $M_{1-x}R_xF_{2+x}$ were synthesized for 93% of the binary systems MF_2–RF_3 with M = Ca, Sr, Ba, Cd. The good availability of crystalline materials facilitates a thorough investigation of the physical properties of this family.

Single crystals of nonstoichiometric tysonite phases $R_{1-y}M_yF_{3-y}$ were prepared for half of the possible combinations. Apparently, this number will increase since the principal practical application of tysonite crystals is solid-state electrochemical devices, for which the cationic composition of the crystals is not limited.

Eight representatives (R = Tb–Lu, Y) of the monoclinic BaR_2F_8 binary systems are known. These crystals have mainly been grown from ternary systems BaF_2–RF_3–$R'F_3$. Of the possible combinations, 24% have been prepared. Considering that the current applications (anti-Stokes luminophores and lasers) limit the RE ions required, prospects for further expansion of the BaR_2F_8 crystals depend on finding new applications.

More than 500 chemically different phases exist for the binary and ternary fluorides examined in Fig. 6. These can be prepared as single crystals. Many of these have broad homogeneous regions in both composition and temperature. Many of the two- and three-component fluorides contain heterovalent cations. Heterovalent isomorphism in CaF_2 and LaF_3 structures leads to crystals with highly disrupted stoichiometry. These disruptions are an effective method of radically changing the properties by changing the chemical composition and the atomic (defect) structure of the crystals.

Not more than 200 of the chemical types of crystals, i.e., about 40% of all possible, were synthesized up to 1987. About 2/3 of the worldwide production of such crystals occurred in the USSR. This is known to be due to completion of a multiyear program of physicochemical and crystal chemical investigations of multicomponent fluorides in general and a study of the high nonstoichiometry in a chemical class of fluorides in particular.

The situation is the same with preparation of single crystals of the simplest multicomponent fluorides. We have already seen in part that a change from two- to three-component crystals (cf. Fig. 6) is accompanied by a substantial increase in the number of qualitatively different chemical compositions. Practical needs will force further complication of the chemical composition.

The total number of ternary systems formed by the 27 fluorides selected by us in the beginning of this article is 2925. Hundreds of the phases in these systems represent a huge reserve of practically useful fluorides.

This short review of the status and prospects for preparing multicomponent fluorides clearly confirms that good prospects exist for developing this area of inorganic materials science. It is controlled by practical demands, which are in turn determined by systematic fundamental investigations of the chemistry and physics of these materials. It is satisfying that domestic investigations have been at the forefront. Their prevalence is obvious from the statistical data presented.

Studies of the physics of multicomponent fluorides delineate the principal directions for the practical use of crystalline materials, including, in particular:

— optical materials transparent from the vacuum UV to the mid IR with improved operating characteristics;
— active laser media with a special (partially disordered) structure and the laser characteristics related to it;
— scintillation materials, dosimeters for ionizing radiation, luminophores (including anti-Stokes);
— photoconductive solid electrolytes for solid-state electrochemical instruments (condensers, batteries, chemical sensors, etc.);
— substrates for semiconducting films and others.

Obviously single crystals are not the only form of multicomponent fluorides. Polycrystalline materials (both as ordinary ceramics and hot pressings) are useful for many practical applications. One-phase materials are by far not needed in all cases. Multiphase eutectic composites may be interesting especially if they are ordered during directed crystallization. Films of multicomponent fluorides represent an independent direction. Few examples of the use of these have been published.

The author thanks P. P. Fedorov and V. B. Aleksandrov for fruitful discussions. My colleagues Z. I. Zhmurova and E. A. Krivandina helped tremendously in studying the growth of nonstoichiometric fluoride crystals.

LITERATURE CITED

1. B. P. Sobolev, "Multicomponent fluorides as crystalline matrices for doping rare earth ions," in: Eighth All-Union Conf. on the Chemistry of Inorganic Fluorides [in Russian], Polevskoi, Aug. 25-27, 1987, Nauka, Moscow (1987), p. 19.
2. B. P. Sobolev, "Nonstoichiometry in MF_m–RF_n systems and new multicomponent fluorides," in: Abstracts of Papers of the Seventh All-Union Symp. on the Chemistry of Inorganic Fluorides [in Russian], Leninabad, Oct. 9-11, 1984, Nauka, Moscow (1984), p. 15.
3. B. P. Sobolev and P. P. Fedorov, "Phase diagrams of the CaF_2-(Y, Ln)F_3 systems. 1. Experimental," *J. Less-Common Met.*, 5, No. 1, 33-46 (1978).
4. B. P. Sobolev, K. B. Seiranian, L. S. Garashina, and P. P. Fedorov, "Phase diagrams of the SrF_2-(Y, Ln)F_3 systems. 1, 2," *J. Solid State Chem.*, 28, No. 1, 51-58 (1979); 39, No. 2, 17-24 (1981).
5. B. P. Sobolev and N. L. Tkachenko, "Phase diagrams of the BaF_2-(Y, Ln)F_3 systems," *J. Less-Common Met.*, 85, No. 2, 155-170 (1982).

6. B. P. Sobolev, Z. I. Zhmurova, V. V. Karelin, et al., "Preparation of single crystals of nonstoichiometric fluorite phases $M_{1-x}R_xF_{2+x}$ (M = alkaline earth, R = rare earth) by the Bridgman—Stockbarger method," in: *Growth of Crystals*, Vol. 16, Consultants Bureau, New York (1991).

7. P. P. Fedorov, T. M. Turkina, V. A. Meleshina, and B. P. Sobolev, "Cellular substructures in single crystalline solid solutions of inorganic fluorides having the fluorite structure," in: *Growth of Crystals*, Vol. 17, Consultants Bureau, New York (1991).

8. H. Guggenheim, "Growth of highly perfect fluoride single crystals for masers," *J. Appl. Phys.*, **34**, No. 8, 2482-2485 (1963).

9. Yu. K. Voron'ko, V. V. Osiko, V. T. Udovenchik, and M. M. Fursikov, "Optical properties of CaF_2—Dy^{3+} crystals," *Fiz. Tverd. Tela*, **7**, No. 1, 267-273 (1965).

10. S. J. Warshaw and R.E. Jackson, "Hydrofluorination unit for purification of fluoride laser materials," *Rev. Sci. Instrum.*, **36**, No. 2, 1774-1776 (1965).

11. D. C. Stockbarger, "Artificial fluorite," *J. Opt. Soc. Am.*, **39**, No. 9, 731-740 (1949).

12. L. M. Shamovskii, P. M. Stepanukha, and A. D. Shushakov, "Growth of single crystalline fluorite activated by rare earths," in: *Spectroscopy of Crystals* [in Russian], Nauka, Moscow (1970), pp. 160-164.

13. D. A. Jones, J. M. Baker, and D. F. D. Pope, "Electron spin resonance of Gd^{3+} in lanthanum fluoride," *Proc. Phys. Soc.*, **74**, No. 3 (477), 249-256 (1959).

14. V. H. Sirgo, "Growth of lanthanum trifluoride laser crystals," *Bull. Am. Phys. Soc.*, Ser. 2, **8**, No. 6, 475 (1963).

15. L. Esterowitz, A. Schnitzler, J. Noonan, and J. Bahler, "Rare earth infrared quantum counter," *Appl. Opt.*, **7**, No. 10, 2053-2070 (1963).

16. V. V. Azarov, V. G. Bonchkovskii, and B. S. Skorobogatov, "Study of $^4I_{9/2} \rightarrow {}^2P_{1/2}$ Nd^{3+} concentrational line broadening in LaF_3 and crystals with the scheelite structure," in: Abstracts of Papers of the Second Symp. on Spectroscopy of Crystals Activated by Rare Earths and Iron Group Elements [in Russian], Oct. 9-14, 1967, Khar'kov (1967), pp. 3-4.

17. M. Robinson and D. M. Cripe, "Growth of laser-quality rare earth fluoride single crystals in a dynamic hydrogen fluoride atmosphere," *J. Appl. Phys.*, **37**, No. 5, 2072-2074 (1966).

18. J. B. Mooney, "Some properties of single crystal lanthanum trifluoride," *Infrared Phys.*, **6**, No. 2, 153-157 (1966).

19. D. A. Jones and W. A. Shand, "Crystal growth of fluorides in the lanthanide series," *J. Cryst. Growth*, **3**, No. 6, 361-368 (1968).

20. B. P. Sobolev, I. D. Ratnikova, P. P. Fedorov, et al., "Polymorphism of ErF_3 and position of third morphotropic transition in the lanthanide trifluorides series," *Mater. Res. Bull.*, **11**, No. 8, 999-1004 (1976).

21. G. S. Shakhkalamyan, T. V. Uvarova, B. V. Sinitsyn, et al., "Growth of single crystals of lanthanide trifluorides of the yttrium subgroup," in: *Studies of Separation and Purification of Rare Metals* [in Russian], Moscow (1979), pp. 100-104; Scientific Works, State Inst. Rare Met., Vol. 91.

22. B. P. Sobolev, P. P. Fedorov, A. K. Galkin, et al., "Fusibility diagrams of certain binary systems formed by rare earth trifluorides," in: *Growth of Crystals*, Vol. 13, Consultants Bureau, New York (1985).

23. Yu. P. Grigoryan, V. V. Karelin, and V. S. Urusov, "Preparation of single crystals of alkaline earth and rare earth fluorides and analysis of the distribution coefficients of isovalent impurities," in: Abstracts of Papers of the Sixth All-Union Symp. on the Chemistry of Inorganic Fluorides [in Russian], Novosibirsk, July 21-23, 1981, Novosibirsk (1981), p. 136.

24. D. Gabbe and A. L. Harmer, "Scheelite structure fluorides: The growth of pure and rare earth doped $LiYF_4$," *J. Cryst. Growth*, **3/4**, 544 (1968).

25. I. A. Ivanova, L. M. Morozov, M. A. Petrova, et al., "Growth of crystals of binary fluorides of lithium—rare earths and their properties," *Izv. Akad. Nauk SSSR, Neorg. Mater.*, **11**, No. 12, 2175-2179 (1975).

26. J. S. Abell, I. R. Harris, B. Cockayne, and J. G. Plant, "A DTA study of zone-refined $LiRF_4$ (R = Y, Er)," *J. Mater. Sci.*, **11**, No. 10, 1807-1816 (1976).

27. R. Uhrin, R. F. Belt, and V. Rosaty, "Preparation and crystal growth of lithium yttrium fluoride for laser applications," *J. Cryst. Growth*, **38**, No. 1, 38-44 (1977).

28. W. A. Shand, "Single crystal growth and some properties of $LiYF_4$," *J. Cryst. Growth*, **5**, No. 2, 143-146 (1969).

29. R. C. Pastor, M. Robinson, and W. M. Akutagawa, "Congruent melting and crystal growth of $LiRF_4$," *Mater. Res. Bull.*, **10**, No. 6, 501-510 (1975).

30. D. A. Jones, B. Cockayne, R. A. Clay, and P. A. Forrester, "Stockbarger crystal growth, optical assessment and laser performance of holmium-doped yttrium-erbium lithium fluoride," *J. Cryst. Growth*, **30**, No. 1, 21-26 (1975).

31. L. S. Korableva, M. S. Tagirov, and M. A. Teplov, "Growth of single crystals of $LiLnF_4$ and control of their quality by magnetic resonance," in: *Paramagnetic Resonance* [in Russian], Izd. Kazan. Univ., Kazan' (1980), pp. 7-14.

32. R. E. Thoma, C. F. Weaver, H. A. Friedman, et al., "Phase equilibria in the system LiF—YF_3," *J. Phys. Chem.*, **65**, No. 7, 1096-1103 (1961).

33. R. E. Thoma, "Binary systems of the lanthanide trifluorides with the alkali fluorides," *Rev. Chim. Miner.*, **10**, No. 1/2, 363-382 (1973).

34. P. P. Fedorov, L. V. Medvedeva, I. P. Zibrov, et al., "Revised phase diagrams of LiF—RF_3 systems," in: Abstracts of Papers of the Seventh All-Union Conf. on Physical Chemical Analysis [in Russian], Frunze (1988), pp. 280-281.

35. T. Vogt, "The yttrium fluorite group," *Neues Jahrb. Miner.*, **2**, No. 1, 9-15 (1914).

36. Kh. S. Bagdasarov, Yu. K. Voron'ko, A. A. Kaminskii, et al., "Induced emission of yttrium fluorite crystals with Nd^{3+} at room temperature," *Kristallografiya*, **10**, No. 5, 746-747 (1965).

37. Kh. S. Bagdasarov, A. A. Kaminskii, and B. P. Sobolev, "Lasers based on cubic 5 NaF—9 YF_3:Nd^{3+}," *Kristallografiya*, **13**, No. 5, 900 (1968).

38. P. P. Fedorov, L. L. Vistin', A. A. Samokhina, et al., "Refinement of phase diagrams of NaF—RF$_3$ systems (R = Y, Gd-Lu) and new crystalline materials with the fluorite structure," in: Eighth All-Union Conf. on the Chemistry of Inorganic Fluorides [in Russian], Polevskoi, Aug. 25-27, 1987, Nauka, Moscow (1987), p. 381.

39. R. E. Thoma, H. Insley, and G. M. Hebert, "The sodium-fluoride—lanthanide trifluoride systems," *Inorg. Chem.*, **5**, No. 7, 1222-1229 (1966).

40. P. P. Fedorov, A. Rappo, F. M. Spiridonov, and B. P. Sobolev, "Phase diagram of the system NaF—YbF$_3$," *Zh. Neorg. Khim.*, **28**, No. 3, 744-748 (1983).

41. P. P. Fedorov, B. P. Sobolev, and S. F. Belov, "Fusibility diagram of the system NaF—YF$_3$ and the cross section Na$_{0.4}$Y$_{0.5}$F$_{2.2}$—YOF," *Izv. Akad. Nauk SSSR, Neorg. Mater.*, **15**, No. 5, 816-819 (1979).

42. B. P. Sobolev, E. G. Ippolitov, B. M. Zhigarnovskii, and L. S. Garashina, "Phase composition of the systems CaF$_2$—YF$_3$, SrF$_2$—YF$_3$, and BaF$_2$—YF$_3$," *Izv. Akad. Nauk SSSR, Neorg. Mater.*, **1**, No. 3, 362-368 (1965).

43. L. S. Garashina, E. G. Ippolitov, B. M. Zhigarnovskii, and B. P. Sobolev, "Phase composition of the system CaF$_2$—YF$_3$," in: *Natural and Technical Mineral Formation* [in Russian], Nauka, Moscow (1966), pp. 289-294.

44. R. C. Pastor, M. Robinson, and A. G. Hastings, "Congruently-melted compounds CaF$_2$—RF$_3$," *Mater. Res. Bull.*, **9**, No. 6, 781-786 (1974); No. 9, 1253-1260.

45. G. P. Dushatina, L. F. Koryakina, L. G. Morozova, et al., "Growth of single crystals based on phases of variable composition with the LaF$_3$ structure in the systems SrF$_2$—(Y, Ln)F$_3$," in: Abstracts of Papers of the Fifth All-Union Conf. on Growth of Crystals, Vol. 2 [in Russian], Tbilisi, Sept. 16-19, 1977, Tbilisi (1977), pp. 150-151.

46. G. V. Anan'ev, K. N. Baranova, M. N. Zarnitskaya, et al., "Growth and physicochemical studies of single crystals of tysonite solid solutions (Y, Ln)$_{1-x}$Sr$_x$F$_{3-x}$," *Izv. Akad. Nauk SSSR, Neorg. Mater.*, **16**, No. 1, 68-72 (1980).

47. A. A. Kaminskii, S. É. Sarkisov, K. B. Seiranyan, and B. P. Sobolev, "Stimulated emission of Sr$_2$Y$_5$F$_{19}$ crystals with Nd ions," *Kvantovaya Élektron.*, **1**, No. 1, 187-189 (1974).

48. I. A. Ivanova, L. F. Koryakina, M. A. Petrova, et al., "Growth of single crystals of bertollide solid solutions with tysonite structure in the systems MF$_2$—LnF$_3$ (M = Sr, Ba)," in: Sixth International Conf. on Growth of Crystals, Vol. 3 [in Russian], Moscow, Sept. 10-16, 1980, Moscow (1980), pp. 186-187.

49. E. G. Ippolitov, L. S. Garashina, B. M. Zhigarnovskii, et al., "Types of phase diagrams of the systems MF$_2$—LnF$_3$," *Dokl. Akad. Nauk SSSR*, **173**, No. 1, 101-103 (1967).

50. H. J. Guggenheim and L. F. Johnson, "New fluoride compounds for efficient infrared-visible conversion," *Appl. Phys. Lett.*, **15**, No. 2, 51-52 (1969).

51. L. F. Johnson and H. J. Guggenheim, "Infrared-pumped visible laser," *Appl. Phys. Lett.*, **19**, No. 2, 44-47 (1971).

52. L. F. Johnson and H. J. Guggenheim, "Laser emission at 3μ from Dy^{3+} in BaY$_2$F$_8$," *Appl. Phys. Lett.*, **23**, No. 2, 96-98 (1973).

53. A. A. Kaminskii, B. P. Sobolev, S. É. Sarkisov, et al., "Physicochemical aspects of the synthesis, spectroscopy and stimulated emission of single crystals of BaLn$_2$F$_8$—Ln^{3+}," *Izv. Akad. Nauk SSSR, Neorg. Mater.*, **18**, No. 3, 482-497 (1982).

54. I. A. Ivanova, L. F. Koryakina, G. I. Merkulyaeva, et al., "Use of the Czochralski method for preparing single crystals of Ba(Y, Ln)$_2$F$_8$," in: Abstracts of Papers of the First All-Union Conf., "Status and Prospects for Development of Preparation Methods of Artificial Single Crystals" [in Russian], Khar'kov, Sept. 18-19, 1979, All-Union Sci.-Res. Inst. Single Crystals, Khar'kov (1979), pp. 6-7.

55. V. V. Kas'yanov and B. V. Sinitsyn, "Growth of crystals based on barium and rare earth fluorides," in: *Studies of Separation and Purification of Rare Metals* [in Russian], Moscow (1979), pp. 105-107; Scientific Works, State Inst. Rare Met., Vol. 91.

56. B. P. Sobolev, T. V. Uvarova, and P. P. Fedorov, "BaLn$_2$F$_8$ - single crystals - new class of laser," in: Abstracts of the Third Hung. Conf. on Cryst. Growth, Sept. 19-23, 1983, (1983), pp. 50-51.

GPSR Compliance

*The European Union's (EU) General Product Safety Regulation (GPSR)
is a set of rules that requires consumer products to be safe and our
obligations to ensure this.*

*If you have any concerns about our products, you can contact us on
ProductSafety@springernature.com*

In case Publisher is established outside the EU, the EU authorized
representative is:

Springer Nature Customer Service Center GmbH
Europaplatz 3
69115 Heidelberg, Germany

Batch number: 09625694

Printed by Printforce, the Netherlands